T0358586

PHYSICS IN COLLISION (PIC 2013)

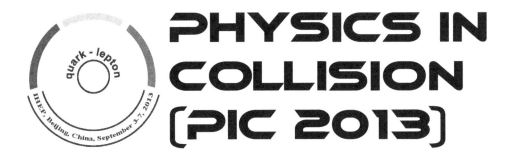

PHYSICS IN COLLISION (PIC 2013)

quark - lepton

IHEP, Beijing, China, September 3-7, 2013

World Scientific

NEW JERSEY · LONDON · SINGAPORE · BEIJING · SHANGHAI · HONG KONG · TAIPEI · CHENNAI

Published by

World Scientific Publishing Co. Pte. Ltd.

5 Toh Tuck Link, Singapore 596224

USA office: 27 Warren Street, Suite 401-402, Hackensack, NJ 07601

UK office: 57 Shelton Street, Covent Garden, London WC2H 9HE

British Library Cataloguing-in-Publication Data
A catalogue record for this book is available from the British Library.

PHYSICS IN COLLISION (PIC 2013)
Proceedings of the XXXIII International Symposium

ISBN 978-981-4618-64-9

Printed in Singapore

CONTENTS

Observation of the $X(1840)$ at BESIII

Energy non-linearity studies at Daya Bay

Physics in Collision (PIC 2013)
International Journal of Modern Physics: Conference Series
Vol. 31 (2014) 1402007 (8 pages)
© The Authors
DOI: 10.1142/S2010194514020078

World Scientific
www.worldscientific.com

Preface

The International Symposium on Physics in Collision is a conference series that began in 1981 in Blacksburg, Virginia, USA. The program of the conference is composed of invited talks and contributions in poster session. The XXXIII international symposium on Physics in Collision was held in Beijing (China), September 04–07, 2013 at Institute of High Energy Physics (IHEP). Invited speakers reviewed and updated key topics in elementary particle physics. Informal discussions on new experimental results and their implications were important parts of the Symposium.

Twenty seven invited talks were presented, covering the main results from all areas of particle physics, ranging from neutrino physics to the most recent results obtained from accelerators and astrophysics results from underground labs, earth-based and space-borne detectors. Twenty three posters were also submitted, out of which eleven were selected to give a short presentation. Copies of almost all the talks are available at the conference website http://pic2013.ihep.ac.cn.

PIC 2013 was organized by IHEP. The meeting was sponsored mainly by Chinese Academy of SciencesNational Natural Science Foundation of China and China Center of Advanced Science and Technology, to which we express our most grateful thanks. We would like to deeply thank all members of the IAC for having collaborated with us to shape a very interesting and complete program for this XXXIII edition of the conference, the members of the Local Committee and the technicians and students of our institute for their help in making PIC2013 a truly unforgettable event.

The electronic proceedings of the XXXIII International Symposium on Physics in Collision also available at SLAC web-site as econf/C1309031.

Physics in Collision

Organized by
Institute of High Energy Physics, Chinese Academy of Sciences

Sponsored by
Chinese Academy of Sciences (CAS)
National Natural Science Foundation of China (NSFC)
China Center of Advanced Science and Technology (CCAST)
Institute of High Energy Physics (IHEP)
Peking University (PKU)
Tsinghua University (THU)
University of Science and Technology of China (USTC)
University of Chinese Academy of Sciences (UCAS)

International Advisory Committee
Alessandro Bettini (Padova Univ., INFN and LSC)
Roy Briere (Carnegie Mellon Univ.)
Jean-Marie Brom (IPHC Strasbourg)
Dusan Bruncko (IEP SAS, Kosice)
Franco L. Fabbri (Lab. di Frascati, INFN)
Ulrich Heintz (Brown Univ.)
Boris Khazin (BINP, Novosibirsk)
Soo-Bong Kim (Seoul National Univ.)
Yoshitaka Kuno (Osaka Univ.)
Wolfgang Lohmann (DESY)
Livia Ludhova (Milano Univ. and INFN)
Giancarlo Mantovani (Perugia Univ. and INFN)
Joao de Mello Neto (Fed. Univ. Rio de Janeiro)
Thomas Müller (KIT)
Helenka Przysieuzniak (LAPP Annecy)
Yoshihide Sakai (KEK)
Xiaoyan Shen (IHEP)
Vladislav Simak (Inst. Phys. ASCR, CTU Prague)
Bernd Stelzer (SFU)
Dong Su (SLAC)
Robert S. Tschirhart (FNAL)

Participants

NAME	Institute	Email
ABLIKIM, M	IHEP	mablikim@ihep.ac.cn
ABT, Iris	MPI	isa@mpp.mpg.de
AI, Xiaocong	IHEP	aixc@ihep.ac.cn
BARSUK, Sergey	LAL	sergey.barsuk@cern.ch
BLACK, Kevin	Boston Univ.	kmblack@physics.bu.edu
BOZZI, Concezio	INFN Sezione di Ferrara	bozzi@fe.infn.it
BRUNCKO, Dusan	IEP SAS Kosice	dusan.bruncko@cern.ch
CHEN, Mingshui	IHEP	chenms@ihep.ac.cn
CHEN, Xun	Shanghai Jiao Tong Univ.	chenxun@pku.edu.cn
CHENG, Yaping	IHEP	chengyp@ihep.ac.cn
DAI, Jianping	IHEP	daijianping@ihep.ac.cn
DONG, Chao	IHEP	dongchao@ihep.ac.cn
FABRIZIO, Margaroli	Sapienza Univ.	Sapienza Univ. of Rome
FANG, Shuangshi	IHEP	fangss@ihep.ac.cn
FANO', Livio	Univeristà degli Studi di Perugia and INFN	livio.fano@pg.infn.it
GAO, Yuanning	Tsinghua Univ.	gaoyn@tsinghua.edu.cn
GAO, Zhen	USTC	gaozhen@mail.ustc.edu.cn
GAO, Qing	IHEP	gaoq@ihep.ac.cn
GARFAGNINI, Alberto	Padova Univ. and INFN-Padova (Italy)	alberto.garfagnini@pd.infn.it
GENG, Cong	USTC	gengcong@mail.ustc.edu.cn
GONG, Bin	IHEP	twain@ihep.ac.cn
GORI, Valentina	INFN Firenze	valentina.gori@cern.ch
GUO, Aiqiang	IHEP	guoaq@ihep.ac.cn
GUO, Yue	IHEP	guoy@ihep.ac.cn
HE, Jibo	CERN	jibo.he@cern.ch
HOU, Zhen	Univ. of California, Davis	hou@ucdavis.edu

HU, Yu	IHEP	huyu@ihep.ac.cn
HU, Chen	IHEP	huchen@ihep.ac.cn
James Gordon Branson	Univ. of California	branson@ucsd.edu
JI, Xiangpan	Nankai Univ.	jixp@mail.nankai.edu.cn
JI, Xiaobin	IHEP	jixb@ihep.ac.cn
KIRYLUK, Joanna	Stony Brook Univ.	joanna.kiryluk@stonybrook.edu
KUNO, Yoshitaka	Osaka Univ.	kuno@phys.sci.osaka-u.ac.jp
LI, Ke	IHEP	like@ihep.ac.cn
LI, Weiguo	IHEP	liwg@ihep.ac.cn
LI, Chunhua	IHEP	chunhuali@ihep.ac.cn
LI, Dengjie	IHEP	lidengjie@ihep.ac.cn
LI, Jingyi	IHEP	liujy@ihep.ac.cn
LIANG, Zhijun	Univ. of Oxford	zhijun.liang@cern.ch
LIU, Beijiang	IHEP	liubj@ihep.ac.cn
LIU, Fang	IHEP	liuf@ihep.ac.cn
LIU, Yanwen	USTC	yanwen@ustc.edu.cn
LIU, Huanhuan	IHEP	huanhuan.liu@ihep.ac.cn
LIU, Jie	IHEP	jieliu@ihep.ac.cn
LOHMANN, Wolfgang	DESY	wolfgang.lohmann@desy.de
LU, Haoqi	IHEP	luhq@ihep.ac.cn
LV, Meng	IHEP	lvmeng@ihep.ac.cn
MA, Jian-Ping	ITP	majp@itp.ac.cn
MADAR, Romain	Freiburg Univ.	romain.madar@cern.ch
MANKEL, Rainer	DESY	rainer.mankel@desy.de
MAO, Zepu	IHEP	maozp@ihep.ac.cn
MARIAZZI, Analisa Gabriela	Instituto de Fisica La Plata	mariazzi@fisica.unlp.edu.ar
MCCUMBER, Michael	Los Alamos National Laboratory	michael.p.mccumber@gmail.com
MITCHELL, Ryan	Indiana Univ. & IHEP	remitche@indiana.edu
MOLANDER, Simon	Stockholm Univ.	simon.molander@fysik.su.se
MONTEIL, Stephane	Univ. Blaise-Pascal — in2p3	monteil@in2p3.fr
NIU, Xunyi	IHEP	niuxy@ihep.ac.cn
NOMURA, Tadashi	KEK	nomurat@post.kek.jp
PAOLONE, Vittorio	Univ. of Pittsburgh	paolonepitt@gmail.com
PIANORI, Elisabetta	Univ. of Warwick	e.pianori@warwick.ac.uk
PIERINI, Maurizio	CERN	maurizio.pierini@cern.ch

POVEDA TORRES, Ximo	Indiana Univ.	ximo.poveda@cern.ch
QI, Hongrong	IHEP	hrqi@ihep.ac.cn
QUINTERO, Amilkar	Kent State Univ.	aquinter@kent.edu
RANUCCI, Gioacchino	Istituto Nazionale di Fisica Nucleare	gioacchino.ranucci@mi.infn.it
SALFELD-NEBGEN, Jakob	DESY	jakob.salfeld@cern.ch
SCHWIENHORST, Reinhard	Michigan State Univ.	schwier@pa.msu.edu
SHEN, Chengping	Beihang Univ.	shencp@ihep.ac.cn
SHEN, Xiaoyan	IHEP	shenxy@ihep.ac.cn
STAL, Oscar	Stockholm Univ.	oscar.stal@fysik.su.se
STEFAN, Hoeche	SLAC National Accelerator Laboratory	shoeche@slac.stanford.edu
SUN, Zhentian	USTC & IHEP	sunzt@ihep.ac.cn
TAKASAKI, Fumihiko	KEK	fumihiko.takasaki@kek.jp
URBAN, Jozef	P.J. Safarik Univ., Kosice, Slovakia	jozef.urban@upjs.sk
VOSS, Rudiger	CERN	rudiger.voss@cern.ch
WANG, Jian	Universite Libre de Bruxelles	jian.wang@cern.ch
WANG, You-kai	Shaanxi Normal Univ.	wangyk@itp.ac.cn
WANG, Zhihong	USTC	wzh1988@mail.ustc.edu.cn
YAN, Liang	USTC	yanl@ihep.ac.cn
YANG, Rongxing	USTC	
YANG, Guang	Argonne National lab/Illinois Institute of Technology	gyang9@hawk.iit.edu
YANG, Hongxun	IHEP	yanghx@ihep.ac.cn
YANG, Masheng	IHEP	yangms@ihep.ac.cn
YIN, Hang	Fermilab	yinh@fnal.gov
YIN, Junhao	IHEP	yinjh@ihep.ac.cn
YUAN, Ye	IHEP	yuany@ihep.ac.cn
ZHANG, Chi	IHEP	zhangchi@ihep.ac.cn
ZHANG, Jianhui	SJTU	zhangjianhui@gmail.com

ZHANG, Jielei	IHEP	zhangjielei@ihep.ac.cn
ZHANG, Jingqing	IHEP	zhangjq@ihep.ac.cn
ZHOU, Xiao rong	USTC	zxrong@mail.ustc.edu.cn
ZHU, Shuai	IHEP	zhushuai@ihep.ac.cn
ZHU, Kai	IHEP	zhuk@ihep.ac.cn

Physics in Collision (PIC 2013)
International Journal of Modern Physics: Conference Series
Vol. 31 (2014) 1460276 (14 pages)
© The Author
DOI: 10.1142/S2010194514602762

World Scientific
www.worldscientific.com

Electroweak measurements from W, Z and photon final states

Hang Yin

(On behalf of ATLAS, CDF, CMS, D0, and LHCb collaborations)

Fermi National Accelerator Laboratory,
Pine Street and Kirk Road, MS 357,
Batavia, Illinois, 50510, USA
yinh@fnal.gov

Received 26 February 2014
Published 15 May 2014

We present the most recent precision electroweak measurements of single W and Z boson cross section and properties from the LHC and Tevatron colliders, analyzing data collected by ATLAS, CDF, CMS, D0, and LHCb detectors. The results include the measurement of the single W and Z boson cross section at LHC, the differential cross section measurements, the measurement of W boson mass, the measurement of W and Z charge asymmetry. These measurements provide precision tests on the electroweak theory, high order predictions and the information can be used to constraint parton distribution functions.

Keywords: LHC; Tevatron; electroweak; W; Z.

1. Introduction

The precision electroweak measurements provide stringent tests the on Standard Model (SM): the measurement of single W and Z boson cross section provides critical tests on the perturbative QCD and the higher order predictions, the measurement of W charge asymmetry can be used to constraint the parton distribution functions (PDFs), and the measurement of W mass and weak mixing angle ($\sin^2 \theta_W$) improves precision of the SM input parameters.

Recently, in both LHC and Tevatron, there are plenty of analyses related to the single W and Z boson have been performed. The LHC is a pp collider, results reviewed in this article used the collision data at center-of-energy of 7 TeV and 8 TeV, collected by the ATLAS,[1] CMS[2] and LHCb[3] detectors. The Tevatron is a $p\bar{p}$ collider, with center-of-energy of 1.96 TeV, using data collected by the CDF[4] and

D0[5] detectors. The W and Z precision measurements have been performed based on these data.

At the Tevatron, the W and Z are produced with the *valence quark*. And at the LHC, in the productions of single W and Z boson, the *sea quark* and *gluon* contributions become larger. Thus, single W and Z measurements at the Tevatron are complementary to that at the LHC. Furthermore, with knowing incoming *quark* direction, the Z forward-backward charge asymmetry at the Tevatron will be more sensitive compared with that measurement at the LHC.

In this article, we review the W and Z cross section measurements from the LHC at first, then review the measurements of boson transverse momentum from both the Tevatron and LHC. In the end, the measurement of the W boson mass, W/Z charge asymmetry measurements, are presented. The physics results are collected from ATLAS, CDF, CMS, D0, and LHCb collaborations.

2. The W and Z Total Cross Section Measurements

Precise determination of the vector boson production cross section and their ratios provide an important test of the SM.

2.1. *Inclusive cross section*

At the LHC, the production of vector boson requires at least one sea quark, and given the high scale of this process, the gluon contribution becomes significant. Furthermore, the inclusive cross section of vector boson production is also sensitive to the PDFs, the cross section predictions depend on the momentum distribution of the gluon. And the theoretical uncertainty is another limitation for the cross section measurement, the current available predictions are at next-to-leading (NLO) and next-to-next-to-leading order (NLO) in perturbative QCD. In Fig. 1, the measured W and Z cross sections from different hadron colliders are compared with NNLO predictions, the predictions agree with measured value well.

Both ATLAS[6] and CMS[7] collaborations presented the inclusive W and Z cross section at $\sqrt{s} = 7$ TeV, with electron and muon decay channels, using a data sample corresponding to 35 pb^{-1} and 36 pb^{-1} of integrated luminosity collected in 2010. And furthermore, CMS[8] using data corresponding to 19 pb^{-1} of integrated luminosity collected in 2012 at $\sqrt{s} = 8$ TeV.

With $\sqrt{s} = 7$ TeV data, the measured inclusive cross sections from CMS are $\sigma(pp \rightarrow WX) \times B(W \rightarrow l\nu) = 10.30 \pm 0.02(stat.) \pm 0.10(syst.) \pm 0.10(th.) \pm 0.41(lumi.)$ nb and $\sigma(pp \rightarrow ZX) \times B(Z \rightarrow l^+l^-) = 0.974 \pm 0.007(stat.) \pm 0.007(syst.) \pm 0.018(th.) \pm 0.039(lumi.)$ nb, and from ATLAS are $\sigma(pp \rightarrow WX) \times B(W \rightarrow l\nu) = 10.207 \pm 0.021(stat.) \pm 0.121(syst.) \pm 0.347(th.) \pm 0.164(lumi.)$ nb and $\sigma(pp \rightarrow ZX) \times B(Z \rightarrow l^+l^-) = 0.937 \pm 0.006(stat.) \pm 0.009(syst.) \pm 0.032(th.) \pm 0.016(lumi.)$ nb.

By presenting the measured cross section as a ratio will result a cancellation of systematic uncertainty, including the luminosity uncertainty. Thus, the ratios

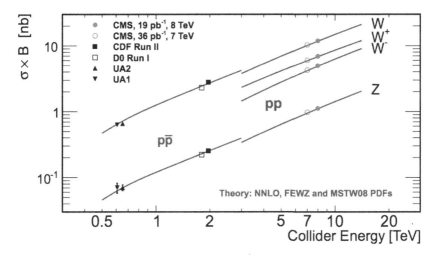

Fig. 1. The inclusive W and Z production cross section times branching ratios as a function of center-of-mass energy measured from CMS and other lower-energy colliders. The blue lines represent the NNLO theory predictions.

between measured $W^{\pm}$ and Z, also the ratio between W^{+} and W^{-}, are shown in Fig. 2 and Fig. 3.

The measurements of vector boson production cross section are consistent between the electron and muon channels, and in agreement with NNLO order predictions.

2.2. *Differential cross section*

The differential cross section of Drell-Yan (DY) process as a function of two leptons invariant mass is sensitive to the PDFs, particular in the high mass region (corresponding to the distribution function of antiquarks with high x value), and also provide stringent test on the high order perturbative QCD calculations. Additionally, this process is a important source of background for other beyond SM searches, the mass spectrum could be changed by new physics phenomena. The ATLAS[9] and CMS[10] report measurement of the high mass DY differential cross section measurement, as shown in Fig. 4. The measured differential cross sections are constant with different NLO (or NNLO) theoretical predictions, with more data in the future, this measurement could be used to constraint PDFs, in particular for antiquarks with large x.

CMS also performed a double-differential cross sections measurement,[10] as functions of boson rapidity (y) and invariant mass. By including additional boson rapidity information (y), the differential cross section provides more information for the PDFs, while the low mass region, the high order effects and final state radiation (FSR) become particularly important. The measured differential cross section as a function of boson y in the mass region of 20 to 30 GeV is shown in Fig. 5. The

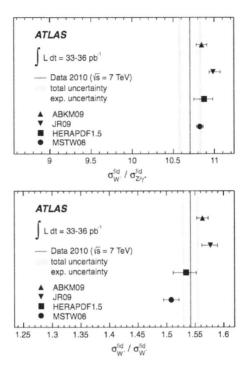

Fig. 2. The ratios between the measured inclusive cross section of $W^{\pm}$ and Z (top panel), and ratio between the measured inclusive cross section of W^+ and W^- (bottom panel). The inner bands represent experimental uncertainty, and outer bands represent total uncertainty. The PDFs uncertainties from the ABKM, JR, and MSTW predictions are considered in 68% C.L. The HERAPDF predictions include all three sources of uncertainty of that set.

double-differential cross section measurement of DY will provides precise inputs to the future PDF sets fitting.

2.3. *Lepton universality*

The coupling of the leptons (e, μ, τ) to the gauge vector boson are independent. The ATLAS performed a measurement of lepton universality,[6] by comparing the cross sections measured using electron channel and muon channel. The ratio between electron and muon channels results is shown in Fig. 6. The result confirms e-μ universality in both W and Z decays.

2.4. *W and Z cross section measurements at LHCb*

In the LHCb, the measurement of the W and Z cross section provide a unique test on the SM with high rapidity W and Z events. Thus, the LHCb results have better precision than ATLAS and CMS in the forward region ($2.0 < \eta < 5.0$). The most recent results[11,12,13,14] are for the inclusive cross section measurements, and the differential cross section measurement in different channels, as functions of rapidity

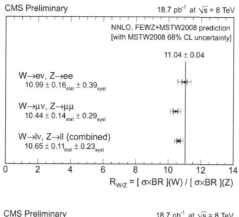

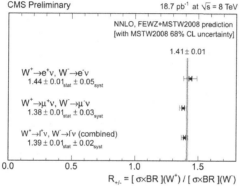

Fig. 3. The ratios between the measured inclusive cross section of $W^{\pm}$ and Z (top panel), and ratio between the measured inclusive cross section of W^+ and W^- (bottom panel). The black bars represent the statistical uncertainties, while the red errors bards include systematic uncertainties. Measurements in the electron and muon channels, and combined, are compared to the theoretical predictions using FEWZ and the MSTW2008 PDF set.

or invariant mass. The LHCb inclusive cross section measurements results are shown in Fig. 7, and one of the differential cross section measurement from LHCb is shown in Fig. 8.

3. The Boson Transverse Momentum Measurements

Initial state QCD radiation from the colliding parton can change the kinematic of the Drell-Yan process system, results Z boson be boosted with a transverse momentum, thus the precision measurement of Z boson p_T can provide a stringent test on the higher order QCD perturbative calculation. And this measurement can be used to reduce the theory uncertainty of the measurement of W mass. In the low p_T region, this process is dominated by soft and collinear gluon emission, with the limitation of standard perturbative calculation, the QCD resummation methods are used in the low p_T region. While in the high p_T region, the process is dominated by

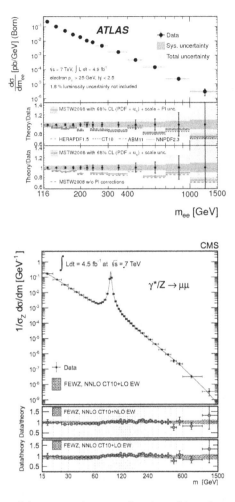

Fig. 4. Measured differential cross section as a function of invariant mass, from ATLAS (top panel) and from CMS (bottom panel). The measured differential cross sections are compared with plenty of high order predictions with different PDF sets.

single parton emission, thus, fixed-order perturbative calculations are used in the high p_T region.

3.1. Traditional method

In general, the boson p_T can be directly measured by using the reconstructed boson p_T. As has been done before,[15] CDF performed a precision measurement of transverse momentum cross section of e^+e^- pairs in the Z−boson mass region of 66–116 GeV/c^2, with 2.1 fb^{-1} of integrated luminosity.[16] Figure 9 shows the ratio between measured data value and RESBOS predictions. And CMS reported a similar measurement using data corresponding to 18.4 pb^{-1} of integrated luminosity,[17] as shown in

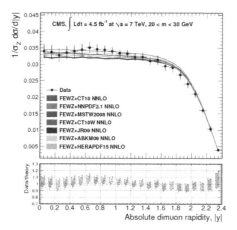

Fig. 5. The measured differential cross section as a function of boson y, in the mass region of 20-30 GeV. The uncertainty bands in the theoretical predictions represent the statistical uncertainty. The bottom plot shows the ratio of measured value to theoretical predictions.

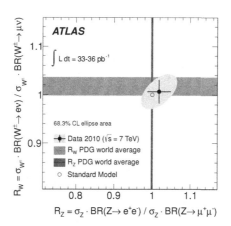

Fig. 6. The correlated measurement of the electron to muon cross section ratios in W and Z events. The vertical band represents the uncertainty from the Z measurement, and the horizontal band represents the uncertainty from the W measurement. The contour illustrates the 68% C.L. for the correlation between W and Z measurements.

Fig. 10. The overall agreement with predictions from the SM is observed, and the results have sufficient precision for the refinements of phenomenology in the future.

3.2. *A novel method*

In the previous Z p_T measurements, the uncertainty is dominated by the experimental resolution and efficiency, and the choice of bin widths is limited by the experimental resolution rather than event statistics. In order to improve the precision of the boson p_T measurements, D0 published measurement of Z boson p_T,[18]

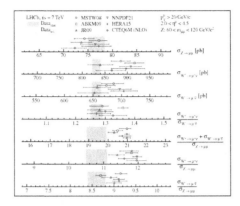

Fig. 7. Measurements of the Z, W^+ and W^- cross-section and ratios from LHCb, data are shown as bands which the statistical (dark shaded/orange) and total (light hatched/yellow) errors. The measurements are compared to NNLO and NLO predictions with different PDF sets for the proton, shown as points with error bars. The PDF uncertainty, evaluated at the 68% confidence level, and the theoretical uncertainties are added in quadrature to obtain the uncertainties of the predictions.

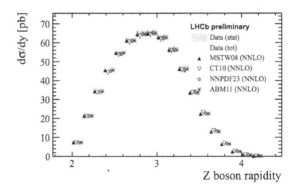

Fig. 8. Differential cross section measurement as a function of Z rapidity at LHCb. The yellow band represent the measured data value, while the NNLO predictions are shown as points with error bars reflecting the uncertainties.

presented with a novel variable ($\phi*$).[19] The ATLAS and LHCb performed same measurements of $\phi*$,[12,20] as shown in Fig. 11 and Fig. 12. The measured results provide sufficient information to the future retuning of theoretical predictions.

4. The W and Z Properties Measurements

There are many other electroweak measurements from Tevatron and LHC, which measure input parameters of the SM with better precision, including the measurement of W mass, the measurement of W charge asymmetry, and the measurement of weak mixing angle.

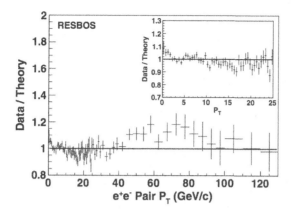

Fig. 9. The ratio of the measured cross section to the RESBOS prediction in the $p_T < 130$ GeV/c region. The RESBOS total cross section is normalized to the data. The insert is an expansion of the low p_T region.

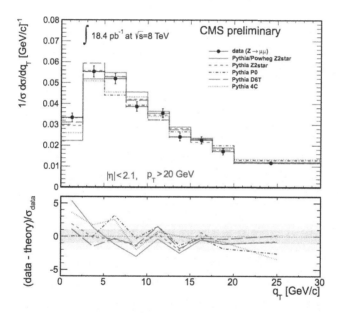

Fig. 10. The transverse momentum distribution of Z boson measurement from CMS. The measure values (black points) are compared with different predictions. In the bottom portion, the difference between data and prediction decided by the uncertainty of data is shown, with the green (inner) and yellow (outer) bands represent the 60% and 90% C.L. experimental uncertainties.

4.1. *Measurement of the W boson mass*

The precision measurement of W mass contribute to the understanding of the electroweak interaction. By combining the more recent measurement of W mass from both CDF,[21] D0,[22] and previous Tevatron measurements, the Tevatron average

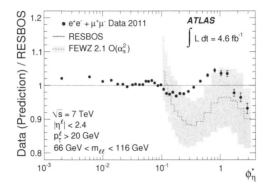

Fig. 11. The ratio of the combined normalized differential cross section to RESBOS predictions as a function of ϕ^*. The inner and outer error bars on the data points represent the statistical and total uncertainties, respectively. The measurements are also compared to predictions, which are represented by a dashed line, from FEWZ 2.1. Uncertainties associated to this calculation are represented by a shaded band.

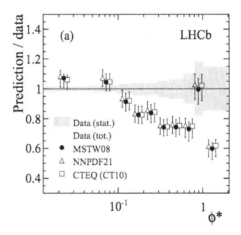

Fig. 12. Ratio of various QCD predictions to the measured data values as a function of ϕ^*, using LHCb data. The NNLO QCD predictions to the measured data are shown in black points.

value is 80387 ± 16 MeV.[23] The combination of W mass with the LEP average further reduces the uncertainty to 15 MeV, as shown in Fig. 13.

4.2. *W charge asymmetry measurements*

The W charge asymmetry is sensitive to PDFs. The Tevatron is a $p\bar{p}$ collider, the u quark tends to carry higher momentum than d quark, thus arise a non-zero asymmetry as a function of W rapidity. The LHC is pp collider, proton has two valance u quarks and one valence d quark, there are more W^+ than W^-. In hadron colliders, the neutrino escapes the detector without producing any measurable signal, without

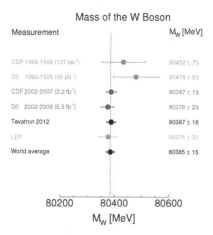

Fig. 13. Measured W boson mass from the CDF and D0 Run I (1989 to 1996) and Run II (2001 to 2009), the new Tevatron average, and the LEP combined results, and the world average obtained by combining the Tevatron and LEP measurements, with assumption that no correlation between them. The shaded band is the new world-average uncertainty (15 MeV).

the neutrino longitudinal momentum, the W charge asymmetry can be measured as lepton charge asymmetry, which is a convolution of the W boson asymmetry and W V-A decay.

The lepton asymmetry has also been measured at the LHC in pp collisions by the ATLAS[24] and CMS,[25][26] Collaborations using data corresponding to 31 pb^{-1} and 840 pb^{-1} of integrated luminosity, respectively. One of CMS results is shown in Fig. 14. With 37 pb^{-1} of integrated luminosity collected with LHCb detector, the lepton charge asymmetry is performed using muon channel,[11] as shown in Fig. 15. These measurements provide useful information for future PDF set fitting.

4.3. *Weak mixing angle measurements*

Weak mixing angle is one of fundamental parameters in the SM. It is one of key input parameters related to the electroweak couples, for both charge current (W) and neutrino current (Z). The weak mixing angle is a running parameter in a wide region of center-of-energy, there are many measurements have been done at the low energy experiments, like atomic parity violation,[27] Milloer scattering,[28] and NuTeV.[29] In the Z peak region, the most precision measurements are from LEP b quark asymmetries, and SLD Left-Right hand asymmetries (A_{LR}).[30] The results from LEP and SLD are deviated by three standard deviations in different directions.

Recently at the hadronic collider, CDF,[31] D0,[32][33] and CMS[34][35] have been performed this measurement, which show reasonable agreement with world average value. Due to the limitations from PDFs and quark fragments, the dominated systematic uncertainty comes from PDFs, which can be possibly suppressed by the update of PDFs sets.

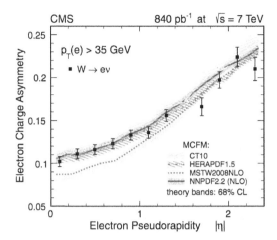

Fig. 14. Lepton charge asymmetry measurement from CMS, using electron channel. With a cut on the electron transverse momentum (35 GeV). The error bars include both statistical and systematic uncertainties.

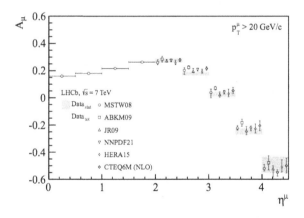

Fig. 15. Lepton charge asymmetry measurement from LHCb. The orange bands correspond to the statistical uncertainties, the yellow bands represent total uncertainty. The measured values are compared with different NNLO predictions.

With data corresponding to 2 fb^{-1} of integrated luminosity, CDF performed a measurement of weak mixing angle[36] using $Z \rightarrow ee$ events. Using data corresponding to 4.8 fb^{-1} of integrated luminosity, ATLAS presented a measurement of weak mixing angle,[37] after combining both electron and muon channels, the measured value at ATLAS is $0.2297 \pm 0.0004(stat.) \pm 0.0009(syst.)$. The summary for weak mixing angle measurement from different experiments is shown in Fig. 16. The PDF uncertainty is the key element for this measurement at hadron colliders, with more

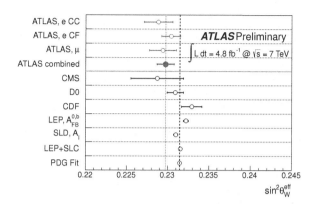

Fig. 16. Comparisons of measured weak mixing angle from different experiments. And the D0 result includes the 90% C.L. PDF uncertainty instead of 68% C.L. PDF uncertainty in other measurements.

data and new PDF set, the precision of weak mixing angle from hadron collider may be comparable to that from LEP.

5. Summary

Precision measurements with single W and Z bosons events provide stringent tests on the SM. With the data collected from the LHC and Tevatron, good consistency between SM and data is observed, these measurements also provide more information for PDF fitting, critical tests on the high order predictions, and more precisely SM input parameters. With good understanding on the detector response, more precision electroweak results with larger data set will come out soon from both LHC and Tevatron.

References

1. ATLAS Collab. (G. Aad *et al.*), *JINST* **3**, S08003 (2008).
2. CMS Collab. (S. Chatrchyan *et al.*), *JINST* **3**, S08004 (2008).
3. LHCb Collab. (A. A. Alves Jr. *et al.*), *JINST* **3**, S08005 (2008).
4. CDF Collab. (A. Abulencia *et al.*), *J. Phys. G*, **34**, 2457 (2007).
5. D0 Collab. (V. M. Abazov *et al.*), *Nucl. Instrum. Meth. Phys. Res. A* **565**, 463 (2006).
6. ATLAS Collab. (G. Aad *et al.*), *Phys. Rev. D* **85**, 072004 (2012).
7. CMS Collab., (S. Chatrchyan *et al.*), *J. High Energy Phys.* **007**, 1110 (2011).
8. CMS Collab., (S. Chatrchyan *et al.*), CMS-PAS-SMP-12-011.
9. ATLAS Collab. (G. Aad *et al.*), *Phys. Lett. B* **725**, 223-242 (2013).
10. CMS Collab., (S. Chatrchyan *et al.*), CERN-PH-EP-2013-168.
11. LHCb Collab., (R. Aaij *et al.*) *J. High Energy Phys.* **1206**, 058 (2012).
12. LHCb Collab., (R. Aaij *et al.*) *J. High Energy Phys.* **1302**, 106 (2013).
13. LHCb Collab., (R. Aaij *et al.*) *J. High Energy Phys.* **1301**, 111 (2013).
14. LHCb Collab., (R. Aaij *et al.*) LHCb-CONF-2012-013.
15. D0 Collab., (V. M. Abazov *et al.*), *Phys. Lett. B* **693**, 522 (2010).

16. CDF Collab., (T. Aaltonen *et al.*), *Phys. Rev. D* **86**, 052010 (2012).
17. CMS Collab., (S. Chatrchyan *et al.*), CMS-PAS-SMP-12-025.
18. D0 Collab., (V. Abazov *et al.*), *Phys. Rev. Lett.* **106**, 122001 (2011).
19. A. Banfi, S. Redford, M. Vesterinen, P. Waller, and T. R. Wyatt, *Eur. Phys. J. C* **71**, 3 (2011).
20. ATLAS Collab., (G. Aad *et al.*) *Phys. Lett. B* **720**, 32-51 (2013).
21. CDF Collab., (T. Aaltonen *et al.*) *Phys. Rev. D* **108**, 151803 (2012).
22. D0 Collab., (V. M. Abazov *et al.*) *Phys. Rev. Lett.* **108**, 151804 (2012).
23. CDF and D0 Collab., (T. Aaltonen *et al.*) *Phys. Rev. D* **88** 052018 (2013).
24. ATLAS Collab., (G. Aad *et al.*), *Phys. Rev. D***85**, 072004 (2012).
25. CMS Collab., (S. Chatrchyan *et al.*), *Phys. Rev. Lett.***109** 111806 (2012).
26. CMS Collab., (S. Chatrchyan *et al.*), *Eur. Phys. J.C***73**, 2318 (2013).
27. S. C. Bennett and C. E. Wieman, *Phys. Rev. Lett.* **82**, 2484 (1999).
28. SLAC E158 Collaboration (P. L. Anthony *et al.*), *Phys. Rev. Lett.* **95**, 081601 (2005).
29. NuTeV Collaboration (G. P. Zeller *et al.*), *Phys. Rev. Lett.* **88**, 091802 (2002) [Erratum-ibid. 90, 239902(2003)].
30. LEP Collaborations ALEPH, DEL-PHI, L3 and OPAL; SLD Collaboration, LEP Electroweak Working Group, SLD Electroweak and Heavy Flavor Groups (G. Abbiendi *et al.*), *Phys. Rep.* **427**, 257 (2006).
31. CDF Collaboration (D. Acosta *et al.*), *Phys. Rev. D* **71**, 052002 (2005).
32. D0 Collaboration (V. M. Abazov *et al.*), *Phys. Rev. Lett.* **101**, 191801 (2008).
33. D0 Collaboration (V. M. Abazov *et al.*), *Phys. Rev. D* **84**, 012007 (2011).
34. CMS Collaboration (S. Chatrchyan *et al.*), *Phys. Rev. D* **84**, 112002 (2011).
35. CMS Collaboration (S. Chatrchyan *et al.*), *Phys. Lett. B* **718**, 752 (2013).
36. CDF Collab., (T. Aaltonen *et al.*), *Phys. Rev. D* **88**, 072002 (2013).
37. ATLAS Collab., (G. Aad *et al.*), ATLAS-CONF-2013-043.

Physics in Collision (PIC 2013)
International Journal of Modern Physics: Conference Series
Vol. 31 (2014) 1460277 (11 pages)
© The Author
DOI: 10.1142/S2010194514602774

World Scientific
www.worldscientific.com

Top cross-sections and single top

Reinhard Schwienhorst
(On behalf of the ATLAS, CMS, CDF, D0 Collaborations)

Department of Physics & Astronomy,
Michigan State University, 567 Wilson Road
East Lansing, Michigan 48823, USA
schwier@pa.msu.edu

Published 15 May 2014

This paper summarizes top quark cross-section measurements at the Tevatron and the LHC. Top quark pair production cross-sections have been measured in all decay modes by the ATLAS and CMS collaborations at the LHC and by the CDF and D0 collaborations at the Tevatron. Single top quark production has been observed at both the Tevatron and the LHC. The t-channel and associated Wt production modes have been observed at the LHC and evidence for s-channel production has been reported by the Tevatron collaborations.

Keywords: Top quark; LHC; Tevatron.

1. Introduction

The top quark is central to understanding physics in the Standard Model (SM) and beyond. This paper summarizes top-quark related measurements from the Tevatron proton-antiproton collider at Fermilab and from the Large Hadron Collider (LHC), the proton-proton collider at CERN. The top quark couplings to the gluon, to the W boson, and now also to the photon and Z boson are all probed in these measurements. Searches for new physics in the top quark final state look for new particles and new interactions.

The Tevatron operation ended in 2011, with CDF and D0 each collecting 10 fb^{-1} of proton-antiproton data[1] at a center-of-mass (CM) energy of 1.96 TeV. The ATLAS[2] and CMS[3] collaborations at the LHC have reported measurements at CM energies of 7 TeV and 8 TeV, with up to 4.9 fb^{-1} and up to 20 fb^{-1}, respectively.

This paper reports recent measurements of top pair production, of single top production, as well as recent searches for new physics in top quark final states. Giving a complete overview of all activities is not possible here, but I will highlight

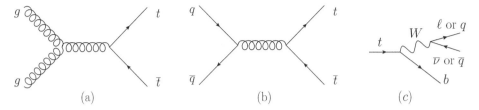

Fig. 1. Feynman diagrams for top pair production (a) via gluon fusion and (b) via quark-antiquark annihilation, and (c) for top quark decay.

relevant measurements and new developments. Section 2 summarizes top quark pair production measurements, Section 4 summarizes single top quark production, Section 5 presents new physics searches in the top quark sector, and Section 6 gives a summary.

2. Top Quark Pair Production

Top quark pair production proceeds mainly via gluon initial states at the LHC, shown in Fig. 1(a), and mainly via quark-antiquark annihilation at the Tevatron, shown in Fig. 1(b).

The production cross-section has been calculated at next-to-next-to leading order (NNLO), including next-to-next-to leading log (NNLL) soft gluon resummation.[4]

The top quark decays to a W boson and a b quark, and the final state topology in top quark pair events is determined by the subsequent decay of the two W bosons, as shown in Fig. 1(c). About a third of top pairs decay to the lepton+jets final state where one W boson decays to an electron or muon and the other to a quark pair. The background to this final state is mainly from W+jets production and QCD multi-jet events where one quark jet is mis-identified as a lepton. This final state topology has reasonable statistics and a manageable background while also allowing for the reconstruction of the two top quarks.

A small fraction of about 6% of top pair events decay to the dilepton (ee, $e\mu$ and $\mu\mu$) final state, which has small backgrounds from Z+jets and diboson production. This topology is attractive for its clean signature, though the individual top quarks can not be reconstructed directly due to the presence of two neutrinos.

About 46% of top pair events decay to an all-hadronic final state which is overwhelmed by a large QCD multi-jet background. Other top pair decays involve τ leptons, and in particular hadronic τ decays are of interest because they provide sensitivity to non-SM top decays. Leptonic τ decays are included in the lepton+jets and dilepton final states, though the lower lepton p_T and the presence of additional neutrinos modifies the event kinematics.

2.1. *Lepton+jets final state*

The lepton+jets final state (where the lepton is an electron or a muon) has backgrounds that can be controlled and higher event statistics than the dilepton final state. Cross-section measurements both at the Tevatron and the LHC by ATLAS[5] and CMS[6] rely on b-quark identification (b-tagging) as well as multivariate analysis techniques to separate the top pair signal from the background sources, mainly W+jets and QCD multi-jet production. The CMS analysis at 7 TeV utilizes the secondary vertex mass to discriminate the top quark pair signal from the backgrounds.[6] This distribution is shown in Fig. 2 for different jet- and b-tag multiplicities. The measured cross-section is 158.1 ± 11.0 pb for an uncertainty of only 7%.

Differential cross-sections have also been measured in top pair production by ATLAS at 7 TeV[7] and by CMS at 7 TeV[8] and 8 TeV.[9,10] The differential cross-section measured by ATLAS at 7 TeV as a function of the top pair transverse momentum is shown in Fig. 3 (left). The differential cross-section is normalized to the total cross-section, thus canceling many systematic uncertainties.

2.2. *Dilepton final state*

The dilepton final state (di-electron, di-muon and electron-muon) is clean with small backgrounds and small uncertainties, hence it provides high-precision measurements

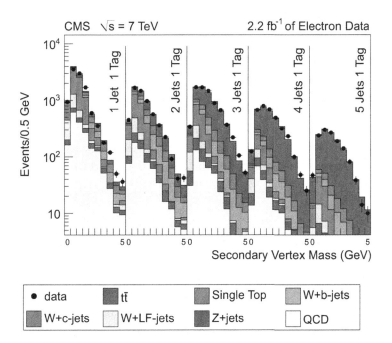

Fig. 2.   Secondary vertex mass distribution in electron+jet events in the measurement of the top pair production cross-section at 7 TeV by the CMS Collaboration.[6]

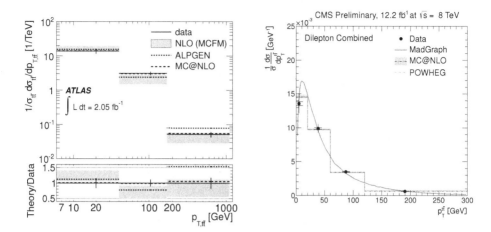

Fig. 3. (Left) ATLAS differential cross-section normalized to the total cross-section versus the transverse momentum of the $t\bar{t}$ system at 7 TeV.[7] (Right) CMS differential cross-section normalized to the total cross-section versus the transverse momentum of the $t\bar{t}$ system at 8 TeV,[10] where data points are shifted horizontally to be directly comparable to the theory prediction.

of the production cross-section. The CMS measurement at 7 TeV has an uncertainty of 4.2%, currently the single most precise measurement.[11]

The differential cross-section has also been measured in the dilepton final state. Figure 3 (right) shows the relative differential cross-section of the transverse momentum of the top quark pair.

2.3. *Pair production summary*

The Tevatron measurements are summarized in Fig. 4. The combined Tevatron top pair cross-section is measured to be 7.60 ± 0.41 pb, an uncertainty of only 5.4%.[12]

The measurements by ATLAS and CMS are shown as a function of the collider energy in Figs. 5 and 6, respectively. Note that the CMS summary figures do not yet include the latest CMS dilepton result.[11] The measured cross-sections are consistent with each other and with the theory predictions.[13, 14] The LHC top pair cross-section combination from Fall 2012[15] is also shown in Fig. 6.

3. Associated Production

Top quark pair production in association with one or more quarks or with a W or Z boson provides a measurement of the top quark strong and weak interactions and is an important background in new physics searches and Higgs boson measurements in $t\bar{t}H$.

The jet multiplicity in top pair events has been measured by both ATLAS[18] and CMS.[19, 20] Figure 7 shows the jet multiplicity in lepton+jets top pair events. At low jet multiplicities, the measurement agrees within the large uncertainty band with the theoretical predictions, while at high jet multiplicities the agreement is poor.

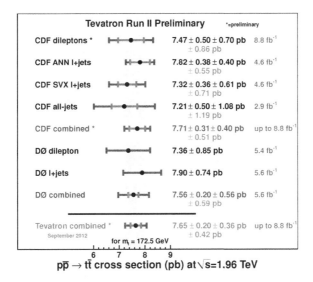

Fig. 4.   Tevatron top quark pair production cross-section measurements and their combination.[12]

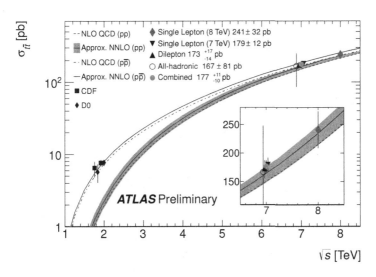

Fig. 5.   ATLAS top quark pair production cross-section measurements as a function of collider energy.[16]

The production of top quark pairs in association with Z bosons has a small cross-section and is difficult to measure. ATLAS had a first search for $t\bar{t}Z$ production at 7 TeV.[22] CMS found evidence for top pair production in association with a boson in two analyses[21] using 7 TeV data: A search for $t\bar{t}Z$ and a search for $t\bar{t}V$, where V

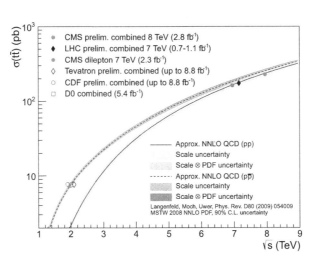

Fig. 6. CMS top quark pair production cross-section measurements as a function of collider energy.[17]

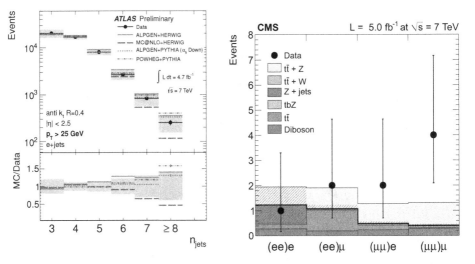

Fig. 7. Multiplicity of jets with $p_T >$ 25 GeV in top quark pair events measured by ATLAS at 7 TeV.[18]

Fig. 8. Event yields in the tri-lepton channels in the CMS $t\bar{t}V$ analysis at 7 TeV.[21]

can be a W boson or a Z boson. Figure 8 shows the two cross-section measurements and their uncertainties.

4. Single Top Production

Single top quark production proceeds via the t-channel exchange of a W boson between a heavy quark line and a light quark line, shown in Fig. 9(a) or via the

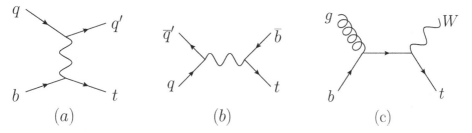

Fig. 9. Feynman diagrams for single top quark production in the (a) *t*-channel, (b) *s*-channel, (c) in association with a *W* boson.

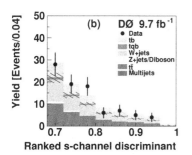

Fig. 10. D0 *s*-channel discriminant signal region.[23]

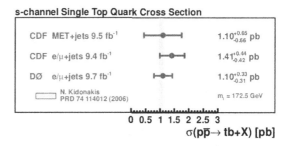

Fig. 11. Summary of Tevatron *s*-channel single top cross-section measurements.

s-channel production and decay of a virtual *W* boson, shown in Fig. 9(b) or as the production of a top quark in association with a *W* boson, shown in Fig. 9(c). At the Tevatron, the *t*-channel cross-section is largest, followed by the *s*-channel, while the *Wt* cross-section is too small to be observed. At the LHC, *t*-channel production benefits from the *qb* initial state, with a large cross-section. The *s*-channel has a smaller cross-section and has not been seen yet. The associated production has a *gb* initial state and can be observed.

4.1. *s-channel production*

Evidence for single top quark production in the *s*-channel was reported recently by the D0[23] and CDF[24] collaborations at the Tevatron. Both collaborations measure a cross-section that is consistent with the SM expectation, and both report an observed significance of 3.8 standard deviations. This is a challenging analysis that relies on multivariate analysis techniques in order to separate the signal from the large backgrounds. The *s*-channel signal region is shown in Fig. 10. A comparison of the CDF and D0 measurements is shown in Fig. 11. CDF also has a measurement using missing transverse energy plus jets events[25] and a combination of the two results.[26] A Tevatron combination of the CDF and D0 results is in progress.

4.2. *Wt associated production*

The production cross-section of a single top quark in association with a W boson has been measured both at 7 TeV and at 8 TeV by both ATLAS[27] and CMS.[28, 29] The final state in Wt events is categorized by lepton multiplicity, similar to top pair production. The difference to top pair production is that Wt production results in exactly one high-p_T b-quark jet. Hence top pair events comprise the largest background in Wt dilepton events, with smaller contributions for Z+jets, dibosons, and events with fake leptons. Multivariate techniques are required to separate the signal from these backgrounds, and systematic uncertainties are large. Nevertheless, the cross-section has now been measured with a relative uncertainty of 25%. The 7 TeV ATLAS analysis measures a cross-section of 16.8 ± 2.9 (stat) ± 4.9 (syst) pb.[27] The multivariate discriminant for the CMS 8 TeV analysis is shown in Fig. 12, CMS cross-section measurement is 23.4 ± 5.5 pb.[29] Both are consistent with the SM expectation.

4.3. *t-channel production*

Single top quark production through the t-channel is sensitive to the parton distribution function (PDF) of the light quarks in the proton. The t-channel final state is comprised of a lepton, neutrino and b quark from the top quark decay, a high-p_T forward jet, and possibly a third jet. The main backgrounds to this signature are from W+jets and top pair production. The ratio of top to antitop quark production in the t-channel is a particularly sensitive variable because many of the experimental uncertainties cancel. ATLAS has measured this ratio in t-channel events at 7 TeV using a neural network to separate the t-channel events from the background.[30] The measured ratio is $1.81^{+0.23}_{-0.22}$ and is compared to several different PDFs in Fig. 13.

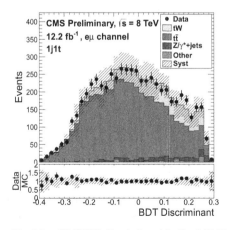

Fig. 12.   CMS Wt discriminant in the 8 TeV analysis.[29]

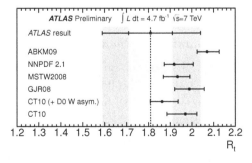

Fig. 13.   Ratio of top to anti-top quark production in t-channel single top measured by ATLAS at 7 TeV.[30]

CMS has measured the same ratio using 8 TeV data to be 1.76 ± 0.27, again consistent with expectations.[31] The precision of these measurements is not yet sufficient to constrain PDFs, but future measurements should be able to improve on this situation. Along with the ratio, ATLAS and CMS have also measured the total t-channel cross section, with an uncertainty of 14% from ATLAS at 8 TeV[30] and an uncertainty of 9% from CMS at 7 TeV.[32]

5. New Physics Searches

Searches for new physics in the top quark sector are particularly sensitive to models that have an enhanced coupling to the third generation of fermions. New heavy bosons Z' and W' appear in many models of new physics, and searches for these heavy resonances are a priority at hadron colliders.

The reconstructed mass of the top quark pair system in the ATLAS 7 TeV analysis is shown in Fig. 14.[33] The transverse momentum of the top quark produced in the decay of a heavy Z' boson is sufficient to collimate quark jets from the top quark decay such that they can no longer be resolved individually. Recent searches for Z' bosons employ algorithms to identify such boosted top quark jets.[33,34] The Z' mass at which such algorithms perform better than resolved algorithms that reconstruct each of the top quark decay jets individually is around 1 TeV as can be seen by the vertical dashed line in Fig. 15. The mass range probed by these searches extends up to masses of 2 TeV. The mass range below 0.75 TeV is also probed at the Tevatron where a CDF search currently provides the best sensitivity.[35]

A new charged heavy boson W' can decay to a top quark together with a b quark, leading to a single top final state. The W' boson may have SM-like left-handed couplings or it may have right-handed couplings to the top quark and the b quark. The ATLAS[36] and CMS[37] analyses probe these couplings separately, with CMS also providing two-dimensional limits as a function of the two couplings.

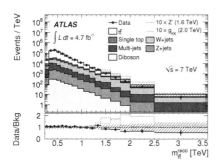

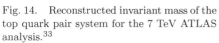

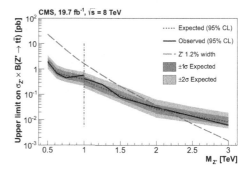

Fig. 14. Reconstructed invariant mass of the top quark pair system for the 7 TeV ATLAS analysis.[33]

Fig. 15. Upper limit on $Z' \rightarrow t\bar{t}$ from CMS at 8 TeV.[34]

6. Summary

The top quark pair production cross-section has been measured in many final states and with high precision by the CDF and D0 collaborations at the Tevatron proton-antiproton collider and by the ATLAS and CMS collaborations at the 7 TeV and 8 TeV LHC proton-proton collider. The single top quark cross-sections have been measured in the *t*-channel and now also in the *s*-channel at the Tevatron, and in the *t*-channel and the *Wt* associated production at the LHC. Many of these measurements are now at the level of precision of the theory predictions, and higher-precision results are yet to come with 8 TeV data. Searches for new physics in top quark final states have reached a sensitivity to high-mass resonances of over 2 TeV. This reach will be extended significantly at the 14 TeV LHC.

References

1. D0 Collaboration, *Nucl. Instrum. Meth.* **A565**, 463 (2006).
2. ATLAS Collaboration, *JINST* **3**, p. S08003 (2008).
3. CMS Collaboration, *JINST* **3**, p. S08004 (2008).
4. M. Czakon, P. Fiedler and A. Mitov, *Phys. Rev. Lett.* **110**, p. 252004 (2013).
5. ATLAS Collaboration, ATLAS-CONF-2012-149, http://cds.cern.ch/record/1493488.
6. CMS Collaboration, *Phys. Lett.* **B720**, 83 (2013).
7. ATLAS Collaboration, *Eur. Phys. J.* **C73**, p. 2261 (2013).
8. CMS Collaboration, *Eur. Phys. J.* **C73**, p. 2339 (2013).
9. CMS Collaboration, CMS-PAS-TOP-12-027, http://cds.cern.ch/record/1523611.
10. CMS Collaboration, CMS-PAS-TOP-12-028, http://cds.cern.ch/record/1523664.
11. CMS Collaboration, *JHEP* **1211**, p. 067 (2012).
12. CDF Collaboration, D0 Collaboration, arXiv:1309.7570 (2013).
13. U. Langenfeld, S. Moch and P. Uwer, *Phys. Rev.* **D80**, p. 054009 (2009).
14. M. Aliev, H. Lacker, U. Langenfeld, S. Moch, P. Uwer *et al.*, *Comput. Phys. Commun.* **182**, 1034 (2011).
15. ATLAS Collaboration and CMS Collaboration, ATLAS-CONF-2012-134, CMS-PAS-TOP-12-003, http://cds.cern.ch/record/1541952.
16. ATLAS Collaboration (2012), http://twiki.cern.ch/twiki/bin/view/AtlasPublic/CombinedSummaryPlots.
17. CMS Collaboration (2013), http://twiki.cern.ch/twiki/bin/view/CMSPublic/PhysicsResultsTOPSummaryPlots.
18. ATLAS Collaboration, ATLAS-CONF-2012-155, http://cds.cern.ch/record/1493494.
19. CMS Collaboration, CMS-PAS-TOP-12-018, http://cds.cern.ch/record/1494576.
20. CMS Collaboration, CMS-PAS-TOP-12-041, http://cds.cern.ch/record/1547532.
21. CMS Collaboration, *Phys. Rev. Lett.* **110**, p. 172002 (2013).
22. ATLAS Collaboration, ATLAS-CONF-2012-126, http://cds.cern.ch/record/1474643.
23. D0 Collaboration, *Phys. Lett.* **B 726**, p. 656 (2013).
24. CDF Collaboration, CDF note 11025, (2013).
25. CDF Collaboration, CDF note 11015, (2013).
26. CDF Collaboration, CDF note 11045, (2013).
27. ATLAS Collaboration, *Phys. Lett.* **B716**, 142 (2012).
28. CMS Collaboration, *Phys. Rev. Lett.* **110**, p. 022003 (2013).
29. CMS Collaboration, CMS-PAS-TOP-12-040, http://cds.cern.ch/record/1563135.

30. ATLAS Collaboration, ATLAS-CONF-2012-056, http://cdsweb.cern.ch/record/1453783.
31. CMS Collaboration, CMS-PAS-TOP-12-038, http://cds.cern.ch/record/1528574.
32. CMS Collaboration, *JHEP* **1212**, p. 035 (2012).
33. ATLAS Collaboration, *Phys. Rev.* **D88**, p. 012004 (2013).
34. CMS Collaboration, *Phys. Rev. Lett.* **111**, p. 211804 (2013).
35. CDF Collaboration, *Phys. Rev. Lett.* **110**, p. 121802 (2013).
36. ATLAS Collaboration, ATLAS-CONF-2013-050, http://cds.cern.ch/record/1547566.
37. CMS Collaboration, CMS-PAS-B2G-12-010, http://cds.cern.ch/record/1525924.

Physics in Collision (PIC 2013)
International Journal of Modern Physics: Conference Series
Vol. 31 (2014) 1460278 (11 pages)
© The Author
DOI: 10.1142/S2010194514602786

World Scientific
www.worldscientific.com

Measurement of top quark properties at the Tevatron and LHC

Fabrizio Margaroli
(On behalf of the CDF, D0, CMS, ATLAS collaborations)

Physics Department, Sapienza University of Rome & INFN,
Piazzale Aldo Moro 5 Rome, 00185, Italy
fabrizio.margaroli@roma1.infn.it

Published 15 May 2014

Almost two decades after its discovery at Fermilab's Tevatron collider experiments, the top quark is still under the spotlight due to its connections to some of the most interesting puzzles in the Standard Model. The Tevatron has been shut down two years ago, yet some interesting results are coming out of the CDF and D0 collaborations. The LHC collider at CERN produced two orders of magnitude more top quarks than Tevatron's, thus giving birth to a new era for top quark physics. While the LHC is also down at the time of this writing, many top quark physics results are being extracted out of the 7 TeV and 8 TeV proton proton collisions by the ATLAS and CMS collaborations, and many more are expected to appear before the LHC will be turned on again sometime in 2015. These proceedings cover a selection of recent results produced by the Tevatron and LHC experiments.

Keywords: Top quark; CKM; FCNC; Higgs.

PACS Numbers: 11.15.Ex, 12.15.−y, 14.65.Ha, 14.80.Ec, 14.80.Fd

1. Introduction

The discovery of the top quark[1,2] during the 1.8 TeV center-of-mass energy collisions of protons and antiprotons at the Tevatron was probably the single major achievement to stem from the effort of the CDF and D0 collaborations. While the existence of top quarks was expected, the community was surprised by the very large value that has been measured for its mass, approximately 175 GeV; the top quark stands as the most massive elementary particle even after the discovery of the Higgs boson.[3,4] This large value prompted a number of questions: why its mass sits at the electroweak scale, being some 40 times larger than the heaviest fermion mass? Why is the Yukawa coupling of the top quark so compatible (by 0.5%) with one and does that mean the top quark plays a special role in the electroweak symmetry breaking (ESB) mechanism? Given the recently known Higgs mass, more puzzles arise: the top quark large mass causes the so-called hierarchy problem, and top quarks participate in two of its most popular solution — SUSY and composite Higgs — where heavy bosonic or fermionic partners of the top quark would appear at energy scales that should be accessible by the LHC. While the large top quark mass poses numerous theoretical challenges to the Standard Model (SM), it also provide the unique opportunity to study a bare quark as the top quark lifetime ($\tau \sim 10^{-25}$ s) is much shorter than the hadronization time ($1/\Lambda_{QCD} \sim 10^{-24}$ s). The top quark is thus the only quark whose properties can be probed directly.

The Tevatron run at 1.96 TeV in its eight years of operations increased the available top quark dataset by two orders of magnitude with respect to the sample needed for discovery, thanks to the increased cross section at higher center-of-mass energy, approximately 7 pb [5] and especially the much larger integrated luminosity $L \simeq 10$ fb^{-1}. In its two years of high energy operations, the proton-proton collisions at the LHC allowed a further increase in the available data on top quarks by two more orders of magnitude, thanks mostly to the much larger cross section, $\sim 160(230)$ pb at 7(8) TeV collisions. The integrated luminosity at the LHC corresponds to 5 fb^{-1} of data during the 7 TeV run and $\simeq 19.6$ fb^{-1} at 8 TeV collisions.

There are notable differences among top quark production at the Tevatron and at the LHC. At the Tevatron, in order to produce top quarks, the incoming partons participating in the collision have to carry a very large fraction x of the proton momentum, where the up and down quark parton density functions (PDF) largely dominate over the others. For this reason, at the Tevatron top quarks are mostly produced through quark-antiquark annihilation. On the other hand, due to the higher LHC beam energies — that translate into a lower x needed to produce top quarks — and the fact that it is harder to extract an antiquark from a proton, top quarks are mostly produced through gluon-dominated initial states at the LHC. Pair production induced by quantocromodynamics (QCD) is more abundant than single top quark production induced by electroweak (EWK) interactions and due to the more striking signature, event selections typically leave clean samples, most properties are thus measured in events where top quarks are pair-produced. Details about top quark production have been amply discussed by the previous speaker so we leave its discussion to the corresponding proceedings.[6] It should be noted that the "historical" distinction of QCD top quark pair production and EWK single top quark production carries a decreasing weight in an era where experimenters are interesting in mixed QCD-EWK processes such as $t\bar{t}W/Z$ or $t\bar{t}H$, and single top quark production is of interest both when it appears together with other quarks, or photons, or heavy bosons.

2. The Crucial Role of the Top Quark Mass

The precise knowledge of the top quark mass is of the utmost importance for the understanding of the SM and the exploration of what lies beyond. In particular, global fits of this and other measurements of electroweak SM parameters provide a stringent prediction for the most likely region for the Higgs mass; the Higgs boson has been discovered at the edge of the 68% confidence level (CL) of these predictions.[7] Now that the Higgs mass is known, its value is fed into the same fit to assess SM self-consistency and constrain beyond SM scenarios (BSM). The knowledge of top quark and Higgs boson masses has recently been used to assess the vacuum stability, with important cosmological consequences.[8] The broad impact the top quark mass has in physics commands the need for ever-increasing precision. The ATLAS collaboration released its most sensitive result in a new analysis that measures simultaneously the top quark mass and both light and b-jet energy scale,[9] the latter two being the two largest systematic uncertainties that affect the mass determination. This measurement is now limited by other systematics, most notably the ones involved in the physics modeling of the signal itself. The top quark mass measurements performed by the Tevatron, CMS and ATLAS are all consistent with the value of 173 GeV within a few hundreds MeV[9-11] and are summarized in Fig. 1. The most precise results make use of the full reconstruction of the top quarks by recombinations of their daughter particles. This technique allows also to probe the hypothesis that the top quark width Γ_t differs from SM predictions, possibly because of exotic top quark decays. The CDF collaboration performed such measurements, setting constraints on generic exotic decays.[12] The main downsides of the mass measurements that make use of full top quark invariant mass reconstruction are the large dependence on the precise knowledge of the jet energy scale, and the intrinsic uncertainty on the definition of the mass of a colored particle. The former point has been addressed by a recent CMS measurement that makes use of the lifetime of the B-hadron stemming from the hadronization of the b-quark decaying from the top quark.[13] In fact, the mean of the B meson boost depends linearly on the mass of the top quark. The measurement's sensitivity to the jet energy scale uncertainty is greatly reduced, although at the cost of increasing the sensitivity to other systematic uncertainties. The second issue is addressed by measuring the properly defined top quark *pole* mass intersecting the measurement of the cross section for $t\bar{t}$ production with its theoretical prediction, as a function of the top quark mass. CMS performs the most precise such measurement finding very good consistency with the results from direct top quark mass measurements.[14] It will be extremely interesting to see such measurements performed using the full 2012 LHC dataset and the most up-to-date theoretical calculations for $t\bar{t}$ cross section.

3. Top Quark and Tests of Conservations Laws

The measurement of the difference between the top quark and the antitop quark mass is an interesting analysis that stems naturally out of the measurement of the top quark mass in top quark pair events. It probes directly the CPT theorem for the first time for "naked" quarks. Semileptonic $t\bar{t}$ system decays are used, so to tag the top or antitop quark according to the charge of the daughter lepton, and

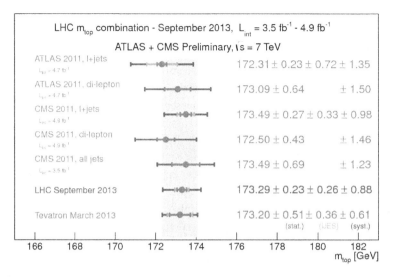

Fig. 1. Summary of the most precise top quark mass measurement from the LHC and Tevatron experiments. The central value as measured by the experiments at both colliders is impressively similar, considering that only a fraction of systematic uncertainties are correlated between the measurements.

reconstruct its mass accordingly. The CMS collaboration produced the single most precise such result as of today.[15] In the measurement of the differences between m_t and $m_{\bar{t}}$ most systematics naturally cancel out, thus enabling vewry high precision: $m_t - m_{\bar{t}} = -0.27 \pm 0.20(stat.) \pm 0.12(syst.)$ GeV; the relative uncertainty corresponds to $\sigma_{\frac{m_t - m_{\bar{t}}}{m_t}} \sim 1.3 \times 10^{-3}$. Recently, the CMS collaboration also produced the first test of baryon number (BN) conservation in the top quark system.[16] In particular, a four-fermion effective lagrangian is assumed that would give rise to decays such as $t \to bc\mu$ and $t \to bue$. Top quark pair events are used for this purpose, where one top quark decays according to the SM, and the other violates BN conservation. The requirement of a charged lepton without the accompanying neutrino greatly suppresses SM backgrounds, thus allowing to set stringent limits on BN violation: $BR(t \to bc\mu) < 16 \times 10^{-4}$.

4. The Last Man Standing: Asymmetry in $t\bar{t}$ Quark Production

At next-to-leading-order (NLO), QCD predicts the top quark to be emitted preferentially in the direction of flight of the proton beam, while the top antiquark in the direction of the antiproton beam. This charge asymmetry comes mainly from the interference between $q\bar{q} \to t\bar{t}$ tree diagram with the NLO box diagram, and from the interference of initial and final state radiations $q\bar{q} \to t\bar{t}g$. The CDF and D0 experiments observed over the years a forward-backward asymmetry (AFB) in top-antitop events that exceeded the SM expectations of a small but non-zero effect. With more data accumulating, CDF observed in 2011 for the first time a 3σ effect in the dependence of AFB as a function of the $t\bar{t}$ invariant mass.[17] A large number of new physics models that would explain such a large effect, while predicting new particles that should have appeared at a scale accessible at the Tevatron and/or at the LHC, have been proposed. The experimentalists proceeded simultaneously in the direction of performing new AFB measurements at the Tevatron, and the related-but-not-identical charge asymmetry measurement at the LHC, while searching for additional new physics manifestations. Theoreticians investigated possible effects whose absence in the current calculations would have lead to an underestimated asymmetry (recalculation of NLO EWK effects, and of perturbative and non-perturbative QCD effects). The latest results produced by the CDF and D0 experiment return a less significant deviation from SM predictions as the new theoretical computation predict larger values and more data and more final states were used.[18–20] Both collaborations investigated various observables and their correlations to establish agreement with the SM. Now that the Tevatron has extracted all possible information related to the top quark forward-backward asymmetry, the next word in the chapter rests in the LHC data. As of today, ATLAS and CMS performed differential measurements of the $t\bar{t}$ asymmetry as a function the invariant mass, the transverse momentum and the rapidity of the $t\bar{t}$ system.[21–24] They both observe consistency between the measurements and the predictions in a broad range of observables and kinematic regimes. Some of these observables and their comparison with theoretical predictions are shown in Fig. 2. Still, more data is needed — and already available — to draw definitive statements.

5. Top and Charged Higgs

A rather minimal extension of the SM, with a limited number of new parameters to adjust to experimental data, consists in the two-Higgs doublets model (2HDM). Different 2HDM alternatives have been suggested, with MSSM the most notable example. The most striking prediction of this set of models is the existence of charged Higgs bosons. In particular, charged Higgs bosons lighter than the top quark, could appear in the top quark decay chain $t \to H^{\pm}b$. The branching ratio of these charged Higgses would be largest when decaying into the heaviest fermions, i.e. a $c\bar{s}$ pair or a $\tau\nu_{\tau}$ pair (charged conjugates processes are implied) where the choice of the former or the latter would be determined by internal parameters of the model. The most recent LHC results use the 7 TeV dataset for these searches. The ATLAS collaboration performed a search of events where the charged Higgs boson would decay favorably to a quark pair.[25] The invariant mass of the charged Higgs boson can thus be completely reconstructed, although at the cost of accepting combinatorial background. A scan of the diet mass compatible with the charged Higgs hypothesis gives back the null results, which can be translated into upper limits on $BR(H^{+} \to c\bar{s}) \sim 2\%$ depending on the assumed charged Higgs mass. Both the ATLAS and CMS collaborations explored the scenario where the Higgs boson decays most favorably to heavy leptons.[25,26] Both leptonic and hadronic decays

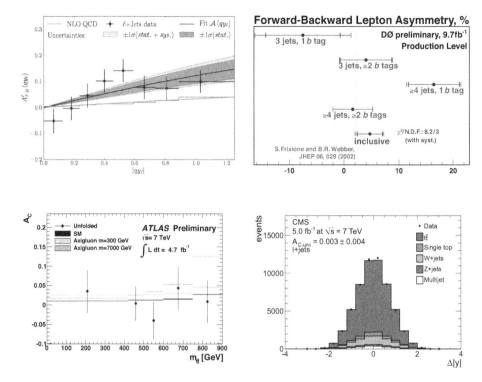

Fig. 2. Distributions of the the lepton (from top quarks) asymmetry in $t\bar{t}$ decays on the top row, and of the charge asymmetry in the bottom row. The collaborations studies also the dependence fo these observables as a function of other event properties.

of the tau lepton are considered; the collaborations use a number of observables sensitive to such exotic decays. In absence of an excess of events consistent with the signal hypothesis, upper limits are set at 95% CL on the charged Higgs branching ratio, that can be translated into limits on 2HDM parameter space. These limits are of the order of $\sim 2\%$ level, about one of order of magnitude stricter than previous results from the Tevatron collider experiments. A charged Higgs heavier than a top quark, could be produced directly and subsequently decay into a top-antibottom pair. This final state resembles closely single top s-channel production, and is very difficult at the LHC due to the very large W+heavy flavor jets background. The most recent result on the subject has been produced by the D0 collaboration several years ago.[28]

6. Top Quark and Flavor

The Cabibbo Kobayashi Maskawa (CKM) matrix elements can be inferred indirectly form the knowledge of the other elements, by exploiting the unitarity of the matrix and its 3×3 structure. New physics could violate these assumptions in several ways, the most obvious being the possible existence of a fourth generation of quarks. Electroweak single top quark production cross section σ_t is proportional to V_{tb}^2, thus allowing indirect extraction of this CKM parameter through comparison with the predicted σ_t. The large number of σ_{t+X} measurements produced at the Tevaton and LHC ($X = q, b, W$) have been directly translated into measurements of V_{tb}. All results are consistent with the SM predictions of $V_{tb} \simeq 1$.

Processes that are induced by flavor-changing neutral currents (FCNC) in top quark production or decay are extremely small according to SM computation, with decay rates of the order of 10^{-10} or below. New physics scenarios such as R-parity violating SUSY, top color technicolor, etc. could enhance FCNC rates by several orders of magnitude, thus making them accessible using current experimental data. Setting stringent limits on the rare decays of a particle — the top quark — that is already rarely produced, requires the wealth of LHC top quark data. FCNC couplings such as ugt or cgt can be probed through the production of a single top quark and no additional particles, through the process $qg \to t \to Wb \to \ell\nu b$ where ($q = u, c$). Decay products are relatively boosted with respect to the dominant $W+b$ background, and the kinematics of the event is fully reconstructed; several such observables are fed into a Boosted Decision Tree (BDT) to enhance sensitivity, and the limits on such FCNC rates are set via a likelihood scan of the the aforementioned multivariate discriminant. The analysis is more sensitive to ugt rather than cgt coupling due to the up quark PDF being larger than the charm quark PDF in the proton. The ATLAS collaboration sets 95% confidence level upper limits are $BR(t \to ug) < 3.1 \times 10^{-5}$ and $BR(t \to cg) < 1.6 \times 10^{-4}$ using $14.2\,\text{fb}^{-1}$ of data.[29] Experimental upper limits on couplings such as tZq can be set analyzing top quark pair production where one top quark decays to a Z boson and a light quark, in events such as $t\bar{t} \to bWqZ \to multileptons$. CMS performed such an analysis on the full 2012 dataset corresponding to $19.5\,\text{fb}^{-1}$ where the kinematics can be fully reconstructed, setting upper limits on this anomalous branching ratio $BR(t \to Zq) < 7 \times 10^{-4}$ which are the most stringent as of today.[30] The tZq coupling can be tested also on single top quark plus Z boson events, where they would appear as an excess of trlepton plus jets events but with different kinematics with respect to the previously discussed analysis. This analysis is performed by the CMS collaboration with the full 2011 dataset corresponding to $4.9\,\text{fb}^{-1}$ using $7\,\text{TeV}$ of data. The corresponding upper limits are $BR(t \to Zq) < 51 \times 10^{-4}$ using single top events;[31] the results could be greatly improved using the full dataset, and combined with the ones extracted from $t\bar{t}$ events, in order to further increase sensitivity. A possible anomalous coupling of special interest is the one of the top quark with the Higgs boson, through tHc coupling. ATLAS analyzes top-antitop quark production where one of the two top quarks decays into an Higgs boson and a quark.[32] By focuing on Higgs decays to photons, the top quark that decays through FCNC can be fully reconstructed with minimal ambiguities. In absence of an excess in the diphoton invariant mass peak, 95% CL upper limits are set on the corresponding branching ratio $BR(t \to Hq) < 17 \times 10^{-4}$. All the above constraints on branching ratios can be translated to constraints on the such anomalous coupling, as shown in the respective papers.

7. The Top Quark and the Newly Discovered Boson

The discovery in 2012 of a new heavy boson [3,4] at the LHC represents an historical milestone in our understanding of nature. To examine whether this particle truly plays the role of the electroweak symmetry breaking (ESB) mechanism agent, it is crucial to study its coupling to all known particles. The Higgs boson has been discovered mainly through its direct coupling with the other known heavy bosons (W/Z) and only indirectly with fermions through loops. A multitude of new physics scenarios could be hiding in those particle loops. The Yukawa structure of the coupling of the Higgs to fermions is largely unexplored: as of today, we only have mild suggestions of the Higgs coupling to b quarks [33] and to τ leptons. [34] Studying the direct coupling of Higgs to fermions through the associated production of top quarks together with the new heavy boson could lead to the direct measurement of the top Yukawa coupling, and thus to possible deviations from the SM predictions in the top-Higgs interaction as foreseen by natural new physics scenarios. In fact, the SM theory appears unnaturaly fine-tuned. This unnatural fine-tuning would be removed by the existence of exotic partners of the top quark that could be either of fermionic nature, such as those predicted by Composite Higgs and Little Higgs, or of bosonic nature, such as the supersymmetric top partner. In particular, both Composite/little Higgs and SUSY would predict final states that would largely overlap with the ones needed for studying top and Higgs production in the SM, i.e. one or more top quarks, in addition to one or more Higgs bosons.

The search for $t\bar{t}H$ production is experimentally very challenging for a multitude of reasons. First, the theoretical NNLO prediction for this process at 8 TeV collisions amounts to only 130fb, [35] i.e. approximately one for every $4 \cdot 10^{11}$ LHC collisions; only about one every 200 Higgs events is produced in association with top quarks. Also, both top quarks and Higgs bosons are extremely short-lived. These searches assume that the Higgs boson decays (and its decay rates) are the ones predicted by the SM. It is thus expected to detect two-to-four particles coming from the Higgs decay, and six particles coming from the top-antitop quark system decay, leading to some of the busiest events under study in high energy physics. In order to reach sensitivity to such a small and complex signal, the search for $t\bar{t}H$ production mandates a careful choice of only a handful of the allowed signatures. Backgrounds are generally very large as the signature is dominated by the presence of the top-antitop pair, and the cross section for $t\bar{t} + X$ production is three orders of magnitude larger.

Both ATLAS and CMS analyzed $t\bar{t}H$ events where $H \to b\bar{b}$; the analyses proceed similarly: first, the sample is split into events with exactly one high p_T lepton — the semileptonic top-antitop decays — or two high p_T leptons — the dileptonic top final state. Both samples are further split into multiple subsamples characterized by different jet multiplicities. The Higgs boson mass is reconstructed within combinatorial ambiguities whenever a sufficient number of objects are identified. The signal is discriminated from the dominant $t\bar{t}$+jets background by mean of BDTs that exploit the peculiar kinematical and topological characteristics of the top-antitop-Higgs production.[36,37] CMS also studied events where the Higgs decays to taus; here BDTs are used again, where the discrimination power comes from combining observables associated to the quality of the reconstructed taus. For all of the above channels, the sensitivity to the signal is estimated by a likelihood fit over the binned BDT distributions.[36]

The excellent resolution of the CMS and ATLAS electromagnetic calorimeters allows the reconstruction of a very narrow Higgs peak. To increase sensitivity and maximize acceptance, all $t\bar{t}$ system decays are collected, split into events with at least five high p_T jets (hadronic channel), and events with at least two high-p_T jets and one high-p_T lepton (leptonic channel). In both instances at least one jet is required to be b-tagged. All jets are required to be sufficiently energetic as to suppress the contamination of the dominant Higgs production process. The contribution of single top plus Higgs production[38] has been estimated to be small. The sensitivity to the signal is estimated by a likelihood fit over the diphoton invariant mass distribution.[39,40]

The most precise current result is obtained by the combination of the several CMS results. The combination of all the above channels is accomplished using the same techniques employed in the global CMS Higgs combination.[3] Figure 1, left side, shows the expected and observed limit from the individual analyses and from their combination, for an assumed Higgs boson mass of 125 GeV. Combining these analyses improves the expected limit by 34% compared to the best individual result for a Higgs mass of 125 GeV. Figure 3, right side, shows the best fit signal strength μ for each channel contributing to the combination. For the combination of all $t\bar{t}H$ channels, the median expected limit for $m_H = 125$ GeV is $2.7 \times \sigma_{SM}$ while the observed limit is $3.4 \times \sigma_{SM}$, and the best-fit value for μ is 0.74 ± 1.34 (68% CL).

The sensitivity of LHC experiments to $t\bar{t}H$ production has quickly increased over the past few months; it is thus possible that the first direct determination of the Yukawa coupling lies just around the corner.

8. Conclusions

An overview of recent results on measurements of top quark properties performed by the CDF, D0, ATLAS and CMS collaborations has been shown at this conference. The Tevatron experiments almost concluded providing their final results using the full Tevatron dataset, while LHC collaborations showed a number of impressive results that take only partly advantage of the much larger integrated dataset. It will thus be exciting to hear about the new top quark results in the next ~ 18 months that separate us from the reopening of the LHC operations. A particular emphasis has been placed on the special relation between the top quark and the Higgs boson,

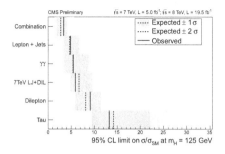 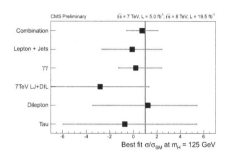

Fig. 3. The observed and expected 95% CL upper limits on $t\bar{t}H$ production (left) and best-fit values (right) of $\mu = \sigma/\sigma_{SM}$ for the lepton + jets (LJ), dilepton (DIL), $\tau\tau$ and $\gamma\gamma$ channels separately from the 2012 8 TeV dataset, the combination of the lepton + jets and dilepton channels from the 2011 7 TeV dataset, and the combination of all of the channels, for $m_H = 125$ GeV.

both in the context of the Standard Model and in complete or effective theories predicting additional Higgs bosons or additional interactions between the neutral Higgs boson and the top quark. The LHC will bring a top quark dataset approximately 1000× larger during the next years, that will allow far deeper understanding of the top quark properties.

Acknowledgments

The author wishes to thank the CDF, D0, ATLAS and CMS collaborations for providing the excellent results discussed in this paper, and Reinhard Schwienhorst for the useful discussions. The authors would like also to thank the conference organizers for the excellent organization and the pleasant atmosphere.

References

1. F. Abe *et al.* [CDF Collaboration], *Phys. Rev. Lett.* **74**, (1995) 2626.
2. S. Abachi *et al.* [D0 Collaboration], *Phys. Rev. Lett.* **74**, 2632 (1995).
3. S. Chatrchyan *et al.* [CMS Collaboration], *Phys. Lett. B* **716**, 30 (2012) arXiv:1207.7235.
4. G. Aad *et al.* [ATLAS Collaboration], *Phys. Lett. B* **716**, 1 (2012) arXiv:1207.7214.
5. P. Baernreuther, M. Czakon and A. Mitov, *Phys. Rev. Lett.* **109**, 132001 (2012) arXiv:1204.5201.
6. R. Schwienhorst, these proceedings.
7. M. Baak, M. Goebel, J. Haller, A. Hoecker, D. Kennedy, R. Kogler, K. Moenig and M. Schott *et al.*, *Eur. Phys. J. C* **72**, 2205 (2012) arXiv:1209.2716.
8. G. Degrassi, S. Di Vita, J. Elias-Miro, J. R. Espinosa, G. F. Giudice, G. Isidori and A. Strumia, *JHEP* **1208**, 098 (2012) arXiv:1205.6497.
9. G. Aad *et al.* [ATLAS Collaboration], ATLAS-CONF-2013-046.
10. Tevatron Electroweak Working Group, CDF and D0 Collaborations, arXiv:1305.3929.
11. S. Chatrchyan *et al.* [CMS Collaboration], arXiv:1307.4617.
12. T. A. Aaltonen *et al.* [CDF Collaboration], *Phys. Rev. Lett.* [*Phys. Rev. Lett.* **111**, 202001 (2013)] [arXiv:1308.4050 [hep-ex]].
13. S. Chatrchyan *et al.* [CMS Collaboration], CMS-PAS-TOP-12-030.

14. S. Chatrchyan *et al.* [CMS Collaboration], arXiv:1307.1907.
15. S. Chatrchyan *et al.* [CMS Collaboration], CMS PAS-12-031.
16. S. Chatrchyan *et al.* [CMS Collaboration], CMS-PAS-B2G-12-023.
17. T. Aaltonen *et al.* [CDF Collaboration], *Phys. Rev. D* **83**, 112003 (2011) arXiv:1101.0034.
18. T. A. Aaltonen *et al.* [CDF Collaboration], *Phys. Rev. D* **88**, 072003 (2013) arXiv:1308.1120.
19. V. M. Abazov *et al.* [D0 Collaboration], D0 CONF-6394.
20. V. M. Abazov *et al.* [D0 Collaboration], arXiv:1308.6690.
21. S. Chatrchyan *et al.* [CMS Collaboration], CMS-PAS-TOP-12-004.
22. S. Chatrchyan *et al.* [CMS Collaboration], Phys. Lett. B **717**, 129 (2012) [arXiv:1207.0065 [hep-ex]].
23. G. Aad *et al.* [ATLAS Collaboration], ATLAS-CONF-2013-078.
24. G. Aad *et al.* [ATLAS Collaboration], ATLAS-CONF-2012-057.
25. G. Aad *et al.* [ATLAS Collaboration], *Eur. Phys. J. C* **73**, 2465 (2013) arXiv:1302.3694.
26. S. Chatrchyan *et al.* [CMS Collaboration], CMS-PAS-HIG-12-052.
27. G. Aad *et al.* [ATLAS Collaboration], *JHEP* **1206**, 039 (2012) arXiv:1204.2760.
28. V. M. Abazov *et al.* [D0 Collaboration], *Phys. Rev. Lett.* **102**, 191802 (2009) [arXiv:0807.0859 [hep-ex]].
29. G. Aad *et al.* [ATLAS Collaboration], ATLAS-CONF-2013-063.
30. S. Chatrchyan *et al.* [CMS Collaboration], CMS PAS TOP-12-037.
31. S. Chatrchyan *et al.* [CMS Collaboration], CMS-PAS-TOP-12-021.
32. G. Aad *et al.* [ATLAS Collaboration], ATLAS-CONF-2013-081.
33. T. Aaltonen *et al.* [CDF and D0 Collaborations], *Phys. Rev. Lett.* **109**, 071804 (2012) arXiv:1207.6436.
34. S. Chatrchyan *et al.* [CMS Collaboration], CMS-HIG-13-004.
35. S. Dawson, C. Jackson, L. H. Orr, L. Reina and D. Wackeroth, *Phys. Rev. D* **68**, 034022 (2003) hep-ph/0305087.
36. S. Chatrchyan *et al.* [CMS Collaboration], CMS-HIG-13-019.
37. G. Aad *et al.* [ATLAS Collaboration], ATLAS-CONF-2012-135.
38. S. Biswas, E. Gabrielli, F. Margaroli and B. Mele, JHEP **07**, 073 (2013) arXiv:1304.1822.
39. S. Chatrchyan *et al.* [CMS Collaboration], CMS-HIG-13-015.
40. G. Aad *et al.* [ATLAS Collaboration], ATLAS-CONF-2013-080.

Physics in Collision (PIC 2013)
International Journal of Modern Physics: Conference Series
Vol. 31 (2014) 1460279 (10 pages)
© The Author
DOI: 10.1142/S2010194514602798

World Scientific
www.worldscientific.com

Diboson production at LHC and Tevatron

Jian Wang
(On Behalf of the ATLAS, CMS, CDF, DØ Collaborations)
Universite Libre de Bruxelles
Brussels, 1050, Belgium
Jian.Wang@cern.ch

Published 15 May 2014

This is a report at the conference Physics In Collision 2013. The experimental results on physics of diboson production are reviewed. The measurements use pp collision at the LHC with center-of-mass energy $\sqrt{s} = 7$ and 8 TeV, and $p\bar{p}$ collision at the Tevatron with $\sqrt{s} = 1.96$ TeV. These include measurements of Wγ, Zγ, WW, WZ and ZZ production. The results are compared with Standard Model predictions, and are interpreted in terms of constraints on charged and neutral anomalous triple gauge couplings.

1. Introduction

The study of diboson production provides an important test of the Standard Model (SM) of particle physics at TeV energy scale. Especially, it is sensitive to the self-interactions among vector bosons via triple gauge couplings. Any significant deviation of the production cross section or kinematic distributions from the SM predictions gives an indication of new physics. In addition, non-resonant diboson production measurements are important to a precise estimation of irreducible backgrounds for the Higgs study.

The experimental results reviewed here, use $p\bar{p}$ collision at the Tevatron at a center-of-mass energy $\sqrt{s} = 1.96$ TeV, with an integrated luminosity up to about $10\,fb^{-1}$, and pp collision at $\sqrt{s} = 7$ and 8 TeV at the LHC, with integrated luminosities up to about 5 fb^{-1} and 20 fb^{-1} respectively. Emphasis is placed on the latest results released in the past year.

2. Cross Section Measurement

2.1. *Wγ and Zγ*

The Wγ and Zγ productions have been studied in $W\gamma \to l\nu\gamma$ and $Z\gamma \to ll\gamma$ decay channels at $\sqrt{s} = 7$ TeV.[1,2] In the Wγ measurement, events are selected by requiring an isolated electron or muon, and missing transverse energy, E_T^{miss}, from the undetected neutrino, in addition to an isolated photon. The dominant background comes from W+jets, where a jet is misidentified as a photon. In the Zγ measurement, events are selected by requiring a same flavor, opposite sign electron or muon pair with an invariant mass close to the Z boson mass, in addition to an isolated photon. The photon is required to be separated from the lepton to suppress the contribution from final state radiation photons.

CMS has measured the cross sections for γ $E_T > 15$ GeV and $m_{ll} > 50$ GeV. The measured Wγ cross section times $W \to l\nu$ branching ratio is 37.0 ± 0.8 (*stat.*) $\pm$ 4.0 (*syst.*) ± 0.8 (*lumi.*)*pb*. The measured Zγ cross section times $Z \to ll$ branching ratio is 5.33 ± 0.08 (*stat.*) ± 0.25 (*syst.*) ± 0.12 (*lumi.*)*pb*. The results are consistent with the SM predictions. The differential cross sections measured by ATLAS, comparing with theoretical predictions, are shown in Fig. 1. In general, the NLO parton-level MC, MCFM,[3,4] agrees with the exclusive (Njet = 0) production cross section measurements, while LO MC (ALPGEN[5] or SHERPA[6]) with multiple parton emission reproduce the γ E_T spectrum.

The $Z\gamma \to \nu\nu\gamma$ decay channel has also been measured.[1,7] The backgrounds are jets misidentified as photons, and instrumental sources such as beam-gas interactions. Very tight photon E_T and E_T^{miss} cuts are applied to suppress such backgrounds. Detector timing is also used to reduce instrumental backgrounds. Both ATLAS and CMS results are in agreement with the SM predictions.

2.2. *WW*

The WW production cross section has been measured in the $WW \to l\nu l'\nu'$ final state.[8–10] Events are selected by requiring two opposite charged isolated leptons,

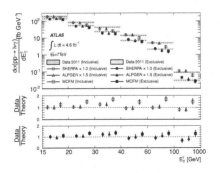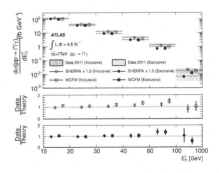

Fig. 1.   ATLAS measured γ E_T differential cross sections of the $l\nu\gamma$ process (left) and the $ll\gamma$ process (right), in the inclusive and exclusive (Njet = 0) extended fiducial regions, at $\sqrt{s} = 7$ TeV

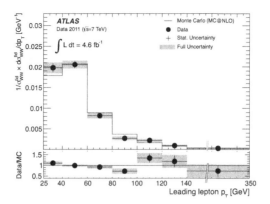

Fig. 2. ATLAS measured normalized differential WW fiducial cross section as a function of the leading lepton p_T compared to the SM prediciton.

electron or muon, accompanied by significant E_T^{miss}. The Z+jets background in the ee and $\mu\mu$ channels is suppressed by a cut on large E_T^{miss} and a Z veto. To minimise the contribution from top-quark background, events containing jets are rejected (jet veto). This leads to a significant theoretical uncertainty in the jet veto efficiency.

The WW production cross sections measured by ATLAS and CMS at $\sqrt{s} = 7$ TeV are 51.9 ± 2.0 (stat.) ± 3.9 (syst.) ± 2.0 (lumi.)pb and 52.4 ± 2.0 (stat.) ± 4.5 (syst.) ± 1.2 (lumi.)pb respectively. The ATLAS measurement of differential cross section as a fuction of the leading lepton p_T is shown in Fig. 2. CMS has also performed a first measurement of WW production at $\sqrt{s} = 8$ TeV. The cross sections is determined to be 69.9 ± 2.8 (stat.) ± 5.6 (syst.) ± 3.1 (lumi.)pb. These measured cross sections are slightly higher than but still compatible with the SM predictions. Systematic uncertainties dominate the total uncertainty.

2.3. WZ

The WZ production cross sections have been measured in the $WZ \to l\nu2l'$ decay channel.[11–13] This final state has very low background after requiring exactly three isolated leptons (electron or muon), a pair of which is of same-flavor and has an invariant mass close to the mass of Z boson, in addition to significant E_T^{miss} from W decay.

In ATLAS measurement at $\sqrt{s} = 7$ TeV, totally 317 candidates are observed with a background expectation of 68 events. In the measurement at $\sqrt{s} = 8$ TeV, 1094 candidate events are observed in total, with a background expectation of 227 events. The WZ cross sections are measured to be $19.0^{+1.4}_{-1.3}$ (stat.)±0.9 (syst.)±0.4 (lumi.)pb at $\sqrt{s} = 7$ TeV, and $20.3^{+0.8}_{-0.7}$ (stat.)$^{+1.2}_{-1.1}$ (syst.)$^{+0.7}_{-0.6}$ (lumi.)pb at $\sqrt{s} = 8$ TeV, for the Z boson mass in the range of 66 to 116 GeV.

CMS has measured the WZ production cross section for the Z boson mass between 71 to 111 GeV. The cross sections are determined to be 20.76 ± 1.32 (stat.)$\pm$

1.13 (syst.)$\pm$0.46 (lumi.)pb at $\sqrt{s} = 7$ TeV, and 24.61$\pm$0.76 (stat.)$\pm$1.13 (syst.)$\pm$ 1.08 (lumi.)pb at $\sqrt{s} = 8$ TeV. Since the LHC is a pp collider, the W^+Z and W^-Z cross sections are not equal. The ratios of production cross sections for W^+Z and W^-Z have also been measured. They are 1.94 $\pm$ 0.25 (stat.) $\pm$ 0.04 (syst.) and 1.81$\pm$0.12 (stat.)$\pm$0.03 (syst.) at $\sqrt{s} = 7$ TeV and 8 TeV respectively, in agreement with the SM predictions.

The total production cross section of WV (WW+WZ) has also been studied in the $WV \rightarrow l\nu qq$ final state at $\sqrt{s} = 7$ TeV.[14,15] The resolution of reconstructed di-jet mass is about 10 GeV, which cannot distinguish W from Z here. This channel has higher branching ratio with respect to the fully leptonic decay mode, at the cost of larger W/Z+jets background. Events are selected by requiring an isolated electron or muon, E_T^{miss} and exactly two high-p_T jets. Signal is extracted by fitting di-jet mass distribution. ATLAS measures a cross section of 72 $\pm$ 9 (stat.) $\pm$ 15 (syst.) $\pm$ 13 (MC stat.)pb and CMS measures a cross section of 68.9$\pm$8.7 (stat.)$\pm$9.7 (syst.)$\pm$ 1.5 (lumi.)pb. Both measurements agree with the SM predictions.

2.4. *ZZ*

The ZZ production cross sections have been measured using the high purity four-lepton ($ZZ \rightarrow 2l2l'$) decay channel..[16–19] Even though the branching ratio to four-lepton final state is small, this process is really clean, with negligible amount of background. Events are selected by requiring two pairs of electrons or muons, with opposite-charge and same-flavour. The invariant mass of each pair is compatible with the Z boson mass. The challenge in the four-lepton analysis is the optimization for lepton efficiencies, especially for low p_T leptons. Some leptons might fall outside the acceptance of the detector while some others may fail the criteria used to select a lepton. With four chances to miss a lepton, even small inefficiencies will add up.

In ATLAS measurement at $\sqrt{s} = 7$ TeV, $ZZ \rightarrow 2l2\nu$ decay mode has also been measured, by applying a tight cut on a E_T^{miss}-related variable in order to suppress the dominant Z+jets background. $2l2l'$ and $2l2\nu$ results are then combined, assuming the SM branching ratios. The ZZ production cross section is determined to be 6.7 $\pm$ 0.7 (stat.)$^{+0.4}_{-0.3}$ (syst.) $\pm$ 0.3 (lumi.)pb at $\sqrt{s} = 7$ TeV, where both Z bosons in the mass range of 66 to 116 GeV. In the measurement at $\sqrt{s} = 8$ TeV, only four-lepton channel is used. Totally 305 candidate events are observed with a background expectation of 20.4. The SM expectation for the number of signal event is 292.5. The ZZ production cross section at $\sqrt{s} = 8$ TeV is measured to be $7.1^{+0.5}_{-0.4}$ (stat.) $\pm$ 0.3 (syst.) $\pm$ 0.2 (lumi.)pb.

The CMS $ZZ \rightarrow 2l2l'$ measurements include $ZZ \rightarrow 2l2\tau$ decay mode as well. The measured cross sections are $6.24^{+0.86}_{-0.80}$ (stat.)$^{+0.41}_{-0.32}$ (syst.)$\pm$0.14 (lumi.)pb at $\sqrt{s} = 7$ TeV and $7.7^{+0.5}_{-0.5}$ (stat.)$^{+0.5}_{-0.4}$ (syst.) $\pm$ 0.4 (theo.) $\pm$0.3 (lumi.)pb at $\sqrt{s} = 8$ TeV, for both Z bosons produced in the mass region of 60 to 120 GeV. Differential cross sections are also measured (as shown in Fig. 3) and well described by the theoretical predictions.

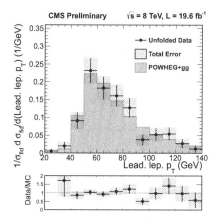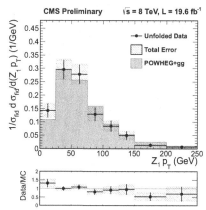

Fig. 3. Differential cross section normalized to the fiducial cross section in the CMS $ZZ \to 2l2l'$ measurement at $\sqrt{s} = 8$ TeV. The differential cross sections in bins of p_T are presented for the leading lepton (left) and the higher-p_T Z (right).

CDF and DØ have measured ZZ cross section in $p\bar{p}$ collision at $\sqrt{s} = 1.96$ TeV.[20] CDF has studied ZZ production through $2l2l'$ and $2l2\nu$ final states, using a dataset corresponding to $9.7 fb^{-1}$ integrated luminosity. The combined measured ZZ cross section is $1.04^{+0.20}_{-0.24}$ (stat.)$^{+0.15}_{-0.08}$ (syst.). DØ has measured the four-lepton channel, and combined it with a previous study of $2l2\nu$ channel, resulting a ZZ cross section of $1.32^{+0.29}_{-0.25}$ (stat.) ± 0.12 (syst.) ± 0.04 (lumi.).

A summary of WW, WZ and ZZ production cross section measurements are listed and compared with the relevant theoretical predictions in Table 1. The theoretical predictions are computed using MCFM to QCD NLO. Please note the theoretical predictions are different for same process measured by different experiments, because the phase spaces are not exactly same defined.

3. Limits on Anomalous Triple Gauge Couplings

The SM describes exactly how vector bosons couple with each other, and diboson productions are sensitive to these couplings. Even if new physics is at very high energy scale, beyond the reach of current colliders (which means direct pair production of new particle is impossible), it could still have indirect effect on triple gauge couplings through virtual corrections. Anomalous triple gauge couplings (aTGC) could be modeled by adding terms to the SM Lagrangian, using a set of parameters, listed in Table 2. All these parameters are equal to zero in the SM. This is a common approach to parameterize low energy effects from high energy scale new physics, which allows for experimental results to be interpreted as model independent constrains on aTGC.

The presence of aTGC would increase diboson production cross sections, in particular at high mass and high p_T regions. Experimental measurements made by

Table 1. Summary of diboson production cross section measurements.

Experiment	$\sqrt{s}$ (TeV)	luminosity (fb^{-1})	Measured cross section (pb)	Theoretical prediction (pb)
WW				
ATLAS	7	4.6	51.9 ± 2.0 (stat.) ± 3.9 (syst.) ± 2.0 (lumi.)	$44.7^{+2.1}_{-1.9}$
CMS	7	4.9	52.4 ± 2.0 (stat.) ± 4.5 (syst.) ± 1.2 (lumi.)	47.0 ± 2.0
CMS	8	3.5	69.9 ± 2.8 (stat.) ± 5.6 (syst.) ± 3.1 (lumi.)	$57.3^{+2.3}_{-1.6}$
WZ				
ATLAS	7	4.6	$19.0^{+1.4}_{-1.3}$ (stat.) ± 0.9 (syst.) ± 0.4 (lumi.)	$17.6^{+1.1}_{-1.0}$
ATLAS	8	13	$20.3^{+0.8}_{-0.7}$ (stat.) $^{+1.2}_{-1.1}$ (syst.) $^{+0.7}_{-0.6}$ (lumi.)	20.3 ± 0.8
CMS	7	4.9	20.76 ± 1.32 (stat.) ± 1.13 (syst.) ± 0.46 (lumi.)	$17.8^{+0.7}_{-0.5}$
CMS	8	19.6	24.61 ± 0.76 (stat.) ± 1.13 (syst.) ± 1.08 (lumi.)	$21.91^{+1.17}_{-0.88}$
WV (V = W or Z)				
ATLAS	7	4.7	72 ± 9 (stat.) ± 15 (syst.) ± 13 (MC stat.)	63.4 ± 2.6
CMS	7	5	68.9 ± 8.7 (stat.) ± 9.7 (syst.) ± 1.5 (lumi.)	65.6 ± 2.2
ZZ				
ATLAS	7	4.6	6.7 ± 0.7 (stat.) $^{+0.4}_{-0.3}$ (syst.) ± 0.3 (lumi.)	$5.89^{+0.22}_{-0.18}$
ATLAS	8	20	$7.1^{+0.5}_{-0.4}$ (stat.) ± 0.3 (syst.) ± 0.2 (lumi.)	$7.2^{+0.3}_{-0.2}$
CMS	7	5.0	$6.24^{+0.86}_{-0.80}$ (stat.) $^{+0.41}_{-0.32}$ (syst.) ± 0.14 (lumi.)	6.3 ± 0.4
CMS	8	19.6	$7.7^{+0.5}_{-0.5}$ (stat.) $^{+0.5}_{-0.4}$ (syst.) ± 0.4 (theo.) ± 0.3 (lumi.)	7.7 ± 0.6
CDF	1.96($p\bar{p}$)	9.7	$1.04^{+0.20}_{-0.24}$ (stat.) $^{+0.15}_{-0.08}$ (syst.)	1.4 ± 0.1
DØ	1.96($p\bar{p}$)	9.8	$1.32^{+0.29}_{-0.25}$ (stat.) ± 0.12 (syst.) ± 0.04 (lumi.)	1.43 ± 0.10

Table 2. Parameterization of aTGC.

Coupling	Parameters	Channels
WWγ	$\Delta\kappa_\gamma, \lambda_\gamma$	WW, Wγ
WWZ	$\Delta g_1^Z, \Delta\kappa_Z, \lambda_Z$	WW, WZ
ZZγ	h_3^Z, h_4^Z	Zγ
Z$\gamma\gamma$	h_3^γ, h_4^γ	Zγ
ZZZ	f_4^Z, f_5^Z	ZZ
ZγZ	f_4^γ, f_5^γ	ZZ

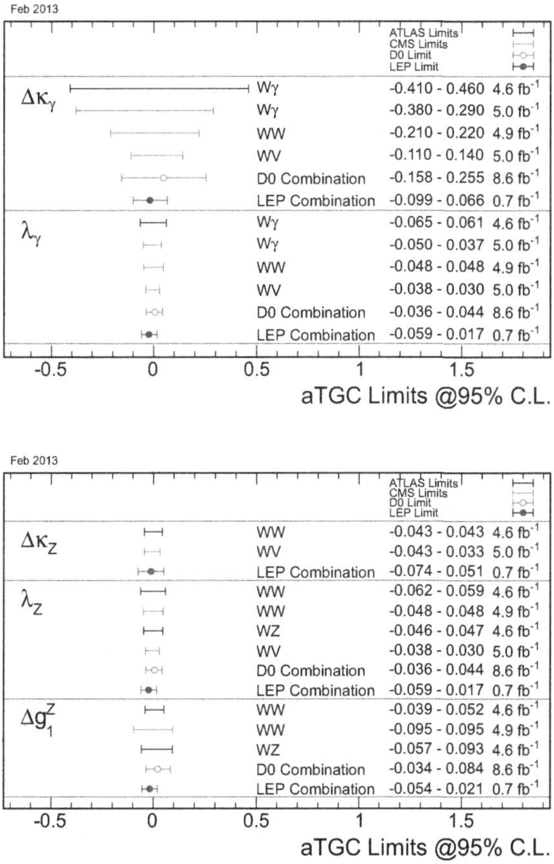

Fig. 4. Limits at 95% C.L. on WWγ (top) and WWZ (bottom) aTGC.

the ATLAS, CMS, CDF and DØ collaborations have found no excess over the SM predictions, leading to limits on charged and neutral aTGC, as shown in Fig. 4 and Fig. 5 respectively, together with previous LEP results. For charged aTGC, LEP results remain competitive while the sensitivity from LHC is approaching. For neutral aTGC, LHC results dominate.

4. First Studies on Quartic Gauge Couplings

CMS has studied the exclusive two-photon production of WW using 5.05 fb^{-1} of data at $\sqrt{s} = 7$ TeV.[21] Events are selected by requiring a $\mu^{\pm}e^{\mp}$ vertex with no associated charged tracks, and $p_T(\mu^{\pm}e^{\mp}) > 30$ GeV. Two events are observed in the data, compared to a SM expectation of 2.2 ± 0.5 signal events with 0.84 ± 0.13 background. The significance of the signal is 1.1σ, with a 95% C.L. upper limit on the SM cross section of 8.4 fb. DØ has studied the exclusive two-photon production

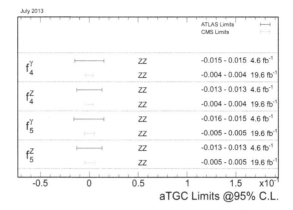

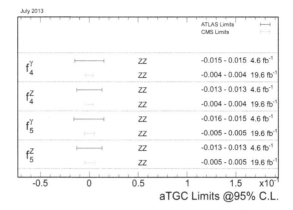

Fig. 5. Limits at 95% C.L. on $Z\gamma Z$, ZZZ (top) and $ZZ\gamma$, $ZZ\gamma$ (bottom) aTGC.

Fig. 6. Limits at 95% C.L. on $WW\gamma\gamma$ aQGC.

of WW in events with an electron, a positron and E_T^{miss}.[22] No excess above the background expectation has been found.

A study of the WVγ, three vector boson production, has also been performed by CMS, using 19.3 fb^{-1} data from pp collision at $\sqrt{s} = 8$ TeV.[23] The analysis selects events containing a W boson decaying to electron or muon, a second V (W or Z) boson decaying to two jets, and a photon. The number of observed events in data is 322, while the estimated background yield is 341.5. This is consistent with the SM predictions, and corresponds to an upper limit of 241 fb at 95% C.L. for WVγ production with photon $p_T > 10$ GeV.

The results of the above analyses are studied for deviations from the SM, and used to constrain anomalous quartic gauge couplings (aQGC). The limits on WW$\gamma\gamma$ aQGC set by these measurements, together with previous results from LEP, are shown in Fig. 6. CMS results have significantly surpassed limits from LEP and DØ.

5. Summary

Latest results of diboson production measurements by ATLAS, CMS, CDF and DØ experiments are reviewed. The measured cross sections are typically consistent with the SM predictions. The results are used to constrain new physics, by setting limits on anomalous triple gauge couplings. First studies on quartic gauge couplings have started as well.

References

1. G. Aad *et al.* [ATLAS Collaboration], *Phys. Rev. D* **87**, 112003 (2013) [arXiv:1302.1283 [hep-ex]].
2. S. Chatrchyan *et al.* [CMS Collaboration], arXiv:1308.6832 [hep-ex].
3. J. M. Campbell and R. K. Ellis, *Nucl. Phys. Proc. Suppl.* **205-206**, 10 (2010) [arXiv:1007.3492 [hep-ph]].
4. J. M. Campbell, R. K. Ellis and C. Williams, *JHEP* **1107**, 018 (2011) [arXiv:1105.0020 [hep-ph]].
5. M. L. Mangano, M. Moretti, F. Piccinini, R. Pittau and A. D. Polosa, *JHEP* **0307**, 001 (2003) [hep-ph/0206293].
6. T. Gleisberg, S. .Hoeche, F. Krauss, M. Schonherr, S. Schumann, F. Siegert and J. Winter, *JHEP* **0902**, 007 (2009) [arXiv:0811.4622 [hep-ph]].
7. S. Chatrchyan *et al.* [CMS Collaboration], *JHEP* **1310**, 164 (2013) [arXiv:1309.1117 [hep-ex]].
8. G. Aad *et al.* [ATLAS Collaboration], *Phys. Rev. D* **87**, 112001 (2013) [arXiv:1210.2979 [hep-ex]].
9. S. Chatrchyan *et al.* [CMS Collaboration], *Eur. Phys. J. C* **73**, 2610 (2013) [arXiv:1306.1126 [hep-ex]].
10. S. Chatrchyan *et al.* [CMS Collaboration], *Phys. Lett. B* **721**, 190 (2013) [arXiv:1301.4698 [hep-ex]].
11. G. Aad *et al.* [ATLAS Collaboration], *Eur. Phys. J. C* **72**, 2173 (2012) [arXiv:1208.1390 [hep-ex]].
12. [ATLAS Collaboration], ATLAS-CONF-2013-021.
13. CMS Collaboration [CMS Collaboration], CMS-PAS-SMP-12-006.

14. [ATLAS Collaboration], ATLAS-CONF-2012-157.
15. S. Chatrchyan *et al.* [CMS Collaboration], *Eur. Phys. J. C* **73**, 2283 (2013) [arXiv:1210.7544 [hep-ex]].
16. G. Aad *et al.* [ATLAS Collaboration], *JHEP* **1303**, 128 (2013) [arXiv:1211.6096 [hep-ex]].
17. [ATLAS Collaboration], ATLAS-CONF-2013-020.
18. S. Chatrchyan *et al.* [CMS Collaboration], *JHEP* **1301**, 063 (2013) [arXiv:1211.4890 [hep-ex]].
19. CMS Collaboration [CMS Collaboration], CMS-PAS-SMP-13-005.
20. V. M. Abazov *et al.* [D0 Collaboration], *Phys. Rev. D* **88**, 032008 (2013) [arXiv:1304.5422 [hep-ex]].
21. S. Chatrchyan *et al.* [CMS Collaboration], *JHEP* **1307**, 116 (2013) [arXiv:1305.5596 [hep-ex]].
22. V. M. Abazov *et al.* [D0 Collaboration], *Phys. Rev. D* **88**, 012005 (2013) [arXiv:1305.1258 [hep-ex]].
23. CMS Collaboration [CMS Collaboration], CMS-PAS-SMP-13-009.

Physics in Collision (PIC 2013)
International Journal of Modern Physics: Conference Series
Vol. 31 (2014) 1460280 (13 pages)
© The Author
DOI: 10.1142/S2010194514602804

World Scientific
www.worldscientific.com

Proton structure functions at HERA

Iris Abt
(On Behalf of the H1 and ZEUS Collaborations)
*Max-Planck Institut für Physik, Föhringer Ring 6,
80805 München, Germany*
isa@mpp.mpg.de

Published 15 May 2014

The "proton structure" is a wide field. Discussed are predominantly the precision measurements of the proton structure functions at HERA and some of their implications for the LHC measurements. In addition, a discussion of what a proton structure function represents is provided. Finally, a connection to nuclear physics is attempted. This contribution is an updated reprint of a contribution to "Deep Inelastic Scattering 2012".[1]

Keywords: Proton; deep inelastic scattering; structure functions; QCD.

PACS Numbers: 13.10+q, 13.40.−f, 13.60.Hb, 24+85.+p

1. Introduction

The proton is quite a fantastic particle. If free, it doesn't decay on any timescale people have been able to explore. It has an immense ability to heal itself, demonstrated in the high rate of diffraction even for interactions with large momentum transfer. What the author really knows, is actually quite limited. The charge was determined to be "+1", the mass was measured to be 1.6×10^{-27} kg and the spin is $1/2$. Spin will not be discussed in this contribution; there are others who will write about it.

If the proton is probed with enough energy, three valance quarks are revealed. If it is probed with even more energy, the QCD affliction of the proton, i.e. the glue and the sea become visible. QCD is always used when the results of one measurement are used to make predictions for another. And one part of this *ansatz* are parton distributions functions, PDFs, of the proton. They are a very successful tool. However, their shape is entirely heuristic; QCD cannot predict them from first principle.

Protons are a vital part of nuclei. Together with neutrons, they provide the rich world of elements that we so dearly love. However, in this environment QCD is generally not the theory of choice to predict what happens. Inside a nucleus, a proton can decay because the energy to become a neutron comes from the nucleus. Nuclei are not spheres; the proton itself is often depicted as one, but that is also too simplistic.

2. Proton Structure Functions at HERA

2.1. *Deep inelastic scattering*

At HERA, the structure of the proton was probed with electrons and positrons. Figure 1 illustrates deep inelastic scattering, DIS, which is generally used to determine the proton structure. The castle is destroyed to learn about its inhabitants, i.e. the quarks and gluons, and their habits. It is interesting to note that the castle rebuilds itself in about 20 % of the interactions. However, this, i.e. diffraction, is not the subject of this contribution.

More commonly used to illustrate DIS is the Feynman diagram depicted in Fig. 2. It should, however, be noted that this Feynman diagram describes the interaction in lowest order while Fig. 1 includes all orders. Anyhow, the process can be described in terms of the kinematic variables x, y, and Q^2. The variable Q^2 is defined as $Q^2 = -q^2 = -(k - k')^2$, where k and k' are the four-momenta of the incoming and scattered lepton, respectively. Bjorken x is defined as $x = Q^2/2P \cdot q$, where P is the four-momentum of the incoming proton. The fraction of the lepton energy transfered to the proton in the rest frame of the proton is given by $y = P \cdot q/P \cdot k = Q^2/sx$, where s is the square of the lepton-proton centre-of-mass energy.

In neutral current, NC, events, most of the information can be deduced from the deflected probe. However, quite often the hadronic system is also used in order

Fig. 1.   Illustration of deep inelastic lepton proton scattering. At HERA the lepton was an electron or positron. In this demonstration, a charged current interaction destroys the proton.

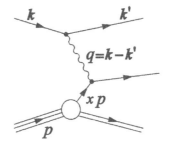

Fig. 2.   Feynman diagram of deep inelastic lepton proton scattering

to optimise the uncertainties. In charged current, CC, events, the outgoing neutrino evades detection and the hadronic system is all we have. This makes things are a lot more difficult.

2.2. *Cross sections*

All our understanding of DIS is connected to the paradigm of factorisation.

In lowest-order QCD, three processes contribute to the NC DIS cross section, namely the Born ($V^*q \to q$, with $V^* = \gamma^*, Z^*$), the boson-gluon-fusion ($V^*g \to \bar{q}q$) and QCD-Compton-scattering ($V^*q \to qg$) processes. The cross section for the production of an observed hadron, H, in the final state in DIS can be expressed in QCD, using the factorisation theorem, as

$$\sigma(ep \to e + H + X) = \sum_{j,j'=q,\bar{q},g} f_{j/p}(x,Q) \otimes \hat{\sigma}_{jj'}(x,Q,z) \otimes F_{H/j'}(z,Q),$$

where the sum runs over all possible initial (final)-state partons j (j'), $f_{j/p}$ are the proton PDFs, which give the probability of finding a parton j with momentum fraction x in the proton, $\hat{\sigma}_{jj'}$ is the partonic cross section, which includes the matrix elements for the three processes mentioned above, and $F_{H/j'}$ are the fragmentation functions, which give the probability that a hadron H with momentum fraction z originates from parton j'.

This contribution concentrates on inclusive measurements, so that the fragmentation "only" shows up in the calculation of the systematic uncertainties connected to the acceptance and the efficiency to reconstruct an event. The cross sections are measured and their description in QCD is used to extract the PDFs which in turn are used to make predictions. This works extremely well as long as the same assumptions are made for the extraction and the predictions.

For a complete overview of neutral current, NC, and charged current, CC, cross sections, please check your favorite textbook. The electroweak Born-level cross section for the $e^{\pm}p$ NC interaction serves as an example here:

$$\frac{d^2\sigma(e^{\pm}p)}{dxdQ^2} = \frac{2\pi\alpha^2}{xQ^4}[Y_+\tilde{F}_2(x,Q^2) \mp Y_-x\tilde{F}_3(x,Q^2) - y^2\tilde{F}_L(x,Q^2)], \tag{1}$$

where α is the fine-structure constant, $Y_\pm = 1 \pm (1-y)^2$ and $\tilde{F}_2(x, Q^2)$, $\tilde{F}_3(x, Q^2)$ and $\tilde{F}_L(x, Q^2)$ are generalised structure functions. The contribution of the longitudinal structure function $\tilde{F}_L$ to $d^2\sigma/dxdQ^2$ is approximately 1%, averaged over the relevant kinematic range; it contributes up to 10% at high y. The $\tilde{F}_3$ term only starts to contribute significantly at Q^2 values of the order of the mass of the Z boson squared.

The reduced cross sections for NC $e^\pm p$ scattering are defined as

$$\tilde{\sigma}^{e^\pm p} = \frac{xQ^4}{2\pi\alpha^2} \frac{1}{Y_+} \frac{d^2\sigma(e^\pm p)}{dx} dQ^2 = \tilde{F}_2(x, Q^2) \mp \frac{Y_-}{Y_+} x\tilde{F}_3(x, Q^2) - \frac{y^2}{Y_+} F_L(x, Q^2). \quad (2)$$

The $x\tilde{F}_3$ can be obtained from the difference of the $e^- p$ and $e^+ p$ cross section.

The different structure functions in the Born-level approximation are directly connected to different combinations of quark momentum distributions. The structure function $\tilde{F}_3$, for example, provides information about the u and d valence quarks.

2.3. *The advent of precision*

The definition of precision is certainly not objective and what it really means in this context is debatable. However, from the viewpoint of HERA, precision structure functions came about when the two experiments, H1 and ZEUS, started to combine their already individually beautiful data.

The two collaboration published their combined results on data taken in the period of 1993 – 2000 in 2010.[2] The 10 years it took indicate that such a combination is difficult. The bins and the kinematic ranges need to be adjusted, the uncertainties evaluated according to their correlation, and, in order to do so, the team has to understand both experiments. The result is depicted in Fig. 3.[2] Due to the careful analysis of the correlations between the systematic uncertainties, the gain is significantly larger than expected if only the statistical precision is considered.

The results of the combination were used to produce the first HERAPDF,[2] entirely deduced from HERA inclusive data, see Fig. 4.

2.4. *The PDF fitting industry*

There is quite a number of groups who perform fits to a wide variety of data sets in order to extract PDFs. They go by acronyms representing names or ideas, in alphabetic order: ABKM, CTEQ, HERAPDF, GJR, NNPDF, MSTW. The acronyms are usual augmented by a version number. HERAPDF was so far restricted to HERA data. Other groups use HERA data, but not exclusively.

Different PDFs can be extracted at varying order in perturbative QCD, using different flavor schemes, different parameters like charm mass and different parametrisations. It is of great importance to use the same schemes and assumptions used for the extraction when making a prediction. The predictions you see in the plots showing cross sections were of course all extracted keeping that in mind.

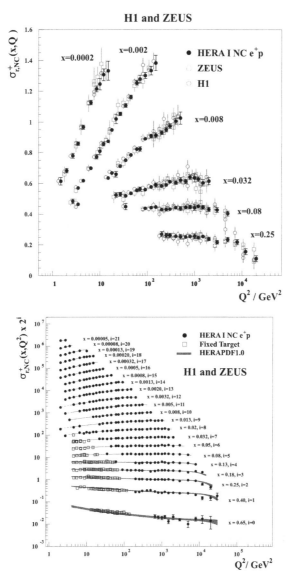

Fig. 3. Combination of reduced cross section. The left panel demonstrates the power of combination for selected values of x while the right panel shows the wide kinematic range covered by HERA.

2.5. *Towards the final HERA precision*

The data taking period that will provide the final precision in cross sections and PDFs from HERA is the HERA II period from 2004 to 2007. The data are still being analysed and some results are only available as preliminary releases so far. It is expected that both the H1 and ZEUS collaborations will publish final results this summer.

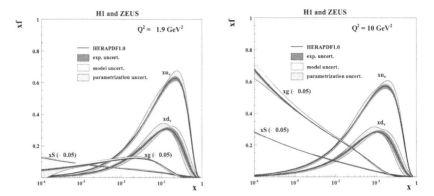

Fig. 4. HERAPDF1.0: The left panel demonstrates the importance of the valence quarks at relatively low Q^2 while the right panel shows the growth of the sea and the glue with Q^2. Please note that sea and glue are scaled down by a factor 20.

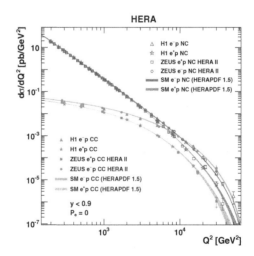

Fig. 5. A demonstration how precision data reveal the unification of the electroweak force and parity violation at $Q^2 > m_{Z^0}^2$.

Figure 5 gives a taste of the precision obtained from HERA II data. At high Q^2, the NC and CC cross sections become equal because the Z^0 starts to dominate over the photon. The difference between electron and positron data also becomes clearly visible at high Q^2.

The CC data also provide other valuable information. As there is no interference from photon exchange, the CC process was used to check the V-A structure of the weak interaction using the polarisation of the lepton beam in HERA II.[3] However, for this contribution the access to the quark structure of the proton is more interesting. As an example, Fig. 6(left) shows the reduced cross sections for CC positron

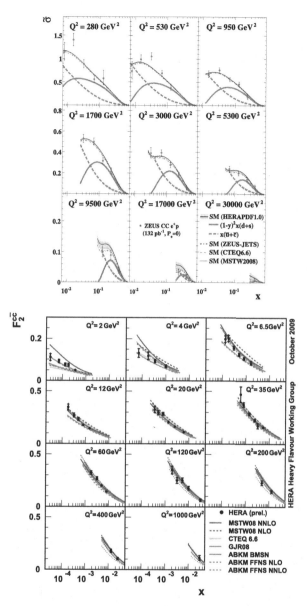

Fig. 6.   Access to the charm content of the proton. Left: through CC interactions, right: through D meson production.

interaction as published by ZEUS[4] . These data give access to the d and s as well as $\bar{u}$ and $\bar{c}$ content of the proton.

While the CC data give a hint of charm, the production of D mesons in all varieties give a direct handle on the c content. This is used to extract the charm structure functions. A preliminary result obtained from combined ZEUS and H1

NAME	NC and CC DIS	NC, lower E(p_beam)	Jets	Charm	Docu	Grids	Data comparison	Date
HERAPDF1.7 NLO	HERAI + partial HERAII	H1+ZEUS	H1 and ZEUS(1)	H1+ZEUS	Figures	N.A.		June 2011
HERAPDF1.6 NLO	HERAI + partial HERAII	---	H1 and ZEUS(1)	---	Writeup and figures	N.A.		March 2011
HERAPDF 1.5 NNLO	HERAI + partial HERAII	---	---	---	Figures	LHAPDF beta 5.8.6		March 2011
HERAPDF 1.5 NLO	HERAI + partial HERAII	---	---	---	Figures	LHAPDF beta 5.8.6		July 2010
Charm mass scan	HERAI	---	---	H1+ZEUS	Writeup and figures	---		August 2010
HERAPDF1.0 NNLO	HERAI	---	---	---	ICHEP2010 writeup and figures	Docu for LHAPDF		April 2010
	HERAI	H1+ZEUS	---	---	Writeup and figures	N.A.		April 2010
	HERAI	---	---	H1+ZEUS	DIS2010 writeup and figures	N.A.		April 2010
HERAPDF1.0 NLO PUBLISHED	HERAI	---	---	---	Paper HERAPDF1.0 page	LHAPDF	Benchmarking HERAPDF1.0	Nov 2009

(left margin: recommended version)

Fig. 7.   Available HERAPDFs as seen on the web.

data is depicted in Fig. 6(right). The charm structure function is an important input to LHC analyses, where the knowledge or lack thereof could provide a dominating systematic uncertainty.

2.6. *The HERAPDF family*

The HERA data was used over the past years to create a family of PDFs. Data were included as results became available. The table depicted in Fig. 7 provides an overview as given on the net at the time of the conference.[5] The youngest member of the family, HERAPDF1.7, is not only based on inclusive measurements, but also on jet and charm data. Charm data was already used previously for a charm mass scan. The currently recommended version is HERAPDF1.5, available at next to leading order, NLO, and at next-to-next to leading order, NNLO, see Fig. 8. It is, as HERAPDF1.0, based on inclusive data only. As soon as the final results for the inclusive measurement of ZEUS and H1 as well as combination paper on charm will be published, a new major version of HERAPDF will be extracted.

   The working group extracting, i.e. fitting, all the HERAPDFs has developed a tool named HERAfitter.[6] This tool allows the extraction of PDFs for a flexible set of input assumptions and input data. This tool has become "open source software" and is a service to the HEP community at large. It is not only used in the HERA, but also the LHC community.

3. Low-x Partons

The physical interpretation of PDFs is not clear to the author. The parametrisations are not predicted by any theory; they are based on common sense arguments like they should be zero at $x = 1$. The variable x itself is generally interpreted as the fraction of the momentum of the proton that a parton carries. At very low x, however, the Heisenberg uncertainty principle teaches us that such a parton cannot

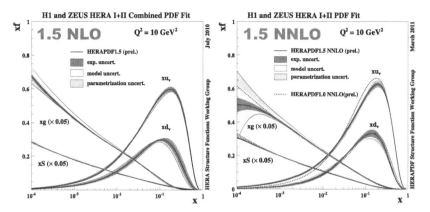

Fig. 8. The currently recommended member of the HERAPDF family. The NNLO extraction has a larger uncertainty on the glue and the sea. Experimental, model and parametrisation uncertainties are given separately.

be confined inside a proton. The interpretation is only valid in a reference frame in which the proton is fast and in the context of a scattering process with a certain momentum transfer, Q^2. In the reference frame where the proton is at rest, the variable x can be interpreted as $x \approx 0.1 \cdot l[\text{fm}]$, where l is the coherence length of fluctuations of the exchanged photon. For $x < 0.1$, l is larger than the size of the proton of about $1\,\text{fm}$. A detailed discussion can be found in.[8] The effect is demonstrated in Fig. 9. The photon fluctuates into a quark-antiquark pair and this fluctuates further into a hadron like object which interacts with the proton.

In general, physics should not be dependent on the rest frame in which it is looked at. Therefore, the two different interpretations should only be seen as something to guide us through perturbation theory. Somehow, what me measure has to be connected to the structure of the underlying interaction. At low x, the PDFs represent more the strong and electroweak field than the proton. It would be nice to base the parametrisations that we use on some understanding of these fields. Another interesting question in this context is, whether the PDFs that could be measured in neutrino proton scattering would be the same as in electron proton scattering. After all, there is no photon to fluctuate in this case. The guess of the author is that at low x the result would be different. The photon probably just looks like a hadron. Thus, probing with a hadron would give the same results as probing with a photon(electron).

4. Test of the Color Dipole Picture

At the very end of the HERA running, data were taken at lower proton energies. This facilitated direct measurements of the longitudinal structure function F_L.[9, 10] These measurement are tricky, because they require the identification of electrons with a relatively low energy, down to $3.4(6.0)\,\text{GeV}$ for H1(ZEUS). Both collaborations published results which clearly show that $F_L > 0$. The results on F_L can be taken

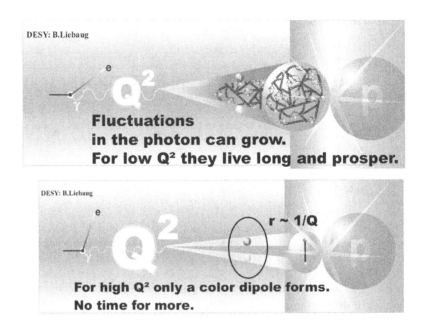

Fig. 9. The *ep* scattering process in the rest frame of the proton. The photon looks like an extended object. At low Q^2, top, it has time to grow into a complicated object. At high Q^2, bottom, there is only time to form a so called "color dipole".

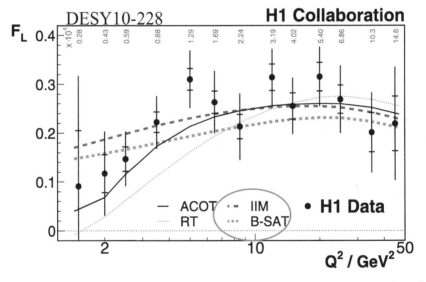

Fig. 10. Measurement of the structure function F_L compared to predictions from colour dipole models, IIM and B-SAT, and others, ACOT and RT.

as a probe of higher orders of perturbative QCD, but they can also be used to test the color dipole picture.[11] The H1 collaboration compared their results to

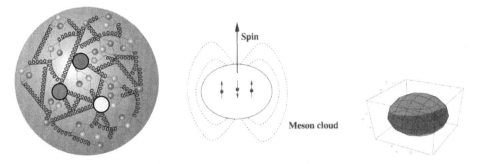

Fig. 11.   Left: Standard illustration of a "HEP-proton". Three valance quarks are imbedded in a cocoon of gluons and sea quarks. Center: A proton as seen by a nuclear physicist. It has a shape to support its magnetic moment and a meson cloud. Right: The shape of a Δ as predicted by lattice QCD.

different model predictions as depicted in Fig. 10.[9] The predictions based on the color dipole model deviate from the PDF-based ones at low Q^2. The data seem to have a preference to fall between the two ways of looking at things. It should be noted that nature does seem to have a sense of humor.

Another test of the color dipole model is provided by Deeply Virtual Compton Scattering, DVCS. The measured dependence of the cross section on the squared momentum transfer at the proton vertex, t, was compared to predictions of the colour dipole model and the model did quite well.[12]

5. The Looks of a Proton

The PDFs as so far discussed in this contribution only give information about what happens in momentum space. It is possible to define generalised parton distribution functions which are used for two-gluon exchanges like in DVCS. The interpretation then is in longitudinal momentum and transverse position space. The aforementioned dependence of the DVCS cross section on t can be parametrised as $d\sigma/dt \sim exp(-b|t|)$ and b can be converted to an average impact parameter. The H1 collaboration has done so[12] and obtained 0.65±0.02 fm for x=0.0012. Is this the transverse expansion of the partons engaged in the interaction? Is this the size of the proton? Or the size of the photon fluctuation? Or the relevant size of the field? Anyhow, there is a quite a number of b-slope measurements available, not only for DVCS, but also for vector meson production. It might be interesting to interpret these measurements with respect to impact parameters.

Most of the time, the "HEP-proton" is depicted as in Fig. 11(left). The charge radius of the proton is not measured in HEP, high energy physics, but in low energy physics. The most precise values come from electronic and muonic hydrogen.[13] The rms values are 0.8768±0.0069 fm and 0.84184±0.00067 fm. The two values disagree by four standard deviations which in itself has triggered some discussion, including discussions about physics beyond the standard model. However, the value of 0.65 fm

I. Abt

is certainly quite different. But that is something that should not surprise too much. The author would assume that the charge radius as measured at low energies is related to the valence quarks. The "radius" measured with DVCS is connected to x=0.0012 and that should not be valence quarks at all.

The picture of Fig. 11(left) is in many ways misleading. Three charged objects as depicted would create an electric dipole moment, but this is known to be less than $0.54\ 10^{-23}$ ecm. And in addition, the proton cannot be spherical at all, because it has a magnetic moment, which was first measured in proton $\to \Delta$ excitations;[14] the most precise measurements come from single trapped protons.[15] The nuclear physicist's view of a proton is shown in Fig. 11(center).[16] Also shown is an indication of a meson cloud. This has to be there; otherwise the proton could not stick together with other baryons. This is again a completely different interpretation than typical HEP view of perturbative QCD. However, the meson cloud was observed in HEP, remember the EMC effect.

The next step in understanding the proton, or perhaps baryon or even nuclei in general, would be to match the pictures of nuclear physics with the ideas of QCD. There is some input on this from lattice QCD where the shape of baryons is actually predicted.[17] Figure 11(right)[17] gives as an example the shape that is the result of a lattice QCD calculation.

It seems a worthwhile enterprise to think about ways how to measure such a shape and how to measure the spatial distribution of what is inside. In Fig. 1 it is shown, how we destroy the castle in order to find out what is inside and who just got married. It would be nice to open the door, have a look inside and just ask. However, many of the dynamics inside the castle might evade our observation, if we look with too much energy, because that means averaging over time; Heisenberg all over again.

6. Summary and Outlook

The proton as we know it still holds a lot of secrets. What is called the proton parton distribution functions, PDFs, was measured with excellent precision at HERA. The final results from HERA will be a very valuable legacy for a long time. They are expected within a year.

These PDFs are and will be one of the inputs to make precise Standard Model predictions for the LHC and other future experiments. The extraction of PDFs is one of the great success stories of HERA. The tools developed in this context will also survive and become part of the HEP daily work.

The interpretation of proton PDFs in the context of understanding the proton itself is not trivial and there is more to the proton than PDFs. There are many questions about size, shape and the spatial distribution of quarks and gluons that should be addressed. In the end, it will take more than perturbative QCD to understand the proton. on energies, respectively.

1460280-12

References

1. I.Abt on behalf of the H1 and ZEUS Collab., Precision Measurement of the Proton Structure, *XX International Workshop on Deep-Inelastic Sacttering and Related Subjects DIS 2012* ed Ian C. Brock (Verlag Deutsche Elektronen Synchrotron), p. 13.
2. H1 and ZEUS Collab., F.D. Aaron *et al.*, JHEP **01** (2010) 109.
3. ZEUS Collab., S. Chekanov *et al.*, Eur.Phys.J. **C61** (2009) 223.
4. ZEUS Collab., H. Abramowicz *et al.*, Eur.Phys.J. **C70** (2010) 945.
5. https://www.desy.de/h1zeus/combined_results/herapdftable.
6. http://herafitter.hepforge.org/.
7. ATLAS Collab., "Determination of the strange quark density of the proton from ATLAS measurements of the $W \rightarrow l\nu$ and $Z \rightarrow ll$ cross sections", 2012. /http://arxiv.org/abs/1203.4051hep-ex/1203.4051.
8. A. Caldwell and G. Grindhammer, Physik Journal **6** (2007) 39.
9. H1 Collab., F.D. Aaron *et al.*, Eur.Phys.J. **C71** (2011) 1579.
10. ZEUS Collab., S. Chekanov *et al.*, Phys.Lett. **B682** (2009) 8.
11. C. Ewerz *et al.*, "The New F_l measurement from HERA and the Dipole model", 2012. /http://arxiv.org/abs/1201.6296hep-ph/1206.6296.
12. H1 Collab., F.D. Aaron *et al.*, Phys.Lett. **B659** (2008) 796.
13. M. O. Distler *et al.*, Phys.Lett. **B696** (2011) 343.
14. R. Beck *et al.*, Phys.Rev.Lett **78** (1997) 606.
15. J. DiScaccia and G. Gabrielse, "Direct Measurement of the Proton Magnetic Moment", 2012. /http://arxiv.org/abs/1201.3038hep-ph/1201.3038.
16. A. Faessler, Progress in Particle and Nuclear Physics **44** (2000) 197.
17. C. N. Papanicolas, Eur.Phys.J. **A18** (2003) 141.

Physics in Collision (PIC 2013)
International Journal of Modern Physics: Conference Series
Vol. 31 (2014) 1460281 (7 pages)
© The Author
DOI: 10.1142/S2010194514602816

www.worldscientific.com

Applications of higher order QCD

Stefan Höche

SLAC National Accelerator Laboratory,
Menlo Park, CA 94025, USA
shoeche@slac.stanford.edu

Published 15 May 2014

In this talk we summarize some recent developments in perturbative QCD and their application to particle physics phenomenology.

1. Introduction

With the discovery of a new boson at the Large Hadron Collider (LHC), particle physics has entered a new era. Since this discovery, the field has quickly moved towards precision measurements on the new particle. In order to further improve these measurements and to find possible small deviations that may hint towards new physics, improved theoretical predictions, including higher-order perturbative QCD corrections for production rates and kinematics are urgently needed. The same is true for other reactions of interest at the LHC, like top quark production and W/Z production. The toolkit used to this end ranges from fixed order calculations at the parton-level over resummation to parton showers and particle-level event generators. Tremendous progress has been made in the field during the past year. Some of the recent developments will be briefly summarized in this talk.

2. Higher-Order Calculations

Fixed-order calculations are available for a large variety of processes. At the tree level, they have long been performed completely automatically using programs like ALPGEN,[1] Amegic++,[2] Comix,[3] CompHEP,[4] HELAC,[5] MadGraph[6] and Whizard.[7] At the next-to-leading order (NLO), automation required two main ingredients: The implementation of known generic methods to perform the subtraction of infrared singularities,[8-10] and the automated computation of one-loop amplitudes. As infrared subtraction terms consist of tree-level matrix elements joined

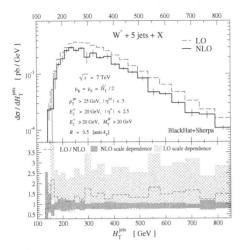
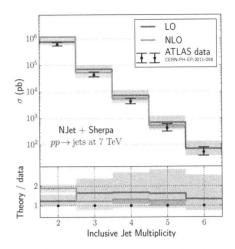

Fig. 1. Distribution of the visible energy in $W + 5$ jet events. Figure taken from Ref. 16.

Fig. 2. Jet multiplicity distribution in pure jet events (right). Figure taken from Ref. 17.

by splitting operators, existing programs for leading order calculations are ideally suited to compute them. Correspondingly, Catani-Seymour dipole subtraction has been implemented in the existing generators Amegic++,[11] Comix, HELAC[12] and MadGraph.[13,14] FKS subtraction is realized in MadGraph only.[15]

The automated computation of virtual corrections has received a boost from generalized unitarity,[18–20] which can be used to determine one-loop amplitudes by decomposing them into known scalar one-loop integrals and rational coefficients determined from tree amplitudes, plus a rational piece.[21–25] Programs like BlackHat,[26] Gosam,[27] HELACNLO,[28] MadLoop,[29] NJet,[30] OpenLoops[31] and Rocket[32,33] implement these techniques and supplement established programs like MCFM[34,35] and dedicated codes based on improved tensor reduction approaches.[36,37] New techniques have also been proposed to accelerate the numerical calculation of the integrand of one-loop amplitudes, independent of the reduction scheme.[31] Figures 1 and 2 show examples from recent NLO calculations for $W+5$ jet production[16] and 5 jets production,[17] both performed using unitarity based techniques. Other recently completed calculations include Higgs boson plus 3 jet production[38] and di-photon plus 2 jet production.[39] The rapid progress in this field is reflected by the fact that all calculations from the experimenter's wishlist for the LHC have now been tackled.[40] Most of the programs used to perform the calculations, or their results, are publicly available.

Driven by the need for higher precision in some selected Standard-Model reactions, the field of next-to-next-to leading order (NNLO) calculations has significantly advanced in the past years. One of the most challenging problems is the regularization of infrared divergences at NNLO. Sector decomposition[41–43] has been used in the past to perform several $2 \to 1$ calculations.[44,45] Antenna subtraction[46,47] was worked out and implemented for $e^+e^- \to 3$ jets.[48,49] q_T subtraction[50] was employed

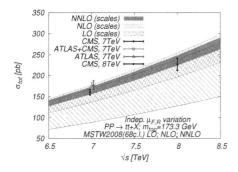
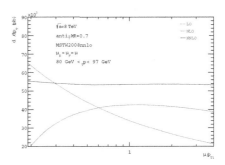

Fig. 3. Energy dependence of the $pp \to t\bar{t}$ total cross. Figure taken from Ref. 58.

Fig. 4. Scale dependence of the $pp \to$ di-jet cross section. Figure taken from Ref. 64.

in several calculations, including Higgs production,[51] W/Z production,[52] associated Higgs production[53] and di-photon production.[54] More recently sector-improved subtraction methods were introduced.[55,56] They have been used to compute cross sections for $pp \to t\bar{t}$[57,58] and $pp \to H$+jet.[59] At the same time, antenna subtraction was extended to initial states[60–63] and employed to compute $pp \to$ di-jets fully differentially at NNLO.[64] Figures 3 and 4 show results from some of these calculations. The calculation of $pp \to t\bar{t}$ has also been combined with higher logarithmic resummation.[65–67] Its theoretical uncertainty is such that uncertainties from scale choices, PDF, strong coupling measurements and top-quark mass measurements are all of the same order.[68]

3. Resummation of Jet Vetoes

The analysis of the Higgs-like particle discovered at the LHC places new demands on resummed calculations. Many of the Higgs analysis channels, most notably $H \to WW^* \to \ell^+\ell^-\nu\bar{\nu}$, veto on the transverse momentum of final state jets to distinguish different Standard Model backgrounds and separate them from the signal. The leading systematic uncertainty is the theoretical uncertainty on the signal cross section in the jet bins. This uncertainty can be reduced by a proper resummation of the logarithms associated with the jet veto. Various groups have investigated this problem, in most cases up to next-to-next-to leading logarithmic accuracy matched to NNLO fixed order, relying either on more traditional resummation methods,[69,70] or on Soft Collinear Effective Theory.[71–75] Higgs plus one jet production was studied at next-to-leading logarithmic order (NLL) and matched to NLO fixed order using SCET.[76]

4. Parton Showers and Matching to NLO Calculations

The interest in parton showers as a means to produce particle-level predictions fully differentially in the phase space of multi-jet events has increased significantly in

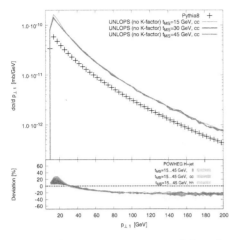

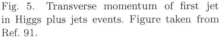

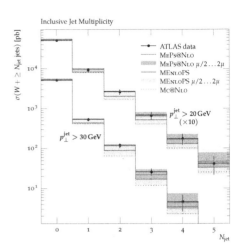

Fig. 5. Transverse momentum of first jet in Higgs plus jets events. Figure taken from Ref. 91.

Fig. 6. Jet multiplicity distribution in W+jets events. Figure taken from Ref. 92.

recent years. New concepts for the construction of parton showers have been proposed, which are based on antenna subtraction[77,78] and/or sectorizing the phase space.[79,80] Efforts were made to include subleading color corrections into showers as a means to improve their logarithmic accuracy.[81,82] However, the crucial development was the proposal of a method to match parton showers to NLO calculations,[83] later extended to eliminate negative weights.[84,85] This matching has been partially or fully automated in several projects,[86–90] such that particle-level predictions at NLO accuracy are now widely available.

The description of multi-jet final states with parton showers can be improved using so-called ME+PS merging methods,[93–97] which, in contrast to matching methods, allow to correct the parton shower for an arbitrary number of emissions with higher-order tree-level calculations. These methods were recently refined and extended, leading to algorithms which can combine multiple NLO calculations of varying multiplicity (like $W + 0$ jet, $W + 1$ jet, $W + 2$ jet, etc.) into a single, inclusive simulation (e.g. of W+jets production).[91,92,98–100] Figures 5 and 6 show examples for the application of ME+PS merging to Higgs boson plus jets production and to W+jets production. A particular scale choice is required for the evaluation of the strong coupling in ME+PS merging, which has also been adopted for the matching to higher-multiplicity NLO calculations on its own in the so-called MINLO approach.[101]

The MINLO method accounts for Sudakov suppression effects in higher-multiplicity final states and allows to extrapolate NLO calculations to zero jet transverse momentum, thus offering the opportunity to match to NNLO calculations for a limited class of processes and observables.[102] A different proposal for a matching to NNLO parton-level calculations was made in,[91,98] which is based on a subtraction method similar to the one used in ME+PS merging at NLO. Both techniques are promising candidates to further increase the precision of event generators for collider physics.

5. Summary

We have presented some of the recent developments in perturbative QCD and applications to particle physics phenomenology. NLO parton-level calculations can nowadays often be provided by fully automated tools. New techniques in event generation allow to also use them for particle-level predictions. NNLO calculations and higher-logarithmic resummation techniques are at the forefront of current research.

References

1. M. L. Mangano, M. Moretti, F. Piccinini, R. Pittau and A. D. Polosa, *JHEP* **07**, p. 001 (2003).
2. F. Krauss, R. Kuhn and G. Soff, *JHEP* **02**, p. 044 (2002).
3. T. Gleisberg and S. Höche, *JHEP* **12**, p. 039 (2008).
4. E. Boos *et al.*, *Nucl. Instrum. Meth.* **A534**, 250 (2004).
5. A. Kanaki and C. G. Papadopoulos, *Comput. Phys. Commun.* **132**, 306 (2000).
6. J. Alwall, M. Herquet, F. Maltoni, O. Mattelaer and T. Stelzer, *JHEP* **06**, p. 128 (2011).
7. W. Kilian, T. Ohl and J. Reuter, *Eur. Phys. J.* **C71**, p. 1742 (2007).
8. S. Frixione, Z. Kunszt and A. Signer, *Nucl. Phys.* **B467**, 399 (1996).
9. S. Catani and M. H. Seymour, *Nucl. Phys.* **B485**, 291 (1997).
10. S. Catani, S. Dittmaier, M. H. Seymour and Z. Trocsanyi, *Nucl. Phys.* **B627**, 189 (2002).
11. T. Gleisberg and F. Krauss, *Eur. Phys. J.* **C53**, 501 (2008).
12. M. Czakon, C. Papadopoulos and M. Worek, *JHEP* **08**, p. 085 (2009).
13. R. Frederix, T. Gehrmann and N. Greiner, *JHEP* **09**, p. 122 (2008).
14. R. Frederix, T. Gehrmann and N. Greiner, *JHEP* **06**, p. 086 (2010).
15. R. Frederix, S. Frixione, F. Maltoni and T. Stelzer, *JHEP* **10**, p. 003 (2009).
16. Z. Bern, L. Dixon, F. Febres Cordero, S. Hoeche, H. Ita *et al.*, *Phys. Rev.* **D88**, p. 014025 (2013).
17. S. Badger, B. Biedermann, P. Uwer and V. Yundin (2013).
18. Z. Bern, L. J. Dixon, D. C. Dunbar and D. A. Kosower, *Nucl. Phys.* **B435**, 59 (1995).
19. Z. Bern, L. J. Dixon, D. C. Dunbar and D. A. Kosower, *Nucl. Phys.* **B425**, 217 (1994).
20. Z. Bern, L. J. Dixon and D. A. Kosower, *Nucl. Phys.* **B513**, 3 (1998).
21. G. Ossola, C. G. Papadopoulos and R. Pittau, *Nucl. Phys.* **B763**, 147 (2007).
22. D. Forde, *Phys. Rev.* **D75**, p. 125019 (2007).
23. R. Ellis, W. Giele and Z. Kunszt, *JHEP* **0803**, p. 003 (2008).
24. G. Ossola, C. G. Papadopoulos and R. Pittau, *JHEP* **05**, p. 004 (2008).
25. R. Ellis, W. T. Giele, Z. Kunszt and K. Melnikov, *Nucl. Phys.* **B822**, 270 (2009).
26. C. F. Berger, Z. Bern, L. J. Dixon, F. Febres-Cordero, D. Forde, H. Ita, D. A. Kosower and D. Maître, *Phys. Rev.* **D78**, p. 036003 (2008).
27. G. Cullen, N. Greiner, G. Heinrich, G. Luisoni, P. Mastrolia, G. Ossola and T. Reiter (2011).
28. G. Bevilacqua, M. Czakon, M. Garzelli, A. van Hameren, A. Kardos *et al.*, *Comput. Phys. Commun.* **184**, 986 (2013).

29. V. Hirschi, R. Frederix, S. Frixione, M. V. Garzelli, F. Maltoni and R. Pittau, *JHEP* **1105**, p. 044 (2011).
30. S. Badger, B. Biedermann and P. Uwer, *Comput. Phys. Commun.* **182**, 1674 (2011).
31. F. Cascioli, P. Maierhofer and S. Pozzorini, *Eur. Phys. J.* **C72**, p. 1889 (2012).
32. R. Ellis, W. Giele, Z. Kunszt, K. Melnikov and G. Zanderighi, *JHEP* **0901**, p. 012 (2009).
33. R. Ellis, K. Melnikov and G. Zanderighi, *JHEP* **0904**, p. 077 (2009).
34. J. Campbell, R. K. Ellis and C. Williams.
35. J. M. Campbell and R. Ellis, *Nucl. Phys. Proc. Suppl.* **205-206**, 10 (2010).
36. A. Denner and S. Dittmaier, *Nucl. Phys.* **B734**, 62 (2006).
37. T. Binoth, J. P. Guillet, G. Heinrich, E. Pilon and C. Schubert, *JHEP* **0510**, p. 015 (2005).
38. G. Cullen, H. van Deurzen, N. Greiner, G. Luisoni, P. Mastrolia *et al.*, *Phys. Rev. Lett.* **111**, p. 131801 (2013).
39. T. Gehrmann, N. Greiner and G. Heinrich (2013).
40. J. Alcaraz Maestre *et al.* (2012).
41. T. Binoth and G. Heinrich, *Nucl. Phys.* **B585**, 741 (2000).
42. C. Anastasiou, K. Melnikov and F. Petriello, *Phys. Rev.* **D69**, p. 076010 (2004).
43. T. Binoth and G. Heinrich, *Nucl. Phys.* **B693**, 134 (2004).
44. C. Anastasiou, K. Melnikov and F. Petriello, *Phys. Rev. Lett.* **93**, p. 262002 (2004).
45. K. Melnikov and F. Petriello, *Phys. Rev. Lett.* **96**, p. 231803 (2006).
46. D. A. Kosower, *Phys. Rev.* **D57**, 5410 (1998).
47. A. Gehrmann-De Ridder, T. Gehrmann and E. W. N. Glover, *JHEP* **09**, p. 056 (2005).
48. A. Gehrmann-De Ridder, T. Gehrmann, E. Glover and G. Heinrich, *Phys. Rev. Lett.* **99**, p. 132002 (2007).
49. A. Gehrmann-De Ridder, T. Gehrmann, E. Glover and G. Heinrich, *JHEP* **0711**, p. 058 (2007).
50. S. Catani and M. Grazzini, *Phys. Rev. Lett.* **98**, p. 222002 (2007).
51. M. Grazzini, *JHEP* **0802**, p. 043 (2008).
52. S. Catani, L. Cieri, G. Ferrera, D. de Florian and M. Grazzini, *Phys. Rev. Lett.* **103**, p. 082001 (2009).
53. G. Ferrera, M. Grazzini and F. Tramontano, *Phys. Rev. Lett.* **107**, p. 152003 (2011).
54. S. Catani, L. Cieri, D. de Florian, G. Ferrera and M. Grazzini (2011).
55. M. Czakon, *Phys. Lett.* **B693**, 259 (2010).
56. R. Boughezal, K. Melnikov and F. Petriello, *Phys. Rev.* **D85**, p. 034025 (2012).
57. P. Bärnreuther, M. Czakon and A. Mitov, *Phys. Rev. Lett.* **109**, p. 132001 (2012).
58. M. Czakon, P. Fiedler and A. Mitov, *Phys. Rev. Lett.* **110**, p. 252004 (2013).
59. R. Boughezal, F. Caola, K. Melnikov, F. Petriello and M. Schulze (2013).
60. A. Daleo, A. Gehrmann-De Ridder, T. Gehrmann and G. Luisoni, *JHEP* **1001**, p. 118 (2010).
61. R. Boughezal, A. Gehrmann-De Ridder and M. Ritzmann, *JHEP* **1102**, p. 098 (2011).
62. T. Gehrmann and P. F. Monni, *JHEP* **1112**, p. 049 (2011).
63. A. Gehrmann-De Ridder, T. Gehrmann and M. Ritzmann, *JHEP* **1210**, p. 047 (2012).
64. A. Gehrmann-De Ridder, T. Gehrmann, E. Glover and J. Pires, *Phys. Rev. Lett.* **110**, p. 162003 (2013).
65. M. Czakon, A. Mitov and G. F. Sterman, *Phys. Rev.* **D80**, p. 074017 (2009).
66. M. Beneke, M. Czakon, P. Falgari, A. Mitov and C. Schwinn, *Phys. Lett.* **B690**, 483 (2010).

67. M. Cacciari, M. Czakon, M. Mangano, A. Mitov and P. Nason, *Phys. Lett.* **B710**, 612 (2012).
68. M. Czakon, M. L. Mangano, A. Mitov and J. Rojo, *JHEP* **1307**, p. 167 (2013).
69. A. Banfi, G. P. Salam and G. Zanderighi, *JHEP* **1206**, p. 159 (2012).
70. A. Banfi, P. F. Monni, G. P. Salam and G. Zanderighi, *Phys. Rev. Lett.* **109**, p. 202001 (2012).
71. F. J. Tackmann, J. R. Walsh and S. Zuberi, *Phys. Rev.* **D86**, p. 053011 (2012).
72. I. W. Stewart, F. J. Tackmann, J. R. Walsh and S. Zuberi (2013).
73. T. Becher and M. Neubert, *JHEP* **1207**, p. 108 (2012).
74. T. Becher, M. Neubert and D. Wilhelm, *JHEP* **1305**, p. 110 (2013).
75. T. Becher, M. Neubert and L. Rothen, *JHEP* **1310**, p. 125 (2013).
76. X. Liu and F. Petriello, *Phys. Rev.* **D87**, p. 094027 (2013).
77. W. T. Giele, D. A. Kosower and P. Z. Skands, *Phys. Rev.* **D78**, p. 014026 (2008).
78. L. Hartgring, E. Laenen and P. Skands, *JHEP* **1310**, p. 127 (2013).
79. A. J. Larkoski and M. E. Peskin, *Phys. Rev.* **D81**, p. 054010 (2010).
80. A. J. Larkoski and M. E. Peskin, *Phys. Rev.* **D84**, p. 034034 (2011).
81. S. Plätzer and M. Sjödahl, *JHEP* **1207**, p. 042 (2012).
82. S. Höche, F. Krauss, M. Schönherr and F. Siegert, *JHEP* **09**, p. 049 (2012).
83. S. Frixione and B. R. Webber, *JHEP* **06**, p. 029 (2002).
84. P. Nason, *JHEP* **11**, p. 040 (2004).
85. S. Frixione, P. Nason and C. Oleari, *JHEP* **11**, p. 070 (2007).
86. S. Alioli, P. Nason, C. Oleari and E. Re, *JHEP* **06**, p. 043 (2010).
87. S. Hoche, F. Krauss, M. Schonherr and F. Siegert, *JHEP* **1104**, p. 024 (2011).
88. R. Frederix, S. Frixione, V. Hirschi, F. Maltoni, R. Pittau *et al.*, *JHEP* **1202**, p. 048 (2012).
89. S. Höche, F. Krauss, M. Schönherr and F. Siegert, *Phys. Rev. Lett.* **110**, p. 052001 (2013).
90. S. Höche and M. Schönherr, *Phys. Rev.* **D86**, p. 094042 (2012).
91. L. Lönnblad and S. Prestel, *JHEP* **1303**, p. 166 (2013).
92. S. Höche, F. Krauss, M. Schönherr and F. Siegert (2012).
93. S. Catani, F. Krauss, R. Kuhn and B. R. Webber, *JHEP* **11**, p. 063 (2001).
94. F. Krauss, *JHEP* **0208**, p. 015 (2002).
95. M. L. Mangano, M. Moretti and R. Pittau, *Nucl. Phys.* **B632**, 343 (2002).
96. J. Alwall *et al.*, *Eur. Phys. J.* **C53**, 473 (2008).
97. L. Lönnblad and S. Prestel, *JHEP* **1302**, p. 094 (2013).
98. N. Lavesson and L. Lönnblad, *JHEP* **12**, p. 070 (2008).
99. T. Gehrmann, S. Höche, F. Krauss, M. Schönherr and F. Siegert, *JHEP* **1301**, p. 144 (2013).
100. R. Frederix and S. Frixione, *JHEP* **1212**, p. 061 (2012).
101. K. Hamilton, P. Nason and G. Zanderighi (2012).
102. K. Hamilton, P. Nason, C. Oleari and G. Zanderighi, *JHEP* **1305**, p. 082 (2013).

Physics in Collision (PIC 2013)
International Journal of Modern Physics: Conference Series
Vol. 31 (2014) 1460282 (12 pages)
© The Author
DOI: 10.1142/S2010194514602828

World Scientific
www.worldscientific.com

Recent results from lattice QCD

Chuan Liu

School of Physics and Center for High Energy Physcis
Peking University, Beijing 100871, China
liuchuan@pku.edu.cn

Published 15 May 2014

Recent Lattice QCD results are reviewed with an emphasis on spectroscopic results concerning the charm quark. It is demonstrated that, with accurate computations from lattice QCD in recent years that can be compared with the existing or upcoming experiments, stringent test of the Standard Model can be performed which will greatly sharpen our knowledge on the strong interaction.

Keywords: Lattice QCD; charm quark; comparison with experiments.

PACS Numbers: 12.38.Gc, 11.15.Ha

1. Introduction

In the past ten years or so, considerable progress has been achieved in lattice Chromodynamics (lattice QCD). Here, I will try to review briefly some selected results, with an emphasis on those related to the charm quark, and compare them with the experiments so far or possibly in the near future. For general recent lattice results, please consult the Lattice 2013 website[1] where the talks are available online.

Lattice QCD is a non-perturbative theoretical method that relies on Monte Carlo estimation of physical quantities using gauge field samples that are generated according to a lattice action. Given a lattice action $S = S_g[U_\mu] + S_f[\bar{\psi}, \psi, U_\mu] = S_g[U_\mu] + \bar{\psi}_x \mathcal{M}_{xy}[U_\mu]\psi_y$, where $\mathcal{M}[U_\mu]$ is called the fermion matrix[a], a physical quantity of interest, $\mathcal{O}[\bar{\psi}, \psi, U_\mu]$, which is built from the basic fields is given by the *ensemble average*:

$$\langle \mathcal{O} \rangle = \frac{1}{\mathcal{Z}} \int \mathcal{D}\bar{\psi}\mathcal{D}\psi\mathcal{D}U_\mu \mathcal{O}[\bar{\psi}, \psi, U_\mu] e^{-S[\bar{\psi}, \psi, U_\mu]}. \tag{1}$$

[a]The explicit form of $\mathcal{M}[U_\mu]$ depends on the type of lattice fermion (staggered, Wilson, etc.) used.

Here the partition function $\mathcal{Z}$ is given by the relevant path integral

$$\mathcal{Z} = \int \mathcal{D}\bar{\psi}\mathcal{D}\psi\mathcal{D}U_\mu e^{-S[\bar{\psi},\psi,U_\mu]} = \int \mathcal{D}U_\mu e^{-S_g[U_\mu]} \det \mathcal{M}[U_\mu]. \tag{2}$$

Following the above equations, a typical lattice calculation therefore consists of two steps: In the first step, also known as the generation step, one generates the gauge field configurations according to the probability distribution: $P[U_\mu] = \mathcal{Z}^{-1}e^{-S_g[U_\mu]} \det \mathcal{M}[U_\mu]$ and stores them for later usage; In the second step, also known as the measurement step, any interested observable $\mathcal{O}[\bar{\psi},\psi,U_\mu]$ is measured from the pre-stored gauge field configurations with the quark and anti-quark fields in the corresponding observable replaced by the corresponding quark propagators, which are relevant matrix elements of $\mathcal{M}^{-1}[U_\mu]$, in a particular gauge field background. Note that the fermion matrix $\mathcal{M}[U_\mu]$ is diagonal in flavor space. Therefore, the determinant of the matrix in Eq. (2) is in fact a product of determinants for each quark flavor. Depending on how many flavors are kept in the generation step, we call them N_f flavor lattice QCD.[b]

2. Spectrum Calculations

2.1. *Hadron masses, conventional computation*

To compute the mass values of hadrons, one starts from a set of interpolating operators $\{\mathcal{O}_\alpha(t) : \alpha = 1, 2, \cdots, N\}$.[2] These operators carry the correct quantum numbers such that the state $\mathcal{O}_\alpha^\dagger(t)|\Omega\rangle$, with $|\Omega\rangle$ being the QCD vacuum state, has the same quantum numbers as those of the hadron in question. In lattice Monte Carlo simulations, the following correlation matrix is measured:

$$\mathcal{C}_{\alpha\beta}(t) = \langle\Omega|\mathcal{O}_\alpha(t)\mathcal{O}_\beta^\dagger(0)|\Omega\rangle. \tag{3}$$

On the other hand, it is known that the same correlation matrix is given by

$$\mathcal{C}_{\alpha\beta}(t) = \sum_n \frac{e^{-E_n t}}{2E_n} Z_\alpha^{(n)} Z_\beta^{(n)*}, \tag{4}$$

where the summation is over all eigenstates (labelled by n) of the Hamiltonian with the E_n's being the corresponding eigenvalues. The measured correlation matrix $\mathcal{C}_{\alpha\beta}(t)$ is then passed through a standard variational calculation, also known as the generalized eigenvalue problem (GEVP),[3,4] for some given t_0:

$$\mathcal{C}_{\alpha\beta}(t)v_\beta^{(n)} = \lambda^{(n)}(t)\mathcal{C}_{\alpha\beta}(t_0)v_\beta^{(n)}. \tag{5}$$

Here $\lambda^{(n)}(t)$'s yield the eigenvalues of the Hamiltonian, namely the E_n's via

$$\lambda^{(n)}(t) \sim e^{-E_n(t-t_0)} \left[1 + O\left(e^{-\Delta E(t-t_0)}\right)\right], \tag{6}$$

while the eigenvectors $v_\beta^{(n)}$ are related to the corresponding overlap $Z_\alpha^{(n)}$.[4]

Standard light hadron spectroscopy has been studied by various groups in recent years, see e.g. Ref. 5. Hadrons containing the heavy quarks have also been studied.

---

[b]Quenched lattice QCD corresponds to $N_f = 0$ in this sense.

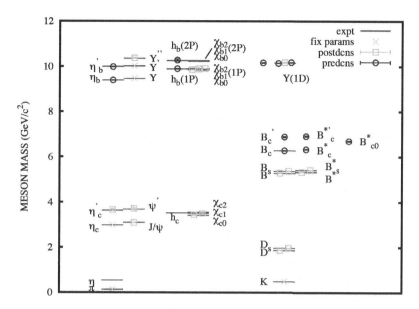

Fig. 1.   Heavy meson spectrum obtained from $N_f = 2 + 1 + 1$ lattice QCD, taken from the paper by HPQCD collaboration.[6]

In Fig. 1, we have shown the heavy meson spectrum from HPQCD collaboration obtained using $N_f = 2 + 1 + 1$ staggered quark configurations.[6] It is seen that, both post-dictions and predictions agree with the experiments, where available, astonishingly well.

2.2. *Hadron masses, multi-hadron scattering effects*

However, one should keep in mind that, in principle, these E_n's are *not* the mass values of the hadrons. The eigenvalue of the QCD Hamiltonian in a particular symmetry sector is only approximately equal to the mass of the hadron if the hadron being studied is a narrow resonance within strong interaction.[c] Therefore, to really study a genuine hadronic resonance, one should study the scattering of hadrons. In fact, Lüscher has established a formalism[7] in which the eigenvalues of the finite-volume Hamiltonian are related to the scattering phases of the two particles.[8]

In the simplest case (single channel, s-wave scattering, neglecting higher l contributions, etc.), this relation reads:

$$\tan \delta_0(E) = \frac{\pi^{3/2} q}{\mathcal{Z}_{00}(1, q^2)}, \tag{7}$$

---

[c]If the hadron is stable in QCD, then the E_n *is indeed* the mass of the hadron. But for resonances, which is true for majority of the hadrons, this is not the case.

where $\mathcal{Z}_{00}(1,q^2)$ is the zeta function that can be computed accurately and q is related to E via

$$E = \sqrt{m_1^2 + \bar{k}^2} + \sqrt{m_2^2 + \bar{k}^2}, \quad q \equiv \frac{\bar{k}L}{2\pi}. \tag{8}$$

The lattice calculation goes the same way as we described above with the exception that the interpolation operators being used, namely the operators $\{\mathcal{O}_\alpha(t) : \alpha = 1, 2, \cdots, N\}$, should also include the two-particle operators. Using Lüscher's formalism, one simply obtains the energy eigenvalues E_n's which are then substituted into Lüscher formula for $\delta_0(E_n)$. For small momentum close to the threshold, one could use the effective range expansion

$$k \cot \delta(E) = \frac{1}{a_0} + \frac{1}{2}r_0 k^2 + \cdots \tag{9}$$

with a_0 being the scattering length and r_0 the effective range.

Lüscher's formula has been generalized to various cases, see e.g. Ref. 10 and references therein. It has also been utilized successfully in lattice studies of hadronic resonances in recent years, see e.g. Ref. 2, 8 and references therein. A very good example is the lattice QCD study of the ρ meson, a typical hadronic resonance in the $I = J = 1$ channel of $\pi\pi$ scattering. In Fig 2, taken from Ref. 9, we have shown the phase shifts of $\pi\pi$ scattering in the $I = 1$, $J = 1$ channel computed within Lüscher's formalism.

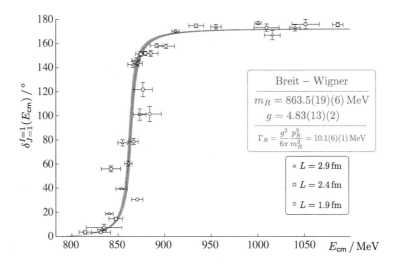

Fig. 2. Phase shifts of $\pi\pi$ scattering in the $I = J = 1$ channel calculated from lattice QCD using Lüscher's formalism, taken from a paper by the Hadron Spectrum Collaboration.[9]

2.3. *Scattering of charmed mesons*

Recently, there have been numerous new hadronic structures observed which contain charm and anti-charm quark. These have been termed the XYZ particles, see Refs 11, 12, 13. What is noticeable is that many of these newly observed states are close to the threshold of two known charmed mesons. For example, the $X(3872)$ is close to the threshold of D and D^* and so is the newly discovered $Z_c^{\pm}(3900)$.[14]

Several years ago, CLQCD has studied the $Z(4430)$ state observed by BELLE collaboration in quenched lattice QCD using Lüscher formalism.[15] Asymmetric volumes were used to investigate the low-momentum behavior of the scattering phase close to the threshold of D^* and a $\bar{D}_1$. The investigation was done in the channel of $J^P = 0^-$ and the interaction between the D^* and a $\bar{D}_1$ was found to be attractive but not strong enough to form a genuine bound state.

Recently, Prelovsek *et al* have studied the case of $X(3872)$ using $N_f = 2$ improved Wilson fermion configurations.[16] They claim that they have found some evidence for the state. With the same set of configurations, they have also studied the case of $Z_c(3900)$ with no signal of the state.[17]

In summary, the study for these XYZ states are still an on-going project. Lattice results obtained so far are still not systematic (most of them are at one lattice spacing, one volume, one pion mass etc.) and therefore one still cannot draw definite conclusions yet. The situation is still somewhat murky and more work needs to be done in the future to clarify the nature of these states from lattice QCD.

2.4. *Decay constants and the story of f_{D_s}*

For a pseudoscalar meson, the decay constant is defined via the matrix elements

$$\begin{cases} \langle \Omega | \bar{s} \gamma_\mu \gamma_5 c | D_s(p) \rangle = i f_{D_s} p_\mu, \\ (m_c + m_s) \langle \Omega | \bar{s} \gamma_5 c | D_s(p) \rangle = -m_{D_s}^2 f_{D_s} \end{cases} \tag{10}$$

with the two definitions related to each other by PCAC.

This quantity can be computed accurately in lattice QCD which then can be compared with the experiments. Around the year of 2008, some puzzling effects occurred that created a 3.8σ difference in the comparison of the lattice results and the experiments. For a more detailed description about this dilemma, the reader is referred to A. Kronfeld's review talk "The f_{D_s} puzzle" in the proceedings of Physics In Collisions in 2009 (PIC2009).[18]

Basically, the puzzle came about because around 2008, the error of the lattice computation dropped significantly due to the HPQCD's new result on f_{D_s}.[19] The tension with the experiments then increased to about 3.8σ around that time, making people contemplating about possible new physics to clarify the puzzle. However, as time goes on, this tension is finally eased, mainly due to the fact that new experimental results[20] came down quite a bit while the lattice result also moves up a little, so that by the end of 2010, the discrepancy is only 1.6σ.[21] Most recent lattice calculations yields compatible results with those in 2009, with the errors

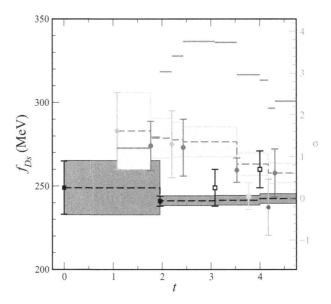

Fig. 3.   The history of f_{D_s} till the end of 2009, taken from A. Kronfeld's review talk "The f_{D_s} puzzle" for PIC2009.

further reduced, see e.g. C. Bernard's talk at Lattice 2013.[22] Note that, although there is no significant "puzzle" right now for f_{D_s}, the story of the the the so-called "f_{D_s} puzzle" has given us a very instructive lesson: precise comparison between the experiments and theory is vital in this game. In the future, should there be more accurate experimental results come about, e.g. at BESIII, there could be further puzzles and by resolving these puzzles, we could sharpen our knowledge about QCD and beyond.

3. Radiative Transitions and Decays of Charmonia

Charmonium states play an important role in our understanding of QCD. For charmonia lying below the open charm threshold, radiative decays are important to illuminate the structure of these states. Recently and in the years to come, more experimental data are being accumulated at BEPCII which will enable us to make precise comparison between the experiments and theory.

To lowest order in QED, the amplitude for $J/\psi \to \gamma H$ is given by

$$M_{r,r_\gamma,r_H} = \epsilon_\mu^*(\vec{q}, r_\gamma)\langle H(\vec{p}_f, r_H)|j^\mu(0)|J/\psi(\vec{p}_i, r)\rangle, \tag{11}$$

where $\vec{p}_i/\vec{p}_f$ is the initial/final three-momentum of the hadron, respectively, while $\vec{q} = \vec{p}_i - \vec{p}_f$ is the three-momentum of the real photon. We use r, r_H, r_γ to denote the polarizations of the relevant particles.[d] $\epsilon_\mu(\vec{q}, r_\gamma)$ is the polarization

---

[d]If the final hadron H is a scalar, then the label r_H is not needed.

four-vector of the real photon and $j^\mu_{e.m.}(0)$ stands for the electromagnetic current operator due to the quarks. We emphasize that the hadronic matrix element $\langle H(\vec{p}_f, r_H)|j^\mu(0)|J/\psi(\vec{p}_i, r)\rangle$ is non-perturbative in nature and consequently should be computed using non-perturbative methods like lattice QCD.

It turns out that the above mentioned matrix element can be obtained from the following three-point correlation function,

$$\Gamma^{(3)}_{i,\mu,j}(\vec{p}_f, \vec{q}; t_f, t) = \frac{1}{T} \sum_{\vec{y}, \tau=0}^{T-1} e^{-i\vec{q}\cdot\vec{y}} \langle \Phi^{(i)}(\vec{p}_f, t_f + \tau) J_\mu(\vec{y}, t+\tau) O_{V,j}(\vec{0}, \tau)\rangle, \quad (12)$$

where $J_\mu = \bar{c}\gamma_\mu c$ is the vector current of the charm quark. By measuring three-point function in Eq. (12) together with relevant two-point functions, one could obtain the desired hadronic matrix element $\langle H(\vec{p}_f, r_H)|j^\mu(0)|J/\psi(\vec{p}_i, r)\rangle$.

For charmonium states various lattice computations, both quenched[23] and unquenched,[24–26] have been performed and the results can be compared with the corresponding experiments. For example, for the radiative transition rate $J/\psi \to \gamma\eta_c$, it is parameterized by a form factor $V(q^2)$:

$$\langle \eta_c(p')|\bar{c}\gamma^\mu c|J/\psi(p)\rangle = \frac{2V(q^2)}{M_{J/\psi} + M_{\eta_c}} \epsilon^{\mu\alpha\beta\gamma} p'_\alpha p_\beta \epsilon(p)_\gamma, \quad (13)$$

and the decay rate is given by the value of $V(0)$. The HPQCD result finally gives:[25]

$$V(0) = 1.90(7)(1), \quad (14)$$

which is larger than the experimental value from CLEO[27] by about 1.7σ. Another lattice calculation using 2 flavors of twisted mass fermions yields even larger (but compatible with that of HPQCD) result.[26] Right now, the experimental error is larger than those in lattice computations. Later on, more accurate experiments at BESIII will surely bring more stringent test to this comparison.

Not only can one computes transitions among charmonium states, one could also compute the radiative transitions of J/ψ to pure-gauge glueballs on the lattice. This has been done recently by CLQCD in quenched lattice QCD.[28, 29] The relevant hadronic matrix element $\langle G(\vec{p}_f, r_G)|J_\mu(0)|V(\vec{p}_i, r)\rangle$ are expanded in terms of form factors and known kinematic functions, see e.g. Ref. 30. Take the tensor glueball as an example, we have

$$\langle G(\vec{p}_f, r_G)|J_\mu(0)|V(\vec{p}_i, r)\rangle = \alpha^\mu_1 E_1(Q^2) + \alpha^\mu_2 M_2(Q^2)$$
$$+ \alpha^\mu_3 E_3(Q^2) + \alpha^\mu_4 C_1(Q^2) + \alpha^\mu_5 C_2(Q^2), \quad (15)$$

where α^μ_i's are kinematic functions and E_1, M_1, E_2 etc. are the form factors. For the scalar glueball, one only needs two form factors $E_1(Q^2)$ and $C_1(Q^2)$ instead.

The physical decay rates for $J/\psi \to \gamma G$ only depend on the values of the form factors at $Q^2 = 0$:

$$\Gamma(J/\psi \to \gamma G_{0^{++}}) = \frac{4\alpha|\vec{p}_\gamma|}{27 M^2_{J/\psi}}|E_1(0)|^2, \quad (16)$$

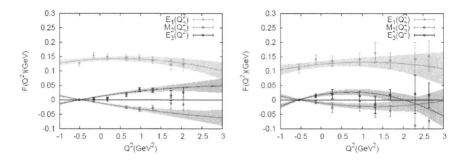

Fig. 4.   The extracted form factors for tensor glueball at two different lattice spacings.[29]

$$\Gamma(J/\psi \to \gamma G_{2^{++}}) = \frac{4\alpha|\vec{p}_\gamma|}{27M_{J/\psi}^2}\left(|E_1(0)|^2 + |M_2(0)|^2 + |E_3(0)|^3\right). \qquad (17)$$

Therefore, the computed matrix elements at various values of Q^2 are then fitted using a polynomial in Q^2 to extract the relevant values at $Q^2 = 0$. In Fig. 4, we show the case for the tensor glueball at two different lattice spacings. After the continuum extrapolation, the physical values for these form factors are obtained which then gives us the prediction of relevant decay rates.

Needless to say, the so-called pure-gauge glueballs are not physical objects that can be measured directly in experiments. In real world they mix with ordinary hadrons with the same quantum numbers. However, the above mentioned lattice calculation can provide us with important information about the pure-gauge glueball component in the measured hadronic states. For example, this calculation helps to clarify, say in the scalar channel, which of the three candidates $f_0(1370)$, $f_0(1500)$ and $f_0(1710)$ contains more glueball component and can thus be regarded as the best candidate for the scalar glueball.[28]

4. The Anomalous Magnetic Moment of the Muon

A precisely measured quantity that brings significant deviation from the prediction of the Standard Model (SM) is the anomalous magnetic moment of the muon. The definition of this quantity is

$$a_\mu = (g_\mu - 2)/2. \qquad (18)$$

This is one of the most accurately measured quantities:

$$a_\mu^{\text{exp}} = 11659208.9(5.4)(3.3) \times 10^{-10}. \qquad (19)$$

The same quantity predicted by the Standard Model is, however,

$$a_\mu^{\text{SM}} = 11659180.2(0.2)(4.2)(2.6) \times 10^{-10}. \qquad (20)$$

The difference of the above two equations is

$$\Delta a_\mu \equiv a_\mu^{\text{exp}} - a_\mu^{\text{SM}} = 287(63)(49) \times 10^{-11}. \tag{21}$$

This is roughly a 3.6σ effect, one of the remaining "puzzles" in SM.

Although the major part of a_μ comes from QED, the major *theoretical uncertainty* comes from the hadronic contributions, denoted as a_μ^{Had}, and is given by

$$a_\mu^{\text{Had}} = a_\mu^{\text{HVP}} + a_\mu^{\text{HLbL}} + \cdots. \tag{22}$$

Here, the leading contribution comes from Hadronic Vacuum Polarisation (HVP), a_μ^{HVP}, that can be measured experimentally with the help of dispersion relations. The next-order correction, the Hadronic Light-by-light scattering (HLbL), a_μ^{HLbL}, cannot be related to experimentally measurable quantities and currently relies on modelling, see e.g. PDG reviews.[31]

There have been several lattice attempts to compute a_μ^{HVP} since past decade,[32–37] which can be related to the current-current correlator in QCD:

$$a_\mu^{\text{HVP}} = \alpha^2 \int_0^\infty \frac{dQ^2}{Q^2} w(Q^2/m_\mu^2)\hat{\Pi}(Q^2), \tag{23}$$

where $w(Q^2/m_\mu^2)$ is a known function;[32] The quantity $\hat{\Pi}(Q^2) = \Pi(Q^2) - \Pi(0)$ is defined via:

$$\begin{cases} \Pi_{\mu\nu}(Q) \equiv (Q_\mu Q_\nu - Q^2 \delta_{\mu\nu})\Pi(Q^2) \\ \Pi_{\mu\nu}(Q) = \int d^4x\, e^{iQ\cdot x} \langle\Omega|T[J_\mu(x)J_\nu(0)]|\Omega\rangle \end{cases}. \tag{24}$$

In principle, the current-current correlator $\langle\Omega|T[J_\mu(x)J_\nu(0)]|\Omega\rangle$ can be computed using standard methods in lattice QCD. However, it does pose several challenges for the current state of the art lattice computations. It turns out that this quantity is dominated by contributions at low Q^2, typically $Q^2 \sim 1\text{GeV}^2$, which is difficult for current realistic lattice calculations. It also requires rather delicate error controlling in lattice extrapolations in order to match the precision of the experimental measurements. Right now, the lattice results are in no comparison with those using dispersion relations yet, as far as the errors are concerned. However, with more and more lattice groups are joining the game, see e.g. Lattice 2013 talks,[38] hopefully we will get a better control in the years to come.

In Fig. 5, I show a summary figure taken partly from Gregory's talk[39] at Lattice 2013. I have added another new point from ETMC collaboration using $2 + 1 + 1$ twisted mass fermion configurations.[40] All available lattice data for $a_\mu^{\text{HVP}} \equiv a_\mu^{\text{had,LO}}$ are summarized together with those obtained using dispersion relations. It is seen that, although some of the lattice results are compatible with those of dispersive analysis, they still need much improvement to reduce the errors in order to match the accuracy for the dispersive analysis which is comparable to that of the experimental ones.

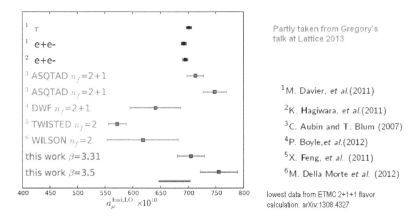

Fig. 5. The current status of lattice computation for a_μ, taken partly from Gregory's talk at Lattice 2013.[39]

5. Conclusions and Outlook

In recent years, as a theoretical tool from first principles, lattice QCD has become an important player in relevant field of physics involving the strong interaction. Due to the crossover from quenched to unquenched calculations, more and more lattice results are available that are both practical and accurate enough to be compared with the existing or upcoming experiments. Some of these lattice results are reviewed here in this talk, with an emphasis on the properties involving the charm quark. I have also try to emphasize the interplay between the experimental results and those obtained from lattice computations.

For the hadron masses and decays, we have seen both post-dictions and pre-dictions that agree rather well with the experiments. Lattice computations nowadays can also handle hadron-hadron scattering processes, which not only helps to obtain important information about hadronic interactions for the resonances, but also to clarify some of the newly discovered exotic hadronic states, the so-called XYZ particles. For the radiative transition and decays of charmonium, lattice computations have matured to a stage that a detailed comparison with experiments is possible. It is also possible to offer us information about glueballs in the radiative decays.

Many of the quantities mentioned above can be obtained rather precisely in lattice calculations, thus providing a precision test for QCD. It is important and instructive to compare the available lattice QCD results with the experiments. Some of the lattice results agree with the experiment perfectly: the hyperfine splitting of the charmonium $M_{J/\psi} - M_{\eta_c}$, for example; some even awaits further more accurate experiments: the charmonium radiative transition rate $\Gamma_{J/\psi \to \gamma\eta_c}$ and the decay constant f_{D_s}. There are also quantities that the lattice cannot compute accurately enough. An example is the muon $g-2$, a quantity showing 3.6σ deviation from the standard model. It is feasible to compute the leading hadronic contributions from

QCD first principles, however, more efforts are needed to bring down the error bars in future lattice calculations. It is in this constant process of comparison between the experiments and theory that we could sharpen our understanding of the theory of strong interaction.

Acknowledgments

This work is supported in part by the National Science Foundation of China (NSFC) under the project No. 11335001 and No.11021092. It is also supported in part by the DFG and the NSFC (No.11261130311) through funds provided to the Sino-Germen CRC 110 "Symmetries and the Emergence of Structure in QCD".

References

1. *Lattice 2013 Homepage.* http://www.lattice2013.uni-mainz.de/37_ENG_HTML.php.
2. *Lattice 2013 webpage.* http://www.lattice2013.uni-mainz.de/presentations/ Plenaries%20Saturday/Thomas.pdf.
3. M. Lüscher and U. Wolff, *Nucl.Phys.* **B339**, 222 (1990).
4. J. J. Dudek, R. G. Edwards, M. J. Peardon, D. G. Richards and C. E. Thomas, *Phys.Rev.* **D82**, p. 034508 (2010).
5. S. Durr, Z. Fodor, J. Frison, C. Hoelbling, R. Hoffmann *et al.*, *Science* **322**, 1224 (2008).
6. C. McNeile, C. Davies, E. Follana, K. Hornbostel and G. Lepage, *Phys.Rev.* **D86**, p. 074503 (2012).
7. M. Lüscher, *Nucl.Phys.* **B354**, 531 (1991).
8. *Lattice 2013 webpage.* http://www.lattice2013.uni-mainz.de/presentations/ Plenaries%20Saturday/Doring.pdf.
9. J. J. Dudek, R. G. Edwards and C. E. Thomas, *Phys.Rev.* **D87**, p. 034505 (2013).
10. N. Li and C. Liu, *Phys.Rev.* **D87**, p. 014502 (2013).
11. S. Choi *et al.*, *Phys.Rev.Lett.* **91**, p. 262001 (2003).
12. B. Aubert *et al.*, *Phys.Rev.Lett.* **95**, p. 142001 (2005).
13. S. Choi *et al.*, *Phys.Rev.Lett.* **100**, p. 142001 (2008).
14. M. Ablikim *et al.*, *Phys.Rev.Lett.* **110**, p. 252001 (2013).
15. G.-Z. Meng *et al.*, *Phys.Rev.* **D80**, p. 034503 (2009).
16. S. Prelovsek and L. Leskovec, *arXiv:1307.5172* (2013).
17. S. Prelovsek and L. Leskovec, *arXiv:1308.2097* (2013).
18. A. S. Kronfeld, *arXiv:0912.0543* (2009).
19. E. Follana, C. Davies, G. Lepage and J. Shigemitsu, *Phys.Rev.Lett.* **100**, p. 062002 (2008).
20. *Heavy Flavor Averaging Group*, (2010). http://www.slac.stanford.edu/xorg/hfag/ charm/index.html.
21. C. Davies, C. McNeile, E. Follana, G. Lepage, H. Na *et al.*, *Phys.Rev.* **D82**, p. 114504 (2010).
22. *Lattice 2013 webpage.* http://www.lattice2013.uni-mainz.de/presentations/6C/ Bernard.pdf.
23. J. J. Dudek, R. G. Edwards and D. G. Richards, *Phys.Rev.* **D73**, p. 074507 (2006).
24. Y. Chen, D.-C. Du, B.-Z. Guo, N. Li, C. Liu *et al.*, *Phys.Rev.* **D84**, p. 034503 (2011).
25. G. Donald, C. Davies, R. Dowdall, E. Follana, K. Hornbostel *et al.*, *Phys.Rev.* **D86**, p. 094501 (2012).

26. D. Becirevic and F. Sanfilippo, *JHEP* **1301**, p. 028 (2013).
27. R. Mitchell *et al.*, *Phys.Rev.Lett.* **102**, p. 011801 (2009).
28. L.-C. Gui, Y. Chen, G. Li, C. Liu, Y.-B. Liu *et al.*, *Phys.Rev.Lett.* **110**, p. 021601 (2013).
29. Y.-B. Yang, L.-C. Gui, Y. Chen, C. Liu, Y.-B. Liu *et al.*, *Phys.Rev.Lett.* **111**, p. 091601 (2013).
30. Y.-B. Yang, Y. Chen, L.-C. Gui, C. Liu, Y.-B. Liu *et al.*, *Phys.Rev.* **D87**, p. 014501 (2013).
31. J. Beringer *et al.*, *Phys.Rev.* **D86**, p. 010001 (2012).
32. T. Blum, *Phys.Rev.Lett.* **91**, p. 052001 (2003).
33. M. Gockeler *et al.*, *Nucl.Phys.* **B688**, 135 (2004).
34. C. Aubin and T. Blum, *Phys.Rev.* **D75**, p. 114502 (2007).
35. X. Feng, K. Jansen, M. Petschlies and D. B. Renner, *Phys.Rev.Lett.* **107**, p. 081802 (2011).
36. P. Boyle, L. Del Debbio, E. Kerrane and J. Zanotti, *Phys.Rev.* **D85**, p. 074504 (2012).
37. M. Della Morte, B. Jager, A. Juttner and H. Wittig, *JHEP* **1203**, p. 055 (2012).
38. *Lattice 2013 webpage.* http://www.lattice2013.uni-mainz.de/static/HStr.html#contrib4120.
39. *Lattice 2013 webpage.* http://www.lattice2013.uni-mainz.de/presentations/9B/Gregory.pdf.
40. F. Burger, X. Feng, G. Hotzel, K. Jansen, M. Petschlies *et al.*, *arXiv:1308.4327* (2013).

Physics in Collision (PIC 2013)
International Journal of Modern Physics: Conference Series
Vol. 31 (2014) 1460283 (13 pages)
© The Author
DOI: 10.1142/S201019451460283X

Reactor antineutrino experiments

Haoqi Lu

Institute of High Energy Physics, Chinese Academy of Sciences
Yuquan Road 19-B, Beijing 100049, P. R. China
Member of Daya Bay collaboration
luhq@ihep.ac.cn

Published 15 May 2014

Neutrinos are elementary particles in the standard model of particle physics. There are three flavors of neutrinos that oscillate among themselves. Their oscillation can be described by a 3×3 unitary matrix, containing three mixing angles θ_{12}, θ_{23}, θ_{13}, and one CP phase. Both θ_{12} and θ_{23} are known from previous experiments. θ_{13} was unknown just two years ago. The Daya Bay experiment gave the first definitive non-zero value in 2012. An improved measurement of the oscillation amplitude $\sin^2 2(\theta_{13}) = 0.090^{+0.008}_{-0.009}$ and the first direct measurement of the $\bar{\nu}_e$ mass-squared difference $|\Delta m^2_{ee}| = (2.59^{+0.19}_{-0.20}) \times 10^{-3}\text{eV}^2$ were obtained recently. The large value of θ_{13} boosts the next generation of reactor antineutrino experiments designed to determine the neutrino mass hierarchy, such as JUNO and RENO-50.

Keywords: Antineutrino; daya bay; θ_{13}; mass hierarchy.

PACS Numbers: 14.60.Pq, 29.40.Mc, 13.15.+g

1. Introduction

The neutrino is a fundamental particle and was first discovered in 1956 by Cowan and Reines.[1,2] In the last few decades, it has been proven that the observed neutrino oscillations can be described in a 3-flavor neutrino framework. A parameterization of the standard Pontecorvo-Maki-Nakagawa-Sakata (PMNS) matrix describing the unitary transformation relating the mass and flavor eigenstates, defines the three mixing angles (θ_{23}, θ_{12}, and θ_{13}) and one charge-parity(CP)-violating phase.[3,4] θ_{12} is about 34° and determined by solar and reactor neutrino experiments. θ_{23} is about 45° and determined by atmospheric and accelerator neutrino experiments. An upper limit of the last unknown angle θ_{13} was given by CHOOZ $\sin^2(2\theta_{13})<0.15$ at 90% confidence level (C.L.).[5] It was hinted to be non-zero by recent results from the T2K,[6] MINOS[7] and Double Chooz experiments.[8] The value of θ_{13} will guide the designs of future experiments for the measurements of the mass hierarchy and CP-violation.

Long-baseline accelerator experiments have limited sensitivity to the θ_{13} mixing angle due to dependencies on the yet unknown mass hierarchy and CP-violating phase. Reactor antineutrino experiments can provide a clear and accurate measurement of θ_{13}, due to their pure antineutrino source, clear signal and independence of the CP phase and matter effects. For reactor-based antineutrino experiments, θ_{13} can be determined in terms of the survival probability of $\bar{\nu}_e$ at certain distances from the reactors,

$$P(\bar{\nu}_e \to \bar{\nu}_e) = 1 - \cos^4 \theta_{13} \sin^2 2\theta_{12} \sin^2 \Delta_{21}$$

$$- \sin^2 2\theta_{13}(\cos^2 \theta_{12} \sin^2 \Delta_{31} + \sin^2 \theta_{12} \sin^2 \Delta_{32}) \tag{1}$$

where $\Delta_{ji} \equiv 1.267\Delta_{ji}^2(\text{eV}^2)$ L(m)/E(MeV), E is the $\bar{\nu}_e$ energy in MeV and L is the distance in meters between the $\bar{\nu}_e$ source and the detector. Δm_{ji}^2 is the difference between the mass-squares of the mass eigenstates ν_j and ν_i ($\Delta m_{ji}^2 = \Delta m_j^2 - \Delta m_i^2$). Since $\Delta m_{21}^2 \ll |\Delta m_{31}^2| \approx |\Delta m_{32}^2|$,[9] the short distance ($\sim$ km) reactor $\bar{\nu}_e$ oscillation is mainly determined by the Δ_{3i} terms.

Most reactor antineutrino oscillation experiments measure antineutrino events via the inverse beta decay(IBD) reaction $\bar{\nu} + p \to e^+ + n$. The IBD reaction is characterized by two time correlated events, the prompt signal coming from the production and subsequent annihilation of the positron, and the delayed signal from the capture of the neutron in the liquid scintillator. For the Daya Bay experiment, 0.1% gadolinium (Gd)-doped liquid scintillator is used to increase the capture cross section of thermal neutrons on Gd and reduce the capture time (about 30 μs) to suppress accidental coincidence backgrounds. The total energy of gammas from neutron capture on Gd is about 8 MeV, which is much higher than the energy range of background radioactivity. The observable antineutrino spectrum is shown in Fig. 1.

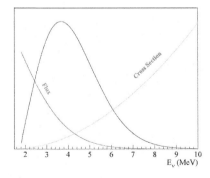

Fig. 1.   The observable $\bar{\nu}$ spectrum, a product of the antineutrino flux from reactors and cross section of IBD reaction.

Table 1.   Comparison of three reactor experiments.

Exp.	Power (GW)	Det.(t) (N/F)	Overburden (m.w.e)N/F	3y Sens. (90%CL)
Double Chooz	8.5	8/8	120/300	0.03
RENO	16.5	16/16	120/450	0.02
DYB	17.4	80/80	250/860	0.008

Due to the small value of θ_{13}, precise measurements are required to reduce the experimental uncertainties. In previous reactor antineutrino experiments, such as Palo Verde[10] and CHOOZ,[5] single detectors were utilized for antineutrino detection. Their measurements had large uncertainty from errors related to detector response and antineutrino flux predictions. There are three ongoing reactor experiments : Daya Bay, Double Chooz, and RENO. They use multiple detectors at different baselines to reduce correlated uncertainties by a relative measurement. A comparison of the three experiments is shown in Table 1. The baselines of their far detectors are 1.65, 1.05 and 1.44 km, respectively.[11]

The following content of this paper mainly focuses on the Daya Bay experiment and future prospects of reactor antineutrino experiments.

2. Daya Bay Experiment

The Daya Bay experiment is designed to explore the unknown value of θ_{13} by measuring the survival probability of electron antineutrinos from the nuclear reactors in Daya Bay, China. The Daya Bay Nuclear Power Plant complex, one of the 5 most prolific sources of reactor neutrinos in the world, consists of 6 reactors (see Fig. 2) producing 17.4 GW of total thermal power. The goal of the experiment is to

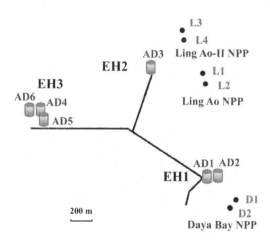

Fig. 2.   Layout of the Daya Bay experiment. There are six reactor cores (D1, D2, L1, L2, L3, L4) and 3 experimental halls (EH1, EH2, EH3).

measure θ_{13} with a sensitivity of $\sin^2 2\theta_{13} < 0.01$ at a 90% C.L. In order to achieve such high sensitivity, the experiment is optimized in several aspects. Multiple sites (one far experimental hall (EH3) and two near experimental halls (EH1,EH2)) are used to effectively cancel the flux uncertainty by relative measurements. The experiment employs 8 identically designed detectors to decrease detector related errors. As shown in Fig. 2, only 6 detectors were deployed before August. 2012. All 8 detectors were installed by October, 2012. The detectors are installed underground to reduce the cosmic-ray muon flux. Each site has redundant muon detectors (water Cerenkov detectors and resistive plate chambers (RPCs)) for muon identification.

2.1. *Detector*

2.1.1. *Antineutrino detector*

The antineutrino detectors (ADs) are filled with Gd doped liquid scintillator for antineutrino event detection. The experiment uses 8 functionally identical detectors (2 at EH1, 2 at EH2 and 4 at EH3), which are cylindrical stainless steel vessels(SSV) with a 5 m diameter and 5 meter height (Fig. 3). Each detector is instrumented with 192 8-inch photomultiplier tubes (PMTs). The detector has a three-zone structure, including a Gd-doped liquid scintillator (GdLS) zone, liquid scintillator (LS) zone and mineral oil (MO) zone. The inner region is the primary target volume filled with 0.1% Gd-LS. The middle layer is LS and act as a gamma catcher, and the outer layer is filled with MO which reduces the impact of radioactivity from PMT glass and the SSV. Two acrylic tanks are used to separate each layer. There are two reflectors at the top and bottom of an AD, which are laminated with ESR (Vikuiti^{TM} Enhanced Specular Reflector Film) film sealed between two 1-cm thick acrylic panels.The reflectors improve light collection and uniformity of detector response.

Three automated calibration units (ACUs) are located on the top of each AD (Fig. 3). The ADs are calibrated periodically by deploying LEDs and radioactive

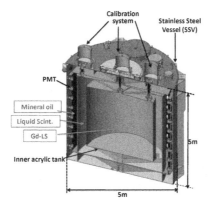

Fig. 3.   Antineutrino detector of Daya Bay experiment.

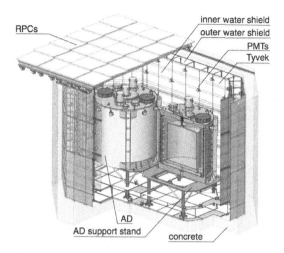

Fig. 4. Veto system of Daya Bay experiment. It includes the inner water shield, outer water shield and top RPC detector.

sources inside the detectors. LEDs are used for timing and PMT gain calibration. Energy calibration is performed with radioactive sources. Based on the first three months of data, the first pair of ADs in EH1 are shown to be functionally identical.[12]

2.1.2. *Muon veto system*

The muon veto system of the Daya Bay experiment is shown in Fig. 4. The antineutrino detectors are immersed in an octagonal pool of ultrapure water. The pool is divided into outer and inner volumes by Tyvek sheets. Both volumes are instrumented with PMTs as active muon detectors via muon Cherenkov light. The Tyvek sheet with very high reflectivity (>95%) can increase light collection efficiency. The outer layer of the water pool is 1 m thick and the thickness of the inner layer is >1.5 m. At least 2.5 m of water surrounds each AD to shield against ambient radioactivity. 288 8-inch PMTs are installed in each near hall pool and 384 in the Far Hall. There is a water circulation and purification system in each hall to maintain water quality. The tops of the water Cherenkov detectors are covered by 4 layers of RPCs. RPCs are gaseous detectors with resistive electrodes operating in streamer mode at Daya Bay.[13] RPC signals are read out by external strips, with a position resolution of about 8 cm. There are 54 modules in each near hall and 81 modules in the Far Hall. The designed efficiency is >99.5% with uncertainty less than 0.25% by combining the water Cherenkov and RPC detectors. From muon data analysis, water Cherenkov detector efficiency is >99.7% for long track muons,[12] which is better than the design requirement.

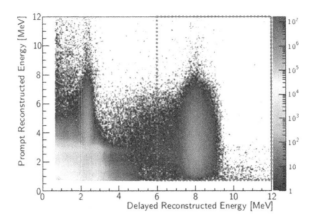

Fig. 5. IBD candidate events are shown within the dashed lines.

2.2. Data analysis

2.2.1. Event selection

After energy calibration and event reconstruction, we select the IBD events under the following criteria:

- PMT light emission events will be rejected by the PMT flasher cut.[12]
- The energy of prompt and delayed candidates satisfy 0.7 MeV < Ep < 12 MeV and 6.0 MeV < Ed < 12 MeV, where Ep is the prompt signal energy and Ed is the delayed signal energy.
- The delayed signal (neutron capture) time satisfies: $1\mu s <t < 200\mu s$, where t is the time between the prompt and delayed signal.
- If a muon goes through the water pool and fires >12 PMTs, it will be treated as a 'water pool muon'. The rejection time window of water pool muons is $[-2\mu s, 600\mu s]$.
- If the visible energy in the AD is (>20 MeV), it will be treated as an 'AD muon'. The rejection time window of AD muon is $[-2\mu s, 1400\mu s]$.
- If the visible energy in the AD is >2.5 GeV, the rejection time window is $[-2\mu s, 0.4s]$
- There must be no additional prompt-like signals $400\mu s$ before the delayed neutron signal and no additional delayed-like signal $400\mu s$ after the delayed neutron signal.

As Fig. 5 shows, the events within the dash lines are the IBD candidate events after selection.

2.2.2. Background

Background control is crucial for a precise measurement in reactor antineutrino experiments. Li9/He8 is the main background from cosmic-ray muons. It is a correlated background that can mimic prompt signals with a beta-decay signal and delayed signals with neutron capture. This background is directly measured by fitting the distribution of IBD candidates versus time since the last muon. The background to signal ratio (B/S) of Li9/He8 is $\sim 0.3 \pm 0.1\%$. Another background from muons is the fast neutron background. Energetic neutrons from muons can give a prompt signal from proton recoil and a delayed signal from the neutron's capture on Gd. By relaxing the prompt energy range from 12 MeV to 50 MeV in the IBD events selection, the fast neutron spectrum in the range of <12 MeV can be extrapolated from the higher energy distribution. The fast neutron spectrum and rate is cross-checked with fast-neutron samples tagged by the muon veto system. The B/S ratio of fast neutrons is 0.1% (0.06%) for the near(far) sites.

Two uncorrelated signals can accidentally mimic an IBD signal. This accident background can be precisely measured from data. The B/S of accidentals is 1.5% (4%) for the near(far) sites. The error in the accidental rate is 1% (4%).

The ^{238}U, ^{232}Th, ^{227}Ac decay chains and ^{210}Po produce a ^{13}C(α n)^{16}O backgrounds. This background is estimated by Monte Carlo(MC). The B/S of this background is 0.01% (0.05%) for the near(far) sites. Am-C calibration sources inside the ACUs on top of each AD constantly emit neutrons, which would occasionally mimic IBD signals by scattering inelastically with nuclei in the shielding material (emitting gamma rays) before being captured on a metal nucleus, such as Fe, Cr, Mn or Ni (releasing more gamma rays). MC is also used to estimate the rate of this background. The background rate and shape are constrained by taking data with a temporary intense source on top of the AD. The B/S ratio of Am-C sources is 0.04% (0.35%) for the near(far) sites.

2.2.3. *Recent results*

The Daya Bay experiment first presented the discovery of a non-zero $\theta_{13}(5.2\sigma)$ in 2012.[14] The most recent analysis[16] is based on the full six-AD data sample from Dec. 24, 2011 to July 28, 2012, which includes more than 300,000 IBD interactions. The rate-only analysis yields $\sin^2(2\theta_{13}) = 0.090 \pm 0.010$ with $\chi^2/\text{NDF} = 0.6/4$. Spectral information is also used after applying an energy nonlinearity correction to the positron spectrum. The oscillation parameters can be extracted from a fit taking into account both rate and spectral information. In Eq. 1, the short baseline ($\sim$ km) reactor $\bar{\nu}_e$ oscillation is mainly determined by the Δ_{3i} terms. We use the following definition[15] of the effective mass-squared difference $\sin^2 \Delta ee \equiv \cos^2 \theta_{12} \sin^2 \Delta_{31} + \sin^2 \theta_{12} \sin^2 \Delta_{32}$. The best-fit values are $\sin^2(2\theta_{13}) = 0.090^{+0.008}_{-0.009}$ and $|\Delta m^2_{ee}| = (2.59^{+0.19}_{-0.20}) \times 10^{-3} \text{eV}^2$ with $\chi^2/\text{NDF} = 163/153$ (68.3% C.L.). The 68.3%, 95.5%, and 99.7% C.L. allowed regions in the $|\Delta m^2_{ee}|$ vs. $\sin^2(2\theta_{13})$ plane are shown in Fig. 6.

The relative deficit and spectral distortion observed between far and near ADs at Daya Bay give the first independent measurement of $|\Delta m^2_{ee}| = (2.59^{+0.19}_{-0.20}) \times 10^{-3} \text{eV}^2$ and the most precise measurement of $\sin^2(2\theta_{13}) = 0.090^{+0.008}_{-0.009}$ to date.

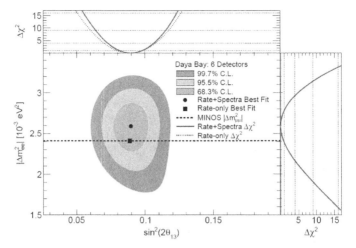

Fig. 6.   Allowed regions for the neutrino oscillation parameters $\sin^2(2\theta_{13})$ and $|\Delta m^2_{ee}|$ at different confidence levels(solid regions). The best estimates of the oscillation parameters are shown by the black dot. The adjoining panels show the dependence of $\Delta\chi^2$ on $|\Delta m^2_{ee}|$(right) and $\sin^2(2\theta_{13})$ (top). The black square and dashed curve represent the rate-only result. The dashed horizontal line represents the MINOS $|\Delta m^2_{\mu\mu}|$ measurement[17].

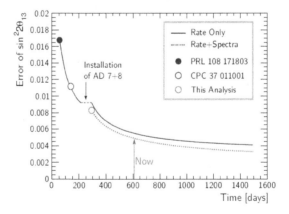

Fig. 7.   Error of $\sin^2(2\theta_{13})$ vs time. The red circle is this analysis result. The current data-taking time is above 600 days.

2.3. Future plan

The sensitivity of $\sin^2(2\theta_{13})$ versus time for Daya Bay is shown in Fig. 7. The projected uncertainty of $\sin^2(2\theta_{13})$ is less than 4% after 3 years of data taking. Daya Bay is also expecting to measure $|\Delta m^2_{ee}|$ in complementary precision to accelerator neutrino experiments. Fig. 8 shows the $|\Delta m^2_{ee}|$ error versus data-taking time. The error of $|\Delta m^2_{ee}|$ will be $< 0.10\times10^{-3}\text{eV}^2$ after 3 years of data-taking.

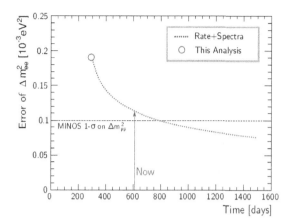

Fig. 8. Error of $|\Delta m^2_{ee}|$ vs time. The red circle is this analysis result. The current data-taking time is above 600 days. The horizontal dashed line is the error of $|\Delta m^2_{\mu\mu}|$ from MINOS.

2.4. *RENO and double chooz status*

RENO updated their result in March, 2013.[18] They gave $\sin^2(2\theta_{13}) = 0.100 \pm 0.010(\text{stat}) \pm 0.015(\text{syst})$ and observed a clear rate deficit (7.1% reduction) between far and near sites. RENO aims to achieve 7% uncertainty in $\sin^2(2\theta_{13})$ and suppress their systemic error to 0.5%. Double Chooz[19] provided their new results from a combined fit of Gd capture and Hydrogen capture with $\sin^2(2\theta_{13}) = 0.109 \pm 0.035$. For Double Chooz, the near detector is under construction and will start running in 2014.

3. Future Prospects

The next generation of neutrino oscillation experiments will mainly focus on the mass hierarchy and CP phase. The mass hierarchy can be determined by a precisely energy spectrum measurement in reactor neutrino oscillations, which is independent of the CP-violating phase. Figure 9 shows the L/E distribution of reactor antineutrino oscillations. It is found that frequencies of oscillation between the normal hierarchy and inverted hierarchy have some difference. The two oscillation frequency components are driven by Δm^2_{31} and Δm^2_{32}, respectively.[20–23] The difference is very sensitive to detector energy resolution. According to the study in Ref. 24, for a single detector at a baseline of 58 km and with a 35 GW reactor, the probability to determine the sign of the hierarchy has a significant difference between 3% and 4% energy resolution as shown in Fig. 10. Another big effect on mass hierarchy sensitivity is potentially from multiple reactor baselines, which are required to be within 500 m.[25] A large detector with good energy resolution (3%) and equal baselines from powerful reactor cores is required for a mass hierarchy determination.

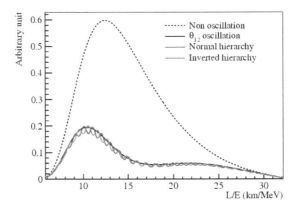

Fig. 9.   Neutrino oscillation difference between normal hierarchy and inverted hierarchy.

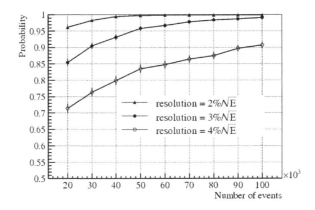

Fig. 10.   The probability to determine the sign of the hierarchy vs. the number of detected events. Different energy resolutions are used to estimate the effect on the probability.

3.1. *The Jiangmen Underground Neutrino Observatory*

The Jiangmen Underground Neutrino Observatory (JUNO) is located at Kaiping, Jiangmen City, in Guangdong Province, China, as shown in Fig. 11, 53 km from the Yangjiang and Taishan nuclear power plants. The total thermal power of the reactors is 36 GW. The detector will be constructed deep underground to reduce the cosmic-ray muon flux, with an overburden of about 700 m of rock. The detector will be filled with 20 ktons of liquid scintillator with 3% energy resolution (at 1MeV). The major goal of JUNO is to determine the mass hierarchy by precisely

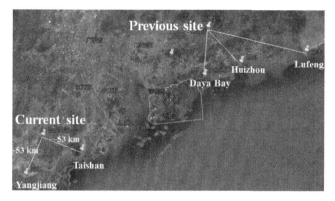

Fig. 11. Experiment site of JUNO. The Taishan and Yangjiang reactor complexes are used for JUNO. The previous site is not considered in that the third reactor complex (Lufeng) is being planed.

measuring the energy spectrum of reactor antineutrinos. However, it is a multi-purpose experiment that can also measure neutrino oscillation parameters, study atmospheric, solar and geo-neutrinos, and perform other exotic searches.

Figure 12 is the detector concept of JUNO. There are around 15,000 20-inch PMTs installed in the central detector for the 3% energy resolution. 6 ktons of mineral oil is used to shield the scintillator from PMT glass radioactivity. The muon veto system includes an outer water Cherenkov detector and a top tracking system. As for the physics prospects of JUNO,[25] if we take into account the spread of reactor cores, uncertainties from energy non-linearity, etc, the mass hierarchy sensitivity with 6 years of data-taking can reach $\Delta\chi^2 > 9$ with a relative measurement and $\Delta\chi^2 > 16$ with an absolute $\Delta m^2_{\mu\mu}$ measurement from accelerator neutrino experiments. Civil construction has begun and will complete in 2019. Liquid scintillator filling and data taking will begin around 2020.

3.2. *RENO 50*

RENO-50 was proposed in South Korea.[26] The detector will be constructed underground and consist of 18 ktons of ultra-low-radioactivity liquid scintillator and 15,000 20-inch PMTs, about 50 km away from the Hanbit nuclear power plant. The scientific goals of the experiment include a high precision measurement of θ_{12} and $|\Delta m^2_{21}|$, determination of the mass hierarchy, the observation of neutrinos from reactors, the Sun, the Earth, supernovae, and any possible stellar objects. The total budget is 100 million dollars for 6 years of construction. Facility and detector construction will be started from 2013 and finished in 2018. The experiment data taking will start in 2019.

4. Summary

Reactor-based antineutrino experiments have obtained many excellent achievements in recent years. Daya Bay measured the non-zero θ_{13} with great precision, together

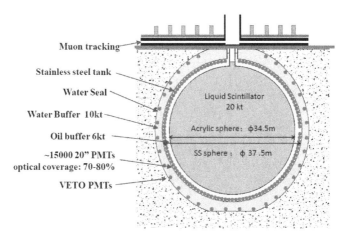

Fig. 12. Detector concept of JUNO, including the central liquid scintillator detector and outer muon veto systems.

with other experiments. The most recent results from Daya Bay provide the first independent measurement of $|\Delta m_{ee}^2|$ and the most precise measurement of $\sin^2(2\theta_{13})$ to date. The precision on $\sin^2(2\theta_{13})$ will be improved to 4% in the future. The large value of θ_{13} boosts the next generation of neutrino oscillation experiments to determine the neutrino mass hierarchy and measure the CP violation phase. Reactor-based antineutrino experiments will continue to play an important role in the mass hierarchy determination and precise measurements of oscillation parameters.

Acknowledgments

The article is supported by the National Natural Science Foundation of China (Y3118G005C). I would like to acknowledge my Daya Bay collaborators for useful suggestion and comments, especially Dr. Yufeng Li, Logan Lebanowski and Viktor Pec who helped me a lot to improve this article.

References

1. C. L. Cowan *et al. Science* **124**, 103 (1956).
2. F. Reines and C. L. Cowan, Jr., *Nature* **178**, 446 (1956).
3. B. Pontecorvo, Sov. Phys. JETP **6**, 429 (1957) and **26**, 984 (1968).
4. Z. Maki, M. Nakagawa and S. Sakata, Prog. Theor. Phys. **28**, 870 (1962).
5. CHOOZ Collab. (M. Apollonio *et al.*), *Phys. Lett. B* **466**, 415 (1999).
6. T2K Collab. (K. Abe *et al.*), *Phys. Rev. Lett.* **107**, 041801 (2011).
7. MINOS Collab. (P. Adamson *et al.*), *Phys. Rev. Lett.* **107**, 181802 (2011).
8. Double Chooz Collab. (Y. Abe *et al.*), *Phys. Rev. Lett.* **108**, 131801 (2012).
9. J. Beringer *et al.* (Particle Data Group), *Phys. Rev. D* **86**, 010001 (2012), Section 13.
10. F. Boehm *et al.*, *Phys. Rev. Lett.* **84**, 400 3764(2000).
11. J. Cao, plenary talk at the 36th International Conference on High Energy Physics, Melbourne, Jul. 2012.

12. DayaBay Collab. (F. An *et al.*), *NIM A* **685**, 78–97 (2012).
13. L. Ma *et al. NIM A* **659**, 154(2011); J. Xu *et al.*, *Chin. Phys. C* **35**, 844 (2011).
14. Daya Bay Collab. (F. An*et al.*), *Phys. Rev. Lett.* **108**, 171803 (2012).
15. The Daya Bay $|\Delta m_{ee}^2|$ definition is consistent with H. Minakata *et al.*, *Phys. Rev. D* bf 74, 053008 (2006) and H. Minakata *et al.*, *Phys. Rev. D* **76**, 053005 (2007).
16. Daya Bay Collab. (F. An *et al.*), http://arxiv.org/abs/1310.6732.
17. (MINOS Collaboration), (P. Adamson *et al.*) *Phys. Rev. Lett.* **110**, 51801 (2013).
18. Seon-Hee Seo, talk on NuTel2013, March 2013.
19. Ralitsa Sharankova, talk on NuFact2013, Aug, 2013, Beijing.
20. S. T. Petcov and M. Piai *Phys. Lett. B* **533**, 94 (2002).
21. S. Choubey, S. T. Petcov and M. Piai *Phys. Rev. D* **68**, 113006 (2003).
22. J. Learned *et al.*, hep-ex/0612022.
23. L. Zhan, Y. Wang, J. Cao, L. Wen,*Phys. Rev. D* **78**, 111103 (2008); *Phys. Rev. D* **79**, 073007 2009.
24. L. Zhan, Neutrino Oscillation Workshop 2012, September 9–16, 2012 Italy.
25. Yu-Feng Li, Jun Cao, Yifang Wang and Liang Zhan, *Phys. Rev. D* **88**, 013008 (2013).
26. Soo-Bong Kim, talk at the international workshop on RENO-50, June 13–14, 2013, Seoul.

Physics in Collision (PIC 2013)
International Journal of Modern Physics: Conference Series
Vol. 31 (2014) 1460284 (10 pages)
© The Author
DOI: 10.1142/S2010194514602841

Review of electron neutrino appearance oscillation experiments

V. Paolone

Department of Physics and Astronomy,
University of Pittsburgh Pittsburgh, Pennsylvania, USA
vipres@pitt.edu

Received 25 November 2013
Published 15 May 2014

A review of ν_e appearance oscillation experiments is presented. The measurement of
the appearance signal is of particular interest because this mode is sensitive to both
θ_{13}, δ_{CP} and other mixing parameters. Using the value of θ_{13} measured by reactor
neutrino experiments (disappearance studies) present and next generation appearance
experiments can explore CP violation.

Keywords: Nue appearance.

1. Introduction

With the establishment of atmospheric neutrino oscillations in the 1990's[1] the first
evidence of physics beyond the standard model was established and the era of
neutrino masses and mixing measurements began. There are fundamental questions
about the nature of this new physics. How are these masses generated? Is it related to
standard model physics: Higgs Mechanism? Seesaw mechanism? What implications
does it have for the early universe and the matter−antimatter asymmetry? At
present Neutrino Physics is an experimentally driven field. Study of neutrino masses
and mixings is our only known window into this new physics.

From both reactor and accelerator experiments we have direct proof that three
active neutrinos exist: ν_e,[2]ν_μ,[3] and ν_τ.[4] Within this three flavor framework neutrino oscillations are described by the PMNS matrix[5,6] and parameterized by three
mixing angles $\theta_{12}(\sim 34°)$, $\theta_{23}(\sim 45°)$, $\theta_{13}(9.1° \pm 0.6°)$, two mass differences Δm_{31}^2
and Δm_{21}^2, and a CP violating phase δ_{CP}. Neutrino oscillations can be studied both
through the disappearance and appearance channels. In the disappearance channel
the initial neutrino flavor flux is reduced and the energy distribution is altered when
comparing the neutrino energy spectra at the near and far locations. In the case of

the appearance channel an excess of a neutrino flavor, usually a small component in the near location flux, is observed in the far detector location. The $\nu_\mu \to \nu_e$ appearance oscillation probability can be written as follows

$$P(\nu_\mu \to \nu_e) \simeq \sin^2 \theta_{23} \sin^2 2\theta_{13} \sin^2 \frac{\Delta m_{31}^2 L}{4E}$$

$$- \frac{\sin 2\theta_{12} \sin 2\theta_{23}}{2 \sin \theta_{13}} \sin \frac{\Delta m_{21}^2 L}{4E} \sin^2 2\theta_{13} \sin^2 \frac{\Delta m_{31}^2 L}{4E} \sin \delta_{\rm CP}$$

$$+ \text{(CP even term, solar term, matter effect term)}, \qquad (1)$$

In this model the ν_e appearance probability is a function not only of θ_{13} but also $\delta_{\rm CP}$ allowing one to explore CP violation in the lepton sector. Other unknowns are the mass hierarchy (is $\Delta m_{31}^2 > 0$ or < 0), and is θ_{23} maximal ($= 45°$) or if not which octant ($<$ or $> 45°$).

2. $\nu_\mu \to \nu_e$ Appearance Experiments

The present experimental efforts studying the $\nu_\mu \to \nu_e$ appearance channel with published results are MINOS[7] and T2K.[8] Their original goal was to observe the appearance channel and therefore determine that θ_{13} is non-zero. In 2011, the T2K collaboration published the first indication[9] of electron neutrino appearance and therefore a non-zero θ_{13}. With the defintive establishment of a non-zero θ_{13} from the reactor experiments the efforts of T2K have shifted to exploration of CP violation. Future efforts, NOνA[10] and LBNE,[11] are being designed for improved sensitivity to CP violation and pinning down the the mass hierarchy.

2.1. MINOS

MINOS is a long-baseline neutrino oscillation experiment, which combines Fermilab's high-rate NuMI beam with a 735 km neutrino path-length (see Fig. 1), was designed to explore the region of the atmospheric neutrino oscillation signal. MINOS presently has produced the most precise determination to date of the neutrino squared mass difference Δm_{23}^2 and has a precision measurement of θ_{23}. MINOS has recently extended its searches to electron neutrino appearance to constrain θ_{13}.

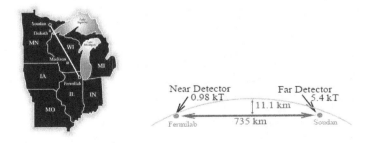

Fig. 1.   A schematic illustration of the MINOS baseline.

Both the far and near detectors are constructed from iron planes interleaved with segmented scintillator tracking planes. The near detector (ND) is 980 tons and the far (FD) 5.4 kilotons in mass. In addition The MINOS detectors are magnetized and can distinguish event by event muon neutrino charged-current (ν_μCC) from muon anti-neutrino charged-current ($\overline{\nu}_\mu$CC) interactions by measuring the sign of the charged-muon track. This MINOS feature is unique among accelerator long-baseline experiments. Minos completed data taking in 2012 and accumulated $> 10^{21}$ protons on target (PoT) in neutrino mode and 3.3×10^{20} PoT in anti-neutrino mode. MINOS distinguishes between Charged Current (CC) ν_μ, Neutral Current (NC), and CC ν_e neutrino interactions using general event shape characteristics. Penetrating/long (beam direction) and small transverse shaped events are predominately CC ν_μ. Short and moderately wide transverse event shapes are predominately NC, while short and wide events are predominately CC ν_e.

In 2012[12] based on 10.6×10^{20} PoT for neutrino running the best fit for $2\sin^2(2\theta_{13})\sin^2(\theta_{23})$ as a function of δ for normal and inverted hierarchy are shown in Fig. 2. MINOS sets a limit of $sin^2 2\theta_{13} < 0.265$ at the 90% confidence limit for the CP-violating phase $\delta = 0$.

MINOS has also preformed a combined fit of both appearance and disappearance data in the PMNS 3-flavor model. The fit used a 4D likelihood surface in θ_{13}, δ_{CP}, θ_{23}, and Δm_{32}^2 and the result is shown in Fig. 3. Constraint on θ_{13} is from the combination of Daya Bay, RENO, and Double Chooz measurements.[14] All others

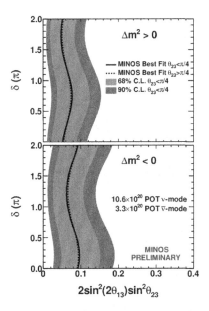

Fig. 2. The plots show the allowed ranges and best fits for $2\sin^2(2\theta_{13})\sin^2(\theta_{23})$ as a function of the CP-violating phase δ. The upper(lower) plot assumes the normal (inverted) neutrino mass hierarchy.

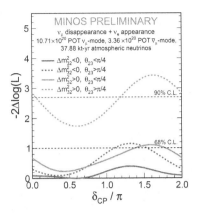

Fig. 3. $-2\Delta\ln(L)$ as a function of δ_{CP} for differing values of mass hierarchy and sign of $\theta_{23} - \frac{\pi}{4}$ relative to the best-fitting solution. Constraints on θ_{13} are used from the reactor experiments. The horizontal dashed lines correspond to 68% or 90% C.L.

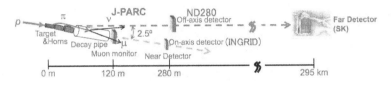

Fig. 4.   T2K beam-line schematic.

variables are fixed to their respective global average. Normal hierarchy and upper octant are disfavored at 81% C.L.

2.2. *T2K*

T2K uses the J-PARC facility operating in Tokai Japan. The beam is aimed to illuminate the off-axis detectors located at an angle of 2.5° off the beam axis (see Fig. 4). The resulting peaked beam spectrum will reduce the intrinsic ν_e beam and NC feed-down background from high energy neutrinos. The existing SuperKamiokande 50 kiloton water Cerenkov detector situated 295 km away is used to detect the oscillated neutrinos. The ND280 site, 280m from the beam source, houses a suite of detectors needed to measure the beam flux and ν_e appearance backgrounds. ND280 consists of the on-axis detector(INGRID) and the off-axis detectors as shown in Fig. 4. The INGRID is composed of seven vertical and horizontal modules interleaved with planes of iron and segmented scintillator. Its primary purpose is to measure and monitor the beam profile and stability using neutrino interactions. The ND280 off-axis detector is a hybrid detector designed to provide measurements of the ν_μ spectrum and the dominant backgrounds from the ν_e beam contamination and neutral current π^o production to improve experimental sensitivity to the $\nu_\mu \rightarrow \nu_e$ appearance signal.

Super-Kamiokande is a 22.5-kt fiducial mass ring-imaging water Cherenkov detector. It is located in the Kamioka mine at a depth of 2,700-m water equivalent. The detector consists of two optically separated concentric cylindrical regions. The inner detector is instrumented with 11,129 20-inch photo-multiplier tubes (PMT), the outer detector with 1,885 8-inch PMTs. A non-showering track such as a muon produces a sharp edged ring while an electron generates a fuzzy one due to electromagnetic showers.

T2K has been taking neutrino data since January 2010 for four running periods: Run 1 (January-June 2010), Run 2 (November 2010-March 2011), Run 3 (January-June 2012) and Run 4 (October 2012-May 2013). The Run 2 period was terminated by the March 11, 2011 Tohoku earthquake, but the accelerator and detector resumed data-taking operations in January 2012. During this period the proton beam power steadily increased and reached 220 kW continuous operation with a world record of 1.2×10^{14} protons per pulse. The present total neutrino beam exposure at SK corresponds to 6.57×10^{20} PoT.

The ν_e selection cuts used are:

- Number of veto hits < 16
- Vertex within fiducial volume
- Number of rings = 1
- Ring is e-like
- $E_{visible} > 100$ MeV
- No Michel electrons (eliminate electrons from stopping muons)
- π^0 rejection cut
- $0 <$ Reconstructed $E_\nu < 1250$ MeV

After applying these cuts 28 ν_e candidates are observed for 6.4×10^{20} PoT. Only 4.64 events are expected if $\sin^2 2\theta_{13} = 0$ as shown in Table 1. This corresponds to a significance of greater than 7 sigma excluding $\theta_{13} = 0$. Table 2 lists the contributions to the systematic errors. These errors are significantly reduced when including the constraints from the near detector(ND280) as shown in Fig. 5.

Two analysis methods were used to extract the oscillation parameters, a maximum likelihood fit using a 2D-distribution of electron momentum versus angle and a fit using reconstructed neutrino energy. Both methods produced consistent results.

Table 1. Expected number of events for different values of θ_{13} and $\sin^2\theta_{12} = 0.306$, $\Delta m^2_{21} = 7.6 \times 10^{-5}$ eV^2, $\sin^2\theta_{23} = 0.5$, $|\Delta m^2_{32}| = 2.4 \times 10^{-3}$ eV^2 and $\delta_{CP} = 0$.

	$\sin^2 2\theta_{13} = 0.0$	$\sin^2 2\theta_{13} = 0.1$
ν_e Signal	0.38	16.42
ν_e Background	3.17	2.93
ν_μ Background (mainly NC π^O)	0.89	0.89
[14] $\nu_\mu + \nu_e$ Background	0.20	0.19
Total	4.64	20.44

Table 2. Systematic uncertainties for different values of θ_{13} and $\sin^2\theta_{12} = 0.306$, $\Delta m_{21}^2 = 7.6 \times 10^{-5}$ eV^2, $\sin^2\theta_{23} = 0.5$, $|\Delta m_{32}^2| = 2.4 \times 10^{-3}$ eV^2 and $\delta_{CP} = 0$.

	$\sin^2 2\theta_{13} = 0.0$	$\sin^2 2\theta_{13} = 0.1$
Beam flux + ν int. in T2K fit	4.9%	3.0%
ν int (from other experiments)	6.7%	7.5%
Far detector	7.3%	3.5%
Total	11.1%	8.8%

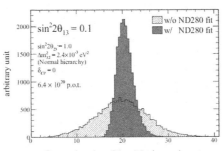

Fig. 5. Effect of ND280 constraints on systematic errors.

The 2D-distribution result is presented here since differences in electron momentum and angle distributions allow better signal and background separation and exploits detector measured variables. The 2D-distribution is shown in Fig. 6. The best fit for $\sin^2 2\theta_{13}$ with normal hierarchy (inverted hierarchy) is $0.150^{+0.039}_{-0.034}$ ($0.182^{+0.046}_{-0.040}$) and $\sin^2\theta_{12} = 0.306$, $\Delta m_{21}^2 = 7.6 \times 10^{-5}$ eV^2, $\sin^2\theta_{23} = 0.5$, $|\Delta m_{32}^2| = 2.4 \times 10^{-3}$ eV^2 and $\delta_{CP} = 0$.

As mentioned in the introduction the appearance oscillation probability depends not only on θ_{13} but many other parameters including δ_{CP}. In Fig. 7 the 68% and 90% CL allowed regions for θ_{13} are shown as a function of δ_{CP}. In addition, the variation of the CL bands as a function of allowed values for θ_{23} using the T2K disappearance result,[13] are also shown. A significant dependency on the value θ_{23} is observed. Improved measurements of θ_{23} will be important when extracting information about oscillation parameters (including δ_{CP}) in present and future LBL appearance oscillation experiments. Also shown, the vertical band in Fig. 7, is the best fit for θ_{13} from the reactor experiments (PDG2012[14]). The band is not centered, but clips the CL bands from the T2K fit which may show a preference to some values of δ_{CP} or some tension with the PMNS model. More data is needed to further reduce the width of the CL bands before a more definitive statement can be made. Beam is expected to resume in the late Spring of 2014 and continuing studies of CP violation (*i.e.* run in anti-neutrino mode) and constrain the allowed regions of δ_{CP} will be T2K's the highest priority.

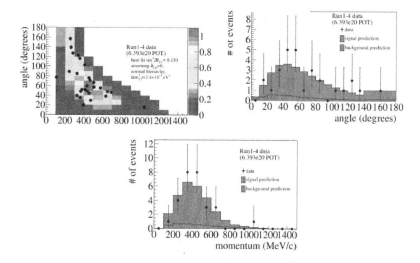

Fig. 6.   Electron momentum versus angle (including projections) for the 28 signal candidates.

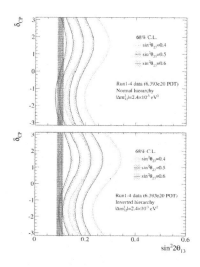

Fig. 7.   The 68% and 90% CL allowed regions for $\sin^2 2\theta_{13}$, as a function of δ_{CP} assuming normal hierarchy (top) and inverted hierarchy (bottom). The solid line represents the best fit $\sin^2 2\theta_{13}$ value for given δ_{CP} values. The values of $\sin^2 \theta_{23}$ and Δm_{32}^2 are varied in the fit with the constraint from the T2K disappearance result. The shaded region shows the average θ_{13} value from the reactor experiments.

2.3. *NOν A*

When ν's travel long baselines through matter (*i.e. the earth*) the energy levels of the propagating eigenstates are altered. It raises the effective mass of ν_e's since only this flavor of neutrino can interact with atomic electronics through the charged

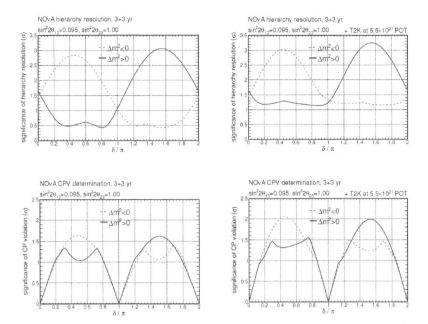

Fig. 8. NOνA mass hierarchy and δ_{CP} sensitivity curves. The left plots are NOνA alone, the right are a combined fit between NOνA and T2K at full expected exposures.

current interaction. Therefore the $\nu_\mu \to \nu_e$ appearance channel is sensitive to the sign of Δm^2 if the baseline is long enough and traveling through the earth. The NOνA experiment using a 500 mile baseline will be sensitive to the sign of Δm^2_{32}. NOνA will consist of a 200-ton near detector located at Fermilab and a 14,000-ton, 65% active far detector using 15.6 m plastic cells filled with liquid scintillator in Ash River, Minnesota. It is using the same NuMI beam-line as MINOS but with its detectors located off-axis with respect to the beam-line direction to exploit the narrow neutrino energy spectrum that is produced. NOνA started taking data in the late summer of 2013 with a partially completed far detector. As shown in Fig. 8 the best sensitivities[15] for both mass hierarchy and δ_{CP} will be obtained when combining the results from NOνA and T2K.

2.4. LBNE

The next improvement in neutrino oscillation sensitivity will be probed by LBNE (Long-Baseline Neutrino Experiment) which is currently in the planning phase. LBNE will use a high intensity "super-beam" (0.7-2.3MW) produced at Fermilab and a massive detector in South Dakota about 1300 km from the source. The 1300km baseline is optimized to observe multiple oscillation peaks. The value of θ_{13} determines the size of the event sample, while δ_{CP} controls the amplitude of the oscillations and the mass differences controls the frequency of oscillations while

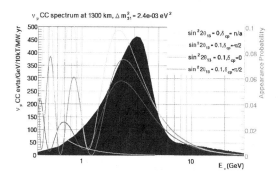

Fig. 9. Effect of different oscillation parameters on the expected LBNE appearance spectrum at a 1300 km baseline.

the mass hierarchy effects both amplitude and frequency. To resolve these degeneracies one needs a long enough baseline but short enough to record a statistically significant sample of events. These effects[11] are shown in Fig. 9.

3. Conclusions and Summary

Neutrino oscillation physics has entered an era of precision measurements. It is a window into new physics and measuring in detail the elements of the mixing/mass matrix is the only window we have to probe this new physics. At this point neutrino physics is experimentally driven and therefore we need to over constrain the parameters in the mixing and mass matrix to check if there are any inconsistencies in the model. Neutrino measurements in general are hard to make, expensive and will require improved statistics. The community will need to combine results from different experiments to improve the physics reach of present and future measurements.

References

1. SK Collab. (Fukuda, Y. *et al.*), *Phys.Rev.Lett.* **81**, 1562 (1998).
2. C. L. Cowan, F. Reines, F. B. Harrison, H. W. Kruse and A. D. McGuire, *Science.* **124**, 103 (1956).
3. G. Danby, J.-M. Gaillard, K. Goulianos, L. M. Lederman, N. B. Mistry, M. Schwartz, J. Steinberger, *Phys. Rev. Lett.* **9**, 36 (1962).
4. DONUT Collab. (K. Kodama *et al.*), *Phys. Lett. B* **504**, 218 (2001)
5. Z. Maki, M. Nakagawa and S. Sakata, *Prog. Theor. Phys.* **28**, 870 (1962).
6. B. Pontecorvo, *Sov. Phys. JETP* **26**, 984 (1968)
7. MINOS Collab. (I. Ambats *et al.*), *NUMI-L-337,FERMILAB-DESIGN-1998-02* (1998).
8. T2K Collab. (K. Abe *et al.*), *Nucl. Instrum. Meth. A* **659**, 106 (2011)
9. T2K Collab. (K. Abe *et al.*), *Phys. Rev. Lett.* **107**, 514 (2011).
10. NOνA Collab. (D. S. Ayres *et al.*), *http://nova-docdb.fnal.gov/cgi-bin/Show Document?docid=593*
11. LBNE Collab., *http://lbne2-docdb.fnal.gov/cgi-bin/ShowDocument?docid=5235&asof=2012-11-06*

V. Paolone

12. MINOS Collab. (P. Adamson *et al.*), *Phys. Rev. Lett.* **110**, 171801 (2013)
13. T2K Collab. (K. Abe *et al.*), *Phys. Rev. D* **85**, 031103 (2012)
14. Particle Data Group (J. Beringer *et al.*), *Phys. Rev. D* **86**, 2012
15. A. Waldron (Representing the NOνA Collab.), *http://indico.cern.ch/conference Display.py?confId=218030*

Physics in Collision (PIC 2013)
International Journal of Modern Physics: Conference Series
Vol. 31 (2014) 1460285 (13 pages)
© The Authors
DOI: 10.1142/S2010194514602853

World Scientific
www.worldscientific.com

Low energy neutrinos

G. Ranucci

Istituto Nazionale di Fisica Nucleare
Via Celoria 16, 20133 Milano, Italy
gioacchino.ranucci@mi.infn.it

G. Bellini, J. Benziger, D. Bick, G. Bonfini, D. Bravo, M. Buizza Avanzini, B. Caccianiga,
L. Cadonati, F. Calaprice, P. Cavalcante, A. Chavarria, A. Chepurnov, D. D'angelo, S. Davini,
A. Derbin, A. Empl, A. Etenko, K. Fomenko, D. Franco, C. Galbiati, S. Gazzana, C. Ghiano,
M. Giammarchi, M. Goeger-Neff, A. Goretti, L. Grandi, C. Hagner, E. Hungenford, Al. Ianni,
An. Ianni, V. Kobychev, D. Korablev, G. Korga, D. Kryn, M. Laubenstein, T. Lewke,
E. Litvinovich, B. Loer, F. Lombardi, P. Lombardi, L. Ludhova, G. Lukyanchenko, I. Machulin,
S. Manecki, W. Maneschg, G. Manuzio, Q. Meindl, E. Meroni, L. Miramonti, M. Misiaszek,
R. MLlenberg, P. Mosteiro, V. Muratova, L. Oberauer, M. Obolensky, F. Ortica, K. Otis,
M. Pallavicini, L. Papp, L. Perasso, S. Perasso, A. Pocar, R. S. Raghavan, G. Ranucci,
A. Razeto, A. Re, A. Romani, N. Rossi, R. Saldanha, C. Salvo, S. SchÖnert, H. Simgen,
M. Skorokhvatov, O. Smirnov, A. Sotnikov, S. Sukhotin, Y. Suvorov, R. Tartaglia, G. Testera,
D. Vignaud, R. B. Vogelaar, F. Von Feilitzsch, J. Winter, M. Wojcik, A. Wright, M. Wurm,
J. Xu, O. Zaimidoroga, S. Zavatarelli and G. Zuzel

(Borexino Collaboration)

Published 15 May 2014

Low energy neutrino investigation has been one of the most active fields of particle
physics research over the past decades, accumulating important and sometimes unex-
pected achievements. In this work some of the most recent impressive successes will be
reviewed, as well as the future perspectives of this exciting area of study.

Keywords: Solar neutrinos; geoneutrinos; supernova neutrinos.

1. Introduction

Under the broad characterization of the title of this talk three very active areas
of neutrino physics are considered: solar neutrinos, geo-neutrinos and supernova
neutrino detection. In the following I will characterize the current status and past
achievements of each field, sketching, as well, the relevant perspectives in the imminent and long term future.

Solar neutrinos research is a mature field of investigation which over the past four decades has accumulated fundamental insights in the particle physics arena. Originally conceived as a powerful tool to deeply investigate the core of our star, solar neutrinos underwent a very successful detour to particle physics, heavily concurring to the successful demonstration of the neutrino oscillation phenomenon, according to the MSW mechanism.[1,2] The solar experiments were also paramount to determine with high accuracy the parameters Δm_{12}^2 and $\sin^2 \theta_{12}$ of the PMNS mixing matrix governing the oscillation in the solar sector.[3] Hence this extremely rich and successful chapter of particle physics produced a large part of the solid foundations upon which the next era of precision measurements of the neutrino oscillation parameters will be built. Furthermore, once completed the mission of unveiling the oscillating nature of neutrinos, solar experiments are now back to the original concept of testing the functioning mechanism of the Sun; in this context important insights are awaited from the running and future experiments, of special relevance to address the current issue regarding the surface metallic content of the Sun.[4]

Geoneutrino science, instead, is a much more recent experimental field, though rich of profound potential implications. The suggestion of the possible detection of antineutrino from the intrinsic radioactivity of the Earth dates back to the early times of the first neutrino detection from Cowan and Reines and is due to Gamow. Only 50 years later the solar neutrino experiment Borexino and the reactor antineutrino experiment KamLAND[5,6] provided the first, and for the moment the only, indications of the observation of antineutrinos from Earth. These pioneering measurements, proving the technical feasibility of geoneutrino detection, demonstrate the practical possibility of the ambitious goal to use the measurements of the antineutrino flux to probe the characteristics of the interior of our planet.

The observation of the neutrino burst from the Supernova 1987A, performed by the Kamiokande,[7] IMB[8] and Baksan[9] detectors, represented in the last century a fundamental breakthrough: for the first time the last moments of a dying star were observed through the copious neutrino flux which accompanies such a cosmic thunderstorm. After that, a great deal of studies showed how the accurate determination of the supernova neutrinos can represent an extraordinarily powerful tool to investigate the evolution of the supernova mechanism; this led to the establishment of an ambitious program for the efficient capture of neutrinos from a galactic star explosion, which not only sees a growing number of experiments potentially able to perform such a task, but also their strategic alliance in a worldwide network, SNEWS, which will greatly enhance the probability of a successful detection of the eagerly awaited "galactic firework", when it will occur.

2. Solar Neutrinos

It is now well known, after several decades of theoretical and experimental investigations, that neutrinos are abundantly produced in the core of the Sun. They originate from the nuclear reactions that power our star, producing the energy required to sustain it over the billions of years of its life. Two different chain reactions occur at the temperatures characteristic of the core of the Sun, the so called *pp* chain and CNO cycle, respectively. Actually, in the case of the Sun the vast majority of the energy (>98%) comes from the pp chain, while the CNO contribution is estimated to less than 1.6%.

The effort to produce a model able to reproduce fairly accurately the solar physical characteristics, as well as the spectra and fluxes of the several produced neutrino components, was led for more than forty years by John Bahcall;[10] this effort culminated in the synthesis of the so called Standard Solar Model (SSM), which represents a true triumph of the physics of XXth century, leading to extraordinary agreements between predictions and observables. Such a beautiful concordance, however, has been somehow recently spoiled as a consequence of the controversy arisen regarding the surface metallic content of the Sun, stemming from a more accurate 3D modeling of the Sun photosphere. Therefore, there are now two versions of the SSM, according to the adoption of the old (high) or revised (low) metallicity of the surface.[11]

¿From the experimental side, solar neutrino experiments also represent a successful 40 years long saga, commenced with the pioneering radiochemical experiments, i.e Homestake, Gallex/GNO and Sage, continued with the Cerenkov detectors Kamiokande/Super-Kamiokande in Japan and SNO in Canada, and with the last player which entered the scene, Borexino at the Gran Sasso Laboratory, which introduced in this field the liquid scintillation detection approach.

It is well known that for more than 30 years the persisting discrepancy between the experimental results and the theoretical predictions of the Solar Model formed the basis of the so called Solar Neutrino Problem, which in the end culminated with a crystal clear proof of the occurrence of the neutrino oscillation phenomenon, via the MSW effect. In particular, the joint analysis of the results from the solar experiments and from the KamLAND antineutrino reactor experiment pin points with high accuracy the values of the oscillations parameter within the LMA (large mixing angle) region of the MSW solution.[3]

So, with this spectacular conclusion in background, which surely makes the solar neutrino study one of the more productive particle physics area over the past two decades, in the following I will illustrate the peculiarity and the specific results of the two still running experiments, Borexino and Super-Kamiokande, plus those of SNO which has released until recently data from its data taking stopped in 2006, emphasizing throughout the description the perspectives of potential further achievements in the field, too.

2.1. *Borexino*

Borexino at the Gran Sasso Laboratory[12] is a scintillator detector which employs as active detection medium 300 tons of pseudocumene-based scintillator. The intrinsic high luminosity of the liquid scintillation technology is the key toward the goal of

Borexino, the real time observation of sub-MeV solar neutrinos through through elastic scattering off the electrons of the scintillator, being the ^{7}Be component the main target. However, the lack of directionality of the method makes it impossible to distinguish neutrino-scattered electrons from electrons due to natural radioactivity, thus leading to the other crucial requirement of the Borexino technology, e.g. an extremely low radioactive contamination of the detection medium, at fantastic unprecedented levels.

The active scintillating volume is observed by 2212 PMTs located on a 13.7 m diameter sphere and is shielded from the external radiation by more than 2500 tons of water and by 1000 tons of hydrocarbon equal to the main compound of the scintillator (pseudocumene), to ensure zero buoyancy on the Inner Vessel containing the scintillator itself. Of paramount importance for the success of the experiment are also the many purification and handling systems, which were designed and installed to ensure the proper manipulation of the fluids at the extraordinary purity level demanded by Borexino.

When data taking started in May 2007, it appeared immediately that the daunting task of the ultralow radioactivity was successfully achieved, representing per se a major technological breakthrough, opening a new era in the field of ultrapure detectors for rare events search.

The exceptional purity obtained implies that, once selected by software analysis the design fiducial volume of 100 tons and upon removal of the muon and muon-induced signals, the recorded experimental spectrum is so clean to show spectacularly the striking feature of the ^{7}Be scattering edge, i.e. the unambiguous signature of the occurrence of solar neutrino detection.

For the quantitative extraction of the ^{7}Be flux, the spectrum is fitted to a global signal-plus-background model. The latest ^{7}Be result has been published in:[13] taking into accounts the systematic errors, stemming essentially from the uncertainty in the energy scale and in the fiducial volume selection, the ^{7}Be evaluation is $46 \pm 1.5(\text{stat})^{+1.5}_{-1.6}(\text{sys})$ counts/day/100 tons, hence, summing quadratically the two errors, a remarkable 5% global precision has been achieved in this critical measurement.

By assuming the MSW-LMA solar neutrino oscillations, the Borexino result can be used to infer the ^{7}Be solar neutrino flux. Using the oscillation parameters from,[14] the detected ^{7}Be count rate corresponds to a total flux of $(4.84 \pm 0.24) \cdot 10^9$ cm^{-2} s^{-1}, very well in agreement with the prediction of the Standard Solar Model.[11] For comparison, the measured count rate in case of absence of oscillations would have been 74 ± 5.2 counts/day/100 tons. The resulting electrons survival probability at the ^{7}Be energy is $P_{ee} = 0.51 \pm 0.07$.

In the low energy regime the unprecedented Borexino background has allowed also the experimental study of the pep and CNO components, a possibility unanticipated during the design of the detector. By far the most important residual background for these solar neutrino fluxes is the ^{11}C decay, a radionuclide continuously produced in the scintillator by the cosmic muons surviving through the rock overburden and interacting in the liquid scintillator. The beta plus decay of ^{11}C originates a continuous spectrum which sits exactly in the middle of the energy region between 1 and 2 MeV, which is just the window for the pep and CNO investigation. Actually, to a less extent also the external background induced by the gammas from the photomultipliers is an obstacle, especially above 1.7 MeV.

In the reference[15] a threefold coincidence strategy encompassing the parent muon, the neutron(s) emitted in the spallation of the muon on a ^{12}C nucleus, and the final ^{11}C signal has been devised and described in detail. Such a strategy applied to the Borexino data led to a pep rate of 3.13 ± 0.23(stat) ± 0.23(sys) counts per day/100 ton,[16] from which the corresponding flux can be calculated, assuming the current MSW-LMA parameters, as $\Phi(pep) = (1.6 \pm 0.3) \cdot 10^8$ cm^{-2} s^{-1}, in agreement with the SSM: indeed the ratio of this result to the SSM predicted value is $f_{pep} = 1.1 \pm 0.2$. The resulting electrons survival probability at the *pep* energy is $P_{ee} = 0.51 \pm 0.07$; it should be underlined that the significance of the *pep* detection is at the 97% C.L.

The same analysis, keeping the *pep* flux fixed at the SSM value, originates a tight upper limit on the CNO flux, i.e. $\Phi(CNO) \leq 7.4 \cdot 10^8$ cm^{-2} s^{-1}, corresponding to a ratio with the SSM prediction less than 1.4.

The experiment has been in condition, as well, to provide important, additional insights for the higher energy ^{8}B component. The distinctive feature of the ^{8}B neutrino flux measurement performed by Borexino[17] is the very low 3 MeV threshold attained, decisively lower than the previous measurements from the Cerenkov experiments.

The measurement is very difficult, since the total background, both of radioactive and cosmogenic origin, in the raw data is overwhelming if compared to the expected signal. The specific background suppression strategy adopted in this case is based on two ingredients: on one hand a careful MC evaluation of the main radioactive contaminants of relevance for this measure, i.e. ^{214}Bi from Radon and the external ^{208}Tl from the nylon wall of the Inner Vessel, and on the other the "in-situ" identification and suppression of the muon and associated cosmogenic signals.

The observed ^{8}B rate in the detector is 0.217 ± 0.038(stat) ± 0.008(sys) cpd/100 ton, corresponding to an equivalent flux $\Phi(^8B) = (2.4 \pm 0.4 \pm 0.1) \cdot 10^6$ cm^{-2} s^{-1}; if, as for the other components, we take into account the oscillation probability, then the ratio with the flux foreseen by the SSM is 0.88 ± 0.19.

Over the next years of running of the detector, the solar neutrino program will be completed: ^{7}Be can be pin pointed to an accuracy of 3% (with respect to the 5% uncertainty of the measurement reported here), providing the final, high accuracy, low energy validation of the MSW-LMA solution; moreover the statistics of the ^{8}B neutrino study will be further increased, the precision of the pep evaluation will be enhanced and an even tighter upper limit on the CNO will be placed. Furthermore, the extremely low ^{14}C level in the liquid scintillator, coupled to the good achieved energy resolution, opens also a possible exploration window between 200 and 240 keV in which the observation of the fundamental pp flux can be attempted: therefore Borexino holds the promise to be the first detector able to perform the complete spectroscopy of the solar neutrino flux.

2.2. *SNO (Solar Neutrino Observatory)*

While SNO has terminated its operations in 2006, the re-analysis of the recorded data has also recently produced interesting outcomes.

Succinctly, it can be reminded that this detector, located underground in Canada, in the Inco mine at Sudbury, employed heavy water to perform concurrently a neutral current all-flavor measurement, and a charged current electron-neutrino specific measurement: the comparison of the two provided the unambiguous demonstration of the occurrence of the neutrino flavour conversion process. This achievement is the collective output of the three phases in which the experiment evolved, characterized by different detection procedures of the neutrons signalling the occurrence of the neutral current reaction: a pure heavy water phase, a salt phase and the final ^{3}He counters stage.

The estimated neutral current, charged current and elastic scattering fluxes over the three phase are statistically in agreement, with the exception of the elastic scattering measure of the ^{3}He stage, lower than the previous results, but consistent with being a downward statistical fluctuation. Also, in the latest ^{3}He measurement procedure the ratio NC/CC resulted equal to 0.301 ± 0.033, slightly lower than the previous phases. While including these updated data in the global solar + KamLAND analysis, the SNO collaboration finds $\Delta m^2 = 7.46^{+0.20}_{-0;19} \cdot 10^{-5} \, \text{eV}^2$, and $\tan^2 \theta_{12} = 0.427^{+0.027}_{-0.024}$.[3] It has to be noted, in particular, that the error on the θ_{12} angle has been reduced of almost a factor two by the ^{3}He measurement, as effect of the different systematic and minimal correlation of the NC and CC measures in this phase.

The recent LETA (Low Energy Threshold Analysis) re-processing of the data has further pushed down the analysis threshold, pursuing (similarly to Super-Kamiokande, see next sub-paragraph) the investigation of the survival probability in the 4-5 MeV region, with the aim to unravel the up-turn, if any, of the ^{8}B spectrum. Such an up-turn is expected as the imprint of the MSW effect in that specific energy region. The result stemming from this analysis is definitively puzzling, since it incredibly seems to point event to a down-turn of the spectrum.

This outcome, coupled to the absence of the up-turn in the Super-Kamiokande results, points to something new lurking in that energy region. Additional high statistic data, however, will be needed to understand what is happening there. In this respect, also the more precise evaluation of the pep rate that Borexino plans to perform within a couple of years will be of fundamental importance.

2.3. *Super-Kamiokande*

The Cerenkov experiment Super-Kamiokande, still taking data, has recently released new results about the ^{8}B flux obtained with threshold lowered to 3.5 MeV.

The long history of this detector started in 1996 and evolved through four phases: the first phase lasted until a major PMT incident in November 2001 and produced the first accurate measurement of the ^{8}B flux via the ES detection reaction. The phase II with reduced number of PMTs, from the end of 2002 to the end 2005, confirmed with larger error the phase I measurement. After the refurbishment of the detector back to the original number of PMTs, the third phase lasted from the middle of 2006 up to the middle of 2008. After that, an upgrade of the electronics brought the detector into its fourth phase, which contemplates also the accelerator neutrino beam experiment T2K. It is important to highlight the evolution of the energy threshold (total electron energy) in all the phases: 5 MeV in phase I, 7 MeV in phase II, 4.5 MeV in phase III and 3.5 MeV in phase IV, thanks to the continuously on-going effort to reduce the radon content in water.

The important result provided by Super-Kamiokande is the value of the equivalent ^{8}B flux, obtained converting the all flavor ES measure into an effective neutrino flux without correcting for the oscillation probability; the Super-Kamiokande IV precise measurement amounts to $2.32\pm0.02(\text{stat})\pm0.04(\text{sys})\cdot10^6\,\text{cm}^2\,\text{s}^{1}$[18] representing a high accuracy confirmation that the electron neutrino flux, which contributes mostly to the result of the experiment, is drastically reduced (about 60%) with respect to the SSM prediction.

The recent release of low threshold data had also the goal to unravel the existence of the low energy, eagerly sought, up-turn in the ^{8}B spectrum. However, the upturn did show up only mildly, the experimental spectrum above 3.5 MeV showing only a 1σ preference for the distortion predicted by standard neutrino flavour oscillation parameters over a flat suppression. In this respect, therefore, the poor definition of the data around 3.5-4 MeV and the contradictory output of SNO call for additional evaluations. More data, hence, are needed to further shed light about this point.

2.4. *Physics implications*

In Fig. 1 from Ref. 19 the MSW predicted P_{ee} (electron neutrino survival probability, grey band) is shown, together with several experimental points, *i.e.* black the ^{8}B from SNO and Super-Kamiokande, green ^{8}B from Borexino and SNO Low Threshold Analysis, blue the ^{7}Be Borexino point, magenta the pp datum as drawn by the comparison of Borexino with the Gallium experiments, and red the pep Borexino point: altogether from this figure we can conclude that the solar data collectively accumulated so far spectacularly confirm over the entire energy range of interest the MSW-LMA solar neutrino oscillation scenario, providing the clear evidence of the transition from the high energy, more suppressed "matter" regime, to the low energy, less suppressed "vacuum" regime.

Furthermore, all the solar and KamLAND data taken together constrain rather precisely the oscillation parameters. Fig. 2, from Ref. 3, shows the allowed region in the parameter space stemming from a two flavor oscillation analysis. In a first

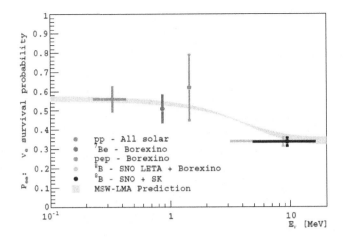

Fig. 1.   Low energy validation of the MSW-LMA solution provided by all solar data.

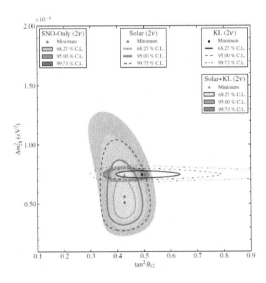

Fig. 2.   Allowed region of the oscillation parameter plane.

approximation, the strongest constraint on the Δm^2 parameter comes from Kam-LAND, while the limit on the mixing angle derives from the solar data.

2.5. *Outlooks*

Despite the high degree of maturity reached by the field, solar neutrino investigation still has interesting future perspectives. The two current generation experiments continuing with data taking, Borexino and Super-Kamiokande, will go ahead along the lines outlined above: Borexino will attempt the high precision, full solar spectroscopy while Super-Kamiokande will accumulate more solar data at lower threshold to try to shed light on the "mystery" of the missing up-turn.

Looking ahead to the near future, SNO+[20] is the new experiment that will replace SNO. It is a liquid scintillator detector, like Borexino, that can go on line soon thanks to the massive re-use of the SNO hardware. In particular the adoption of a new liquid scintillator, linear alkyl benzene, featuring the great advantage of being acrylic-compatible, will allow the re-use of the SNO acrylic vessel. Only the support system had to be built from scratch in order to cope with the different buoyancy condition with respect to the heavy water situation. The other new major piece of equipment that has been realized is the scintillator purification system. The project has been funded for these major items on June 2009; all the planned refurbishments have been completed, and a preliminary operation of water fill is now in progress. The solar program of the experiment will be targeted mainly to pep and CNO detection, profiting of the depth of SNOLAB, which suppress enormously the cosmogenic ^{11}C background. It should be, however, highlighted that the main goal of SNO+ will be the double beta decay with Tellurium dissolved in the scintillator, and currently it is not yet decided whether the solar phase will come before or after the Tellurium phase. The data taking with scintillator will start within 2014.

Far in the future there may be experiments based on completely new techniques like for example LENS,[21] which is the modern version of the old brilliant idea for an Indium experiment. The experimental tool used for the detection of solar neutrinos is the tagged capture of electron-flavor neutrinos on ^{115}In via charged current: $\nu_e + ^{115}\text{In} \rightarrow ^{115}\text{Sn}^* + e^- \rightarrow ^{115}\text{Sn} + 2\gamma$. The low Q of the reaction ($114\,\text{keV}$) allows in principle the full solar neutrino spectroscopy, including the *pp* neutrinos. Time and space coincidence, the former intrinsic to the detection reaction and the latter realized via a granular detector design, will be the key to suppress the radioactive backgrounds as well as the inherent background from the beta decay of ^{115}In.

Finally, it should be mentioned that massive cryogenic detectors for dark matter search based on liquefied noble gases, if realized with the mass of several tens of tons of material, can be perfectly suited for the precise pp spectroscopy. This is true in particular for detector based on Neon and Xenon, which are exempted from low energy radioactive isotopes; this is not the case of Argon due to the presence of the radioactive ^{39}Ar.

3. Geoneutrinos

Geoneutrinos, the antineutrinos from the progenies of U, Th and ^{40}K decays in the Earth, bring to the surface information from the whole planet, concerning its content of radioactive elements. Their detection can shed light on the sources of the terrestrial heat flow, on the present composition and on the origins of the Earth.

Although geo-neutrino detection was conceived very long ago, only recently they have been considered seriously as a new probe of our planet, as a consequence of two fundamental advances that occurred in the last few years: the development of extremely low background neutrino detectors and the progress on understanding neutrino propagation.

Geoneutrinos potentially can unlock many interesting unanswered questions regarding our planet: what is the radiogenic contribution to terrestrial heat production? How much is the content of U and Th in the crust and in the mantle, respectively? What is hidden in the Earth's core, e.g a geo-reactor or ^{40}K, etc.? And finally, is the standard geochemical model (denoted as BSE) consistent with geo-neutrino data?

In particular this last question would ideally require multiple measurements carried out in several locations of the Earth's surface, to be compared with the theoretical predictions of geoneutrino flux (different groups performed such calculations, which are each other in agreement at the 10% level[22–24]).

The detection occurs via the classical inverse beta reaction $\bar{\nu}_e + p \rightarrow e^+ + n$, signaled by the delayed 2.2 MeV gamma stemming from the subsequent neutron capture on protons, for which the most important background is represented by anti-neutrinos from reactors. Exploiting this approach, Borexino and KamLAND[5,6] are the experiments that so far provided the first geo-neutrino measurements.

In particular the latest geoneutrino data from Borexino has been reported in Ref. 5, for a data taking period of 1353 days. With a fiducial exposure of $(3.69\pm0.16)\cdot 10^{31}$ proton × year, after all selection cuts and background subtraction, (14.3 ± 4.4) geo-neutrino events were detected, assuming a fixed chondritic Th/U ratio of 3.9. This result corresponds to a geoneutrino signal of (38.8 ± 12.0) TNU (Terrestrial Neutrino Unit: $= 1$ event / year/ 10^{32} protons), disfavouring also the no-geoneutrino hypothesis with a p-value of 6×10^{-6}.

The reactor neutrino background was evaluated equal to $31.2^{+7}_{-6.1}$, well in agreement with the expectation (33.3 ± 2.4), thus confirming the validity of the entire measurement procedure.

The updated KamLAND results in Ref. 6 are in line with the Borexino output, amounting to (31.1 ± 7.3) TNU.

The main virtue of the pioneering results of Borexino and KamLAND is on one hand the demonstration that geoneutrinos can be detected, paving the way to a completely new branch of geosciences, and on the other that there is a broad agreement with the predictions of the standard BSE model of the composition of the Earth. Obviously, the low statistics of both measures hampers the possibility to distinguish among variants of this general model. This is left for future massive detectors: in particular after the imminent SNO+, which will complement the Borexino and KamLAND data with measurements of similar statistical size, the two planned gigantic detectors JUNO and LENA[25,26] will have the capability of several hundred events per year.

4. Supernova Neutrinos

On February 23, 1987 the Kamiokande,[7] IMB[8] and Baksan[9] detectors recorded for the first (and up to now the only) time a burst of neutrinos from a supernova, the famous SN1987A. Despite the small number of recorded signals, 11, 8 and 5 respectively, plenty of information was squeezed from them, providing substantial confirmations and new insights on the supernova mechanism. On the basis of this success, it can be anticipated that further deep insights and details of the supernova origin and evolution will be gained when the next burst will be observed, especially if the supernova will occur at the typical galactic distance from the Earth of 10 kpc, compared to the 50 kpc distance of the SN1987A, which indeed was located in the Large Magellanic Cloud.

A number of multipurpose detectors, some already running and some still in the construction phase, will detect the next supernova neutrinos. Among the current detectors, Super-Kamiokande is surely the largest with its 32000 tons of water. For a 10 kpc distant supernova it will observe about 7300 $\bar{\nu}_e + p \rightarrow e^+ + n$ events, $\sim$300 $\nu + e \rightarrow \nu + e$ scattering events, $\sim$360 ^{16}O neutral current gamma events and $\sim$100 ^{16}O charged current events. The scattering events will identify the supernova direction with an accuracy of about $5°$.

The large statistics of inverse beta events, instead, will allow the precise energy spectrum measurement, enabling the possibility to discriminate among various proposed models. Besides being the major detection channel of the liquid scintillation detectors, it may take place also in Super-Kamikande, if the current effort to dissolve gadolinium in water will be successful.

At Gran Sasso since 1992 the LVD[27] detector is in operation: it is a segmented liquid scintillator type apparatus with 840 counters and a total target mass of 1000 tons. It can detect about 300 $\bar{\nu}_e + p \rightarrow e^+ + n$ events for the reference 10 kpc supernova. The energy threshold of the counter is 4 MeV and it can tag neutrons with an efficiency of $50 \pm 10\%$. The online trigger efficiency is more than 90% for a distance less than 40 kpc.

KamLAND, Borexino and the future SNO+ with their 1000, 300 and 1000 tons of detection medium, respectively, are perfect supernova observatories. They feature several interactions channels: by normalizing the expected number of events for a 10 kpc supernova to a target mass of 1000 tons, the predictions are $\sim$300 $\bar{\nu}_e + p \rightarrow e^+ + n$ events, several tens of CC and NC events on ^{12}C and about 300 $\nu + p \rightarrow \nu + p$ NC events. This last interaction, being not affected by neutrino oscillations, has the very interesting capability to measure the initial spectrum originated by the supernova explosion. Clearly these numbers will increase of two order of magnitudes in the next generation, huge scintillation detectors JUNO and LENA.

The South Pole IceCube detector has a giga-ton target volume, with the photosensors arranged in a three dimensional structure. The entire inventory of strings (86) has now been deployed. In this detector the observation of a supernova burst will not occur via individual neutrino observation, but through the coherent increase of the PMT dark noise. As a consequence, the detector can measure the time variation of the energy release fairly accurately.[28]

Finally, it has to be highlighted that Super-Kamiokande, LVD, IceCube and Borexino are part of the SNEWS consortium (Supernova Early Warning System), which, upon detection of a neutrino burst, sends prompt alerts to astronomers to trigger the optical observation of the potential identified source. More detailed information on this topic can be found in Ref. 29.

5. Conclusions

Low energy neutrino physics and astrophysics is a field that, despite the many successes already accumulated, is promising for the future many more exciting results.

Solar neutrino investigation is reaching the stage in which the precise spectroscopy of the whole neutrino energy spectrum is a concrete case, up to a level that makes it feasible the check of the SSM, hence paving the path for possibly resolving the discrepancy between high Z and low Z models and understanding the mystery of the up-turn in the ^{8}B spectrum.

An entire new area of investigation is represented by the geoneutrinos study, which exploits the more powerful detectors aimed at solar and reactor neutrino detection to open a completely new window on the characteristics of the interior of our planet.

Supernova neutrino experiments are already a powerful and large family, with new members expected to join in the near future, and hence if Nature will arrange for us a "galactic firework", surely we will not miss the fun!

Acknowledgments

The author wihes to thank the organizers for the invitation to contribute to such an interesting and enlightening conference.

References

1. L. Wolfenstein, *Phys. Rev. D* **17**, 2369 (1978).
2. S.P. Mykheyev and A.Yu. Smirnov, *Sov. J. Nucl. Phys.* **42**, 913 (1985).
3. B. Aharmim *et al.*, arXiv:1109.0763 [nucl-ex] (2008).
4. M. Asplund *et al.*, *Annu. Rev. Astron. Astrophys.* **47**, 481 (2009).
5. G. Bellini *et al.* (Borexino Collaboration), *Phys. Lett. B* **722**, 295 (2013).
6. A. Gando *et al.*, *Phys. Rev. D* **88**, id. 033001 (2013).
7. K. Hirata *et al.*, *Phys. Rev. Lett* **58**, 1490 (1987).
8. R. M. Bionta *et al.*, *Phys Rev. Lett.* **58**, 1494 (1987).
9. E. N. Alexeyev *et al.*, *Phys. Lett. B* **205**, 209 (1988).
10. http://www.sns.ias.edu/ jnb/ John Bahcall's home page.
11. A. M. Serenelli, W. C. Haxton, and C. Pena-Garay, *Astrophys. J.* **743**, 24 (2011).
12. G. Alimonti *et al.*, *Nucl. Instr. and Meth. A* **600**, 568 (2009).
13. G. Bellini *et al.* (Borexino Collaboration), *Phys. Rev. Lett.* **107**, 141302 (2011).
14. Review of Particle Physics, K. Nakamura *et al.* (Particle Data Group), *J. Phys. G* **37**, 075021 (2010).

15. H. Back *et al.* (Borexino Collaboration), *Phys. Rev. C* **74**, 045805 (2006).

16. G. Bellini *et al.* (Borexino Collaboration), *Phys. Rev. Lett.* **108**, 051302 (2012).

17. G. Bellini *et al.* (Borexino Collaboration), *Phys. Rev. D* **82**, 033006 (2010).

18. H. Sekiya, for the Super-Kamiokande Collaboration, proceedings of the 33rd International Cosmic Ray Conference, Rio De Janeiro 2013 (ICRC2013) arXiv:1307.3686 [hep-ex].

19. G. Bellini *et al.* (Borexino Collaboration), arXiv:1308.0443 [hep-ex] (2013).

20. M. Chen, *Nucl. Phys. B (Proc. Suppl.)* **154**, 65 (2005).

21. R. S. Raghavan, *Journal of Physics: Conference Series* **120**, 052014 (2008).

22. S. Enomoto, Ph.D. thesis, Tohoku University (2005).

23. G. Fogli *et al.*, *Phys. Lett. B* **623**, 80 (2005).

24. F. Mantovani *et al.*, *Phys. Rev. D* **69**, 013001 (2004).

25. Y. Li *et al.*, *Phys. Rev. D* **88**, 013008 (2013).

26. M. Wurm *et al.*, *Astropart. Phys.* **35**, 685-732 (2012).

27. N. Yu. Agafonova *et al.*, *Astropart. Phys.* **28**, 516 (2008).

28. R. Abbasi *et al.*, arXiv:1108.0171 [astro-ph.HE] (2011).

29. K. Schoelberg, arXiv:1205.6003 [astro-ph.IM] (2011).

Physics in Collision (PIC 2013)
International Journal of Modern Physics: Conference Series
Vol. 31 (2014) 1460286 (8 pages)
© The Author
DOI: 10.1142/S2010194514602865

World Scientific
www.worldscientific.com

Neutrinoless double beta decay experiments

Alberto Garfagnini

Physics and Astronomy Department, Padova University,
via Marzolo 8 Padova, I-35131, Italy
alberto.garfagnini@pd.infn.it

Published 15 May 2014

Neutrinoles double beta decay is the only process known so far able to test the neutrino intrinsic nature: its experimental observation would imply that the lepton number is violated by two units and prove that neutrinos have a Majorana mass components, being their own anti-particle. While several experiments searching for such a rare decay have been performed in the past, a new generation of experiments using different isotopes and techniques have recently released their results or are raking data and will provide new limits, should no signal be observed, in the next few years to come. The present contribution reviews the latest public results on double beta decay searches and gives an overview on the expected sensitivities of the experiments in construction which will be able to set stronger limits in the near future.

Keywords: Beta-decay; majorana neutrinos.

1. Introduction

Double beta decay is the simultaneous beta decay of two neutrons in a nucleus. The process can be calculated in the standard model of particle physics as a second order process: $(A, Z) \rightarrow (A, Z + 2) + 2e^- + 2\overline{\nu}_e$. The two neutrinos double beta decay $(2\nu\beta\beta)$ process has been observed in 11 nuclei, where single beta decay is energetically forbidden, and very high half-lives, between 7×10^{18} yr and 2×10^{24} yr have been measured.[1,2] Several models extending the standard model predict that a neutrinoless double beta decay $(0\nu\beta\beta)$ should also exists: $(A, Z) \rightarrow (A, Z+2)+2e^-$. It's observation would imply that lepton number is violated by two unit and that neutrino have a Majorana mass component. The standard mechanism for $0\nu\beta\beta$ assumes that the process is mediated by light and massive Majorana neutrinos and that other mechanisms potentially leading to neutrinoless double beta decay are negligible.[3] With a light Majorana neutrino exchange, it would be possible to derive an effective neutrino mass using nuclear matrix element and phase space factor predictions. For recent reviews on the subject, see Refs. 4, 5.

Several experiments searching for neutrinoless double beta decay have been performed in the past going back for at least half a century with increasing sensitivities. Two main approaches have been followed. Indirect methods are based on the measurements of anomalous concentrations of the daughter nuclei in selected samples after very long exposures (i.e. radiochemical methods). Direct methods, on the other hand, try to measure in real time the properties of the two electrons emitted in $\beta\beta$ decay. The detectors can be homogeneous, when the $\beta\beta$ source is the detector medium and in-homogeneous when external $\beta\beta$ sources are inserted in the detector.

In the following sections, the status of the art on $0\nu\beta\beta$ searches will be presented with special emphasis on large scale running experiments.

2. Double Beta Decay in ^{136}Xe

^{136}Xe is a very interesting double beta decay emitter candidate. It has a high $Q_{\beta\beta} = 2457$ keV, in a region which has lower contamination from radioactive background events. It can can be dissolved in liquid scintillators or used as gas allowing to realize a homogeneous detector providing both scintillation and ionization signals. The experiment EXO-200 uses xenon in an homogeneous medium (both as active and as detector), while in KamLAND-Zen it is dissolved as a passive $\beta\beta$ source in a liquid scintillator detector.

2.1. *EXO*

The Enriched Xenon Observatory[6] is an experiment in operation at the Waste Isolation Pilot Plant (WIPP), at a depth of about 1600 m water equivalent near Carlsbad in New Mexico (USA). The experiment is built around a large liquid Xenon Time Projection Chamber filled with about 200 kg of liquid Xenon enriched to about 80.6% in the ^{136}Xe isotope. In contrast to standard TPC the experiment uses liquid xenon which can be concentrated in a smaller volume with the same mass concentration. To overcome the limitation of worse energy resolutions, compared to gaseous TPCs, the experiment exploits the readout of both scintillation and ionization signals produced by interacting particles in xenon. Moreover, by combining both signals the experiment is able to reject background events characterized by different charge to light collection ratio. Finally, by using the difference in the arrival time between the scintillation and ionization signals a z-coordinate of the event is reconstructed. The experiment started data taking in May 2011.

With the first data collected in 752.66 hours (between May 21, 2011 and July 9, 2011), EXO measured for the first time the $2\nu\beta\beta$ life-time of ^{136}Xe[7]

$$T_{1/2}^{2\nu} = 2.11 \pm 0.04(\text{stat.}) \pm 0.21(\text{syst.}) \times 10^{21} \text{ yr} ,$$

the measured value was significantly higher that that previously reported in the literature.

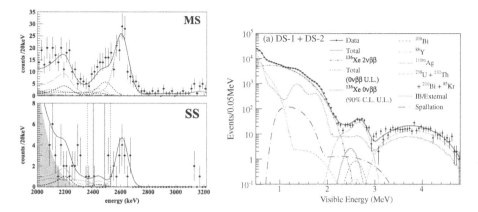

Fig. 1. Left: EXO-200. Energy spectra in the ^{136}Xe $Q_{\beta\beta}$ region for Multiple Site (top) and Single Site (bottom) events. The 1 (2) σ region around $Q_{\beta\beta}$ is shown by solid (dashed) vertical lines. Taken from Ref..[8] Right: KamLAND-Zen. Energy spectra of selected candidate events together with the best-fit backgrounds and $2\nu\beta\beta$ decays. Figure taken from Ref..[13]

In June 2012, the collaboration reported the first results on $0\nu\beta\beta$ decay, analyzing and exposure of 32.5 kg·yr. The $2\nu\beta\beta$ and $0\nu\beta\beta$ signals were extracted simultaneously with a fit to the single-site (SS) and multiple-site (MS) spectra, in an energy range between 0.7 MeV and 3.5 MeV (see left plots of Fig. 1). The fit takes into consideration the main radioactive background sources and a lower limit to the $0\nu\beta\beta$ life-time has been derived:[8]

$$T_{1/2}^{0\nu} > 1.6 \times 10^{25} \text{ yr} \quad @ \text{ 90\% C.L.}$$

Very recently the EXO collaboration has updated the $2\nu\beta\beta$ life-time measurement[9] using 127.6 days of live-time, collected between September 2011 and April 2012, to

$$T_{1/2}^{2\nu} = 2.165 \pm 0.016(\text{stat.}) \pm 0.059(\text{syst.}) \times 10^{21} \text{ yr} ,$$

which is the most precisely measured half-life $2\nu\beta\beta$ decay to date.

A future evolution of EXO[10] is moving in the direction of a tonne scale experiment, with an active mass of few tonnes of ^{136}Xe and improved energy resolution and background suppression.

2.2. *KamLAND-Zen*

The KamLAND-Zen experiment is based on a modification of the existing Kam-LAND[11] detector carried out in the summer of 2011. KamLAND is located at a depth of about 2700 m water equivalent at the Kamioka underground neutrino observatory near Toyama in Japan. The experiment has been equipped with 13 tons of Xe-loaded liquid scintillator (Xe-LS) contained inside a 3.08 m diameter spherical inner balloon. The isotopic abundance of the enriched xenon has been measured to be about 90.9% ^{136}Xe and 8.9% ^{134}Xe. The experiment started data taking in

October 2011 and after and exposure of 30.8 kg yr of ^{136}Xe (77.6 days) it reported a measurement of the $2\nu\beta\beta$ half-life:[12]

$$T_{1/2}^{2\nu} = 2.38 \pm 0.02(\text{stat}) \pm 0.40(\text{syst}).$$

Careful studies have been performed by the collaboration to identify the various background sources contributing to the energy spectra (see right plot of Fig. 1). The spectrum shows a clear peak in the ROI that is compatible to a ^{110m}Ag contamination of the inner vessel. An attempt to purify the Xe-LS has been performed, but unfortunately the filtration did not produce the desired effect: the background counting rate due to ^{110m}Ag decreased only slightly, from 0.19 ± 0.02 cts/(tonne·day) to 0.14 ± 0.03 cts/(tonne·day).

The $0\nu\beta\beta$ limit reported so far by the experiment is:[13]

$$T_{1/2}^{0\nu} > 1.9 \times 10^{25} \text{ yr @ 90\% C.L.}$$

The data taking of the experiment has been completed and a new purification campaign to remove the ^{110m}Ag contamination has been started.

3. Double Beta Decay in ^{76}Ge

Germanium became a warhorse of $0\nu\beta\beta$ decay searches once it was realized that ^{76}Ge emitter could be embedded in solid state detectors using a calorimetric approach with High Purity Germanium (HPGe) diodes. Thanks their excellent energy resolutions (germanium diodes are still the best gamma spectroscopy detectors to date) in the order of 0.1-0.2% FWHM at 2 MeV and to the industrial manufacturing technology, sizable mass detectors can be built. Unfortunately, due to the quite low natural abundance of ^{76}Ge (7.8%), isotopically enriched material has to be procured, before constructing HPGe didoes. Milestone experiments have been performed by the Heidelberg-Moscow[14] (HdM) and IGEX[15] collaborations: they used 11 kg and 8 kg of isotopically enriched (up to 86%) germanium diodes operated in low activity vacuum cryostats located in underground laboratories, Laboratori Nazionali del Gran Sasso (LNGS), in Italy for HdM, and Laboratorio Subterraneo de Canfranc (LSC), in Spain for IGEX.

Germanium $Q_{\beta\beta}$ is not very high, ($Q_{\beta\beta} = 2039.061 \pm 0.007$ keV[16]) and lies in a region where contamination from background sources are possible: apart from the ^{238}U and ^{232}Th decay chains that can contribute to the $0\nu\beta\beta$ region-of-interest (ROI), sizable contamination can arise from long-lived cosmogenically produced isotopes (^{68}Ge and ^{60}Co in copper and germanium activation) and from few anthropogenic radioisotopes. Therefore the experiments have to fight background reduction with careful screening of all the materials close to the detectors and develop pulse shape discrimination techniques to further reduce the background contamination.

Part of the HdM collaboration claimed evidence for a peak at $Q_{\beta\beta}$ which corresponds to a half live central value of $T_{1/2}^{0\nu} = 1.19 \cdot 10^{25}$ yr.[17] The result was later refined with pulse shape analysis (PSA) techniques giving a half life $T^{0\nu}_{1/2} = 2.23^{+0.44}_{-0.31} \cdot 10^{25}$ yr.[18]

Two larger scale experiment, GERDA[19] in Europe and MAJORANA[20] in USA, are exploiting the germanium diodes technology and beside scrutinizing the previous claims[17,18] they will try to push the experimental sensitivity to the limits.

3.1. *GERDA*

The GERmanium Detetor Array (GERDA) experiment operates germanium didodes made of isotopically modified material, enriched to about 88% in ^{76}Ge, without encapsulation in a liquid argon cryogenic bath. The experiment aims to pursue very low backgrounds thanks to ultra-pure shielding against environmental radiation. The germanium detectors are suspended in strings into the cryostat where 64 m^3 of LAr are used both as a coolant and shield. The stainless steel cryostat vessel is covered, from the inside, with copper lining to reduce gamma radiation from the cryostat walls. The vessel is surrounded by a large tank filled with high purity water (590 m^3) which further shields the inner volumes from the experimental hall radiation (absorbing γs and moderating neutrons). Moreover it provides a sensitive medium for the muon system which operates as a Cerenkov muon veto.

The first phase of the experiment has started on November 9, 2011 using eight reprocessed coaxial germanium detectors from the HDH and IGEX experiments together with three natural germanium diodes. In July 2012, two of the coaxial detectors with natural isotopic abundance have been replaced by five new enriched Broad Energy Germanium (BEGe) detectors. The latters are a sub-sample of the thirty new BEGe detectors recently constructed by Canberra for the Phase II of the experiment.

The first 5.04 kg yr data, collected before the insertion of the new BEGe detector, have been used to measure the half-life of the $2\nu\beta\beta$ decay of ^{76}Ge.[21] The extracted half-life is

$$T^{2\nu}_{1/2} = \left(1.84^{+0.09}_{-0.08}\text{ fit }^{+0.11}_{-0.06}\text{ syst}\right) \times 10^{21}\text{yr}.$$

After having studied the background decomposition of the collected energy spectra,[22] keeping the $0\nu\beta\beta$ ROI blinded, Pulse Shape Discrimination techniques have been developed on coax and BEGe detectors.[23]

The $0\nu\beta\beta$ analysis[24] covers the full data taking, from November 9, 2011 to May 21, 2013, for a total exposure of 21.60 kg·yr. Data were grouped into three subsets with similar characteristics: (1) the golden data set contains the major part of the data from coaxial detectors, (2) two short run periods with higher background levels when the BEGe detector were inserted (silver data set) and (3) the BEGe detectors data set. Figure 2 shows the energy spectrum around the region of interest with and without PSD selection. No excess of events was observed over a flat background distribution. Seven events were seen in the range $Q_{\beta\beta} \pm 5$ keV before PSD cuts, while 5.1 were expected. Only 3 events survived (classified as Single-Site Events[24])

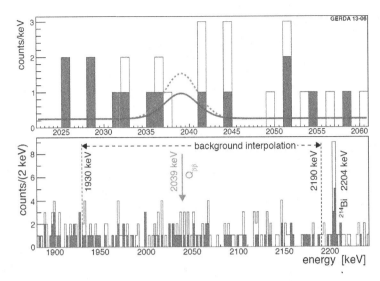

Fig. 2. Top: Energy spectrum for the sum of all ^{enr}Ge detectors in the range 2020-2060 keV, without PSD (empty histogram) and with PSD (black histogram). The red dashed curve is the expectation based on the central value of,[17] the blue curve is instead based on the 90% upper limit derived by GERDA, $T_{1/2}^{0\nu} = 2.1 \times 10^{25}$ yr. Bottom: Energy spectrum in the range 1880-2240 keV, the vertical dashed lines indicate the interval 1930-2190 keV used for the background estimation. Taken from Ref..[24]

after PSD cuts and no event remained in $Q_{\beta\beta} \pm \sigma_E$. To derive the number of signal counts $N^{0\nu}$ a profile likelihood fit of the spectrum was performed. The fit function was given by the sum of a constant term for the background and of a Gaussian for the signal events. The profile likelihood ratio was limited to the region $T_{1/2}^{0\nu} > 0$. The best fit was obtained for $N^{0\nu} = 0$, that is no excess of signal events above background. The lower limit on the half-life is:

$$T_{1/2}^{0\nu} > 2.1 \times 10^{25} \text{yr} \quad @ 90\% \text{ C.L.},$$

including systematic uncertainties. This limit corresponds to $N^{0\nu} < 3.47$ events. A Bayesian calculation using a flat prior distribution for $1/T_{1/2}^{0\nu}$ from 0 to 10^{-24}/yr gives $T_{1/2}^{0\nu} > 1.9 \times 10^{25}$yr @ 90% C.L.

The data not show any peak at $Q_{\beta\beta}$ and the result does not support the claim described in Ref. 18. The GERDA result is consistent with the negative result fron IGEX and HdM. A combined profile likelihood fit including all these three negative results gives a 90% probability limit: $T_{1/2}^{0\nu} > 3.0 \times 10^{25}$yr @ 90% C.L.

A new phase II of the GERDA experiment is planned to start in the middle of 2014, after a six months shutdown prepared to upgrade the experiment infrastructure and install thirty new BEGe diodes. The detectors, which have been recently characterized at the HADES underground laboratory[25] in Belgium, will increase the experiment active mass by about 20 kg (of which 3.6 kg, corresponding to 5 BEGe diodes, were inserted during phase I and their data included in the published results[24]). Thanks to liquid argon instrumentation and enhanced pulse shape discrimination power of the new BEGe detectors, a background reduction from the current BI of about $2 \cdot 10^{-2}$ cts/(keV kg yr) to $0.1 \cdot 10^{-2}$ cts/(keV kg yr) is expected. With the new configuration, the experiment is supposed to collect an exposure of about 100 kg yr and and improve the $0\nu\beta\beta$ sensitivity to $T_{1/2}^{0\nu} > 1.35 \cdot 10^{26}$ yr.

3.2. *MAJORANA*

While GERDA is taking data in Europe, the MAJORANA collaboration in USA is planning to build a large mass germanium experiment using the status of the art technology in diode production with a accurate selection and custom production of radio-pure materials. The proposal is to mount HPGe diodes inside ultra clean electro-formed copper vacuum cryostats and place the whole apparatus in a very deep underground laboratory. Presently, the MAJORANA collaboration is building a prototype, the MAJORANA demonstrator[26] (MJD), to prove the feasibility of the experiment and measure the background conditions. The MJD will be constructed and operated in the Sanford Underground Research Facility (SURF) at a depth of 1500 m in South Dakota, USA. The MJD will use about 40 kg of germanium diodes (with about 30 kg of enriched germanium diodes) and, with performances comparable to those of GERDA, is expected to start data taking in 2014.

Depending on the results of GERDA Phase II and the MAJORANA demonstrator, a next generation germanium experiment using of the order of 1 ton of enriched germanium diodes is under discussion. The effort could be built in stages, starting to merge the GERDA Phase II and MAJORANA diodes in a common effort environment while constructing new diodes from enriched material to enlarge the detector active mass.

4. Double Beta Decay in ^{130}Te

Tellurium is another good candidate suitable for $\beta\beta$ decay searches: due to its high natural abundance (33.8%) isotopic enrichment is not needed and can be used in the form of TeO_2 to build bolometric detectors. Bolometers are calorimeters operated at milli-kelvin temperatures that can measure the energy released in the crystal by interacting particles through their temperature rise. Finally the $Q_{\beta\beta}$ of the decay is relatively high ($Q_{\beta\beta} = 2527$ keV) meaning small background contamination in the ROI.

4.1. *Cuore*

The Cryogenic Underground Detector for Rare Events (CUORE)[27] is an experiment under construction exploiting a large mass of bolometers: its design consists of about 1000 natural TeO_2 crystals grouped in 19 separated towers. Each crystal is a cube with a side of 5 cm, with a mass of 750 g. The small temperature rise originating from nuclear decays in the crystals were read using Neutron Transmutation Doped (NTD) Ge termistors. The array will be operated at about 10 mK in a custom He^3/He^4 dilution refrigerator. The experiment will be located at LNGS in the same experimental hall of the GERDA experiment. The technology has been successfully validated with the Cuoricino[28] experiment. The first installed CUORE tower, CUORE0 has been configured as a stand-alone experiment and is currently taking data to study the background rates and sensitivities expected for CUORE. Recent results on CUORE0[29] show a good energy resolution of 5.6 keV FWHM at the $Q_{\beta\beta}$ and a background counting rate of 0.074 ± 0.012 cts/(keV·kg·yr). The full CUORE experiment will start data taking in about two years with an expected sensitivity on $0\nu\beta\beta$ of 2.1×10^{26} yr.

References

1. J. Beringer *et al.*, Particle Data Group, *Review of Particle Physics, Phys. Rev.* **D86** (2012).
2. A. S. Barabash, *Phys. Rev.* **C81**, 035501 (2010).
3. W. Rodejohann 2011 *Int. J. Mod. Phys.* **E20**, 1833 (2011).
4. B. Schwingenheuer, *Ann. Phys.* **525**, 269 (2013).
5. P. S. Bhupal Dev, et al *Phys. Rev.* **D 88**, 091301(R) (2013).
6. M. Auger et al., *JINST* **7**, P05010 (2012).
7. EXO Collab. (N. Ackerman *et al.*), *Phys. Rev. Lett.* **107**, 212501 (2011).
8. EXO Collab. (M. Auger *at al.*), *Phys. Rev. Lett.* **109**, 032505 (2012).
9. EXO Collab. (J. B. Albert *et al.*, *An improved measurement of the $2\nu\beta\beta$ half-life of ^{136}Xe with EXO-200*, [arXiv/1306.6106v3].
10. G. Gratta, *SNOLAB Future Projects Planning Workshop 2013*, Cryopit Workshop, 21 August 2013, Sudbury, Ontario, Canada.
11. KamLAND Collab. (S. Abe *et al.*), *Phys. Rev.* **C 81**, (2012) 025807.
12. KamLAND-Zen Collab. (A. Gando *et al.*), *Phys. Rev.* **C 85**, 045504 (2012).
13. KamLAND-Zen Collab. (A. Gando *et al.*), *Phys. Rev. Lett.* **110**, 062502 (2013).
14. HdM Collab. (H. V. Klapdor-Kleingrothaus *at al.*), *Eur. Phys. J.* **A 12**, 147 (2001).
15. IGEX Collab. (C. E. Aalseth *et al.*), *Phys. Rev.* **D 65**, 092007 (2002).
16. B. J. Mount, M. Redshaw and E. G. Mayers, *Phys. Rev.* **C 81**, 032501 (2010).
17. H. V. Klapdor-Kleingrothaus, *et al. Phys. Lett.* **B 586**, 198 (2004).
18. H. V. Klapdor-Kleingrothaus and I. V. Krivosheina I. V. *Phys. Lett.* **A 21**, 1547 (2006).
19. GERDA Collab. (K. H. Ackermann *et al.*) *Eur. Phys. J.* **C 73**, 2330 (2013).
20. MAJORANA Collab. (R. Gaitskell *et al.*), *White paper on the Majorana zero-neutrino double-beta decay experiment* [nucl-ex/0311013].
21. GERDA Collab. (M. Agostini *et al.*), *J. Phys. G: Nucl. Part. Phys.* **40**, 035110 (2013).
22. GERDA Collab. (M. Agostini *et al.*), *The background in the neutrinoless double beta decay experiment* GERDA, accepted by *Eur. Phys. Jour.* C [arXiv/1306.5084].
23. GERDA Collab. (M. Agostini *et al.*), *Eur. Phys. J.* **C 73**, 2583 (2013).
24. GERDA Collab. (M. Agostini *et al.*), *Phys. Rev. Lett.* **111**, 122503 (2013).
25. E. Andreotti *et al.*, *JINST* **8**, P06012 (2013).
26. MAJORANA Collab. (R. D. Martin *et al.*), *Status of the MAJORANA demonstrator* [arXiv/1311.3310].
27. CUORE Collab. (C. Arnaboldi *et al.*) *Nucl. Instr. and Meth.* **A 518**, 775 (2004) [hep-ex/0212053].
28. C. Arnaboldi *et al. Phys. Rev.* **C 78**, 035502 (2008).
29. CUORE Collab. (M. Vignati *et al.*), *First data from CUORE0*, talk at TAUP 2013, 3–8 September 2013, Asilomar, CA, USA.

Physics in Collision (PIC 2013)
International Journal of Modern Physics: Conference Series
Vol. 31 (2014) 1460287 (12 pages)
© The Author
DOI: 10.1142/S2010194514602877

World Scientific
www.worldscientific.com

Search for fermionic Higgs boson decays in pp collisions at the ATLAS and CMS experiments

Romain Madar

(On behalf of the ATLAS and CMS Collaborations)

Physikalisches Institut, Albert-Ludwigs-Universität,
Hermann-Herder-Str. 3, Freiburg 79104, Germany
romain.madar@cern.ch

Published 15 May 2014

The newly discovered boson was only observed in diboson final states. The measurement of the fermionic decay modes of this new particle is essential to establish its compatibility with the Standard Model Higgs boson. The two main final states accessible at the LHC are the decay into pairs of b quarks and τ leptons, both of which are experimentally challenging. In some cases, the exploitation of specific Higgs boson production modes allows to significantly increase the experimental sensitivity. One can also extend the search by looking for di-muon final states. This article gives an overview of the Higgs boson searches in these three fermionic final states at the ATLAS and CMS experiments, focusing on the most recent developments and results.

1. Introduction

The Higgs boson (H) is the relic particle of the electroweak symmetry breaking and is of fundamental interest. Since the discovery of the Higgs boson[3,4] by the ATLAS and CMS experiments,[1,2] it is known that this new particle couples to bosons while there is only indirect evidence of couplings to fermions (via loops). The role of direct searches for fermionic Higgs boson decays is to directly measure these couplings. Also, the experimental uncertainty on the Higgs boson branching ratios is highly dominated by the most important decay mode, $H \rightarrow bb$. As a consequence, improving the precision on this branching fraction will reduce the uncertainty on all Higgs boson decay widths. Finally, loop induced production modes involve fermions but might also have contributions due to still unknown particles. Thus, a direct measurement of fermionic couplings would allow to disentangle New Physics (NP) from Standard Model (SM) contributions of these loop diagrams.

The overall analysis strategy is motivated in Section 2. Sections 3, 4 and 5 describe the search for the $b\bar{b}$, $\tau\tau$ and $\mu\mu$ final states respectively, while Section 6 focuses on searches dedicated to the ttH production mode.

2. Overview of the Search Strategy

Assuming a mass of $m_H = 125$ GeV, the total cross section of the $pp \to H + X$ process is of the order of $\sigma_{\text{tot}} \sim 20$ pb, with four main contributing processes. The gluon fusion (GGF), via heavy quark loops, gives the highest contribution (88% of σ_{tot}) but doesn't have any specific event topology allowing to reject the background. The vector boson fusion process (VBF) has a lower contribution (6.6% of σ_{tot}) but is a tree level process with two forward jets in the final state, offering a very specific signature. The production in association with a vector boson has a low contribution (5% of σ_{tot}) but has a clean final state due to lepton(s) from the vector boson decay. The production in association with a $t\bar{t}$ pair has an extremely low contribution (0.4% of σ_{tot}) and a busy environment, but its measurement provides a unique access to the coupling between the top quark and the Higgs fields.

The Higgs boson decay is dominated by the $b\bar{b}$ final state ($\mathcal{B}_{b\bar{b}} = 57\%$) which is very difficult to extract from the large non-resonant $pp \to b\bar{b} + X$ background ($\sigma_{b\bar{b}} \sim 10^7$ pb). However, this decay mode can be identified using the signature of the leptonically decaying vector boson produced in association with H. The second important decay mode is the $\tau\tau$ final state ($\mathcal{B}_{\tau\tau} = 6\%$) which, due to specific τ signatures, can be better separated from the multijet background . This final state also suffers from a high background contribution, especially for hadronic τ decay (τ_{had}) final states. However, the jet topology of the VBF production mode or highly boosted Higgs bosons recoiling against a jet allow to significantly increase the sensitivity of this channel. The Higgs boson can also decay to a $c\bar{c}$ pair with a comparable rate ($\mathcal{B}_{c\bar{c}} = 6\%$), but this final state is extremely difficult to access in hadronic collisions. Finally, the di-muon final state benefits from a very good mass resolution, but suffers from an extremely low branching ratio ($\mathcal{B}_{\mu\mu} = 0.02\%$) together with a large background from the Z production ($\sigma_Z \sim 10^4$ pb).

The overall strategy is then to exploit all possible combinations of Higgs boson production and decay modes. A summary of the presented searches is given in Table 1. The following section gives more details about each search channel.

3. Search for $H \to bb$

The $b\bar{b}$ final state exploits all production modes except GGF, as explained in Section 2. The VBF production mode leads to a four jets final state. The large multijet background can be reasonably reduced exploiting the jet topology of the signal process. The associated production takes advantage of the vector boson decay signature and is clearly the most sensitive channel for this decay mode. Finally, the search for the $ttH(\to b\bar{b})$ process is described Section 6.

Table 1. Overview of the final states and production modes analysed and the integrated luminosity used for both the ATLAS and CMS experiments.

	ATLAS $\mathcal{L}$ [fb^{-1}] (7/8 TeV)	Ref.	CMS $\mathcal{L}$ [fb^{-1}] (7/8 TeV)	Ref.
$H \to bb$				
VBF	-/-	-/-	-/19.0	Ref. 5
VH	4.7/20.3	Ref. 6	5.0/19.0	Ref. 7
ttH	4.7/-	Ref. 6	-/19.0	Ref. 9
$H \to \tau\tau$				
GGF+VBF	4.9/13.0	Ref. 10	4.9/19.4	Ref. 11
VH	-/-	-	-/19.0	Ref. 12
ttH	-/-	-	-/19.0	Ref. 9
$H \to \mu\mu$				
Inclusive	-/20.7	Ref. 13	-/-	-

Independently of the production mode, the key observable for $H \to b\bar{b}$ searches is the invariant mass of the $b\bar{b}$ system ($m_{b\bar{b}}$), since the Higgs boson would appear as an excess of events close to m_H. The energy scale and resolution of b-quarks are then crucial for the experimental sensitivity. Both experiments achieved significant improvements by refining their treatment of the b-jet energy calibration: the kinematic dependence of the b-jet energy response is now taken into account and the energy correction due to B-hadron semi-leptonic decays is also added. These modifications, applied in a slightly different way in the two experiments, lead to a 10 to 15% improvement on the $m_{b\bar{b}}$ resolution, depending on the phase space region.

3.1. *Optimized search for the associated production*

For both experiments, the general strategy is to categorize the events according to the number of reconstructed leptons n_ℓ ($\ell \equiv e, \mu$): the 0-lepton channel is enriched in $Z(\to \nu\nu)H(\to b\bar{b})$ events, the 1-lepton channel is enriched in $W(\to \ell\nu)H(\to b\bar{b})$ events and the 2-lepton channel is enriched in $Z(\to \ell\ell)H(\to b\bar{b})$ events. The neutrinos, coming from W and Z decays, don't interact with the detector and will be indirectly measured via the missing transverse energy E_T^{miss}, *i.e.* as a momentum imbalance in the plane transverse to the beam axis.

The main Standard Model processes contributing to these final states are the $t\bar{t}$ pair production (leading to a $b\bar{b}$ pair and bc pairs with a misidentified c-quark), the W/Z+light quarks production (due to b-jet mis-identification), the $W/Z + b\bar{b}$ production leading to an identical final state, and the multi-jet production with and without real b-quarks (due to fake and/or non isolated lepton, in addition to misidentified b-jet). Several kinematic properties of these backgrounds offer a way to identify signal enhanced phase space regions. In addition to the $b\bar{b}$ system invariant mass, considering events with high transverse momentum of the vector boson (p_T^V) allows to increase the signal-to-background (S/B) ratio. Both experiments follow the same strategy to achieve a reliable and precise background modelling. First, several regions with different background and signal compositions are defined, using n_ℓ, the number of reconstructed jets (n_{jet}) and b-jets (n_b) as well as p_T^V. Then, different parameters affecting the background prediction are defined, such as normalisation factors or continuous parameters modelling the effect of a given systematic uncertainty on the background. Finally, a global fit to the observed data is performed in order to determine the set of parameters which describes the data best, simultaneously in all regions. In this way, correlations of systematic variations across the regions are taken into account, the signal contamination in low S/B regions is properly propagated and the various background compositions of the samples ensure that every single background is constrained.

The strategy to further reduce the background on top of existing categorisation, is to exploit differences in event topology between signal and backgrounds. On top of $b\bar{b}$ system properties, the angle between E_T^{miss} and the $b\bar{b}$ system (expected to be around π for $Z(\to \nu\nu)H(\to b\bar{b})$), or the compatibility of the (ℓ, E_T^{miss}) system with a W decay (as expected for $W(\to \ell\nu)H(\to b\bar{b})$) are also used. The ATLAS search is performed by analysing the m_{bb} distribution after cut-based selections on these variables while the CMS search combines them in a Boosted Descision Tree (BDT) which is used as discriminant observable.

An important proof of principle of any experimental search is to be able to measure a known process having the same signature as the signal. The $VH(\to b\bar{b})$ search can be validated by measuring the SM production of VZ pairs. Both the ATLAS and CMS experiments are able to measure VZ production at the expected rate.

Figure 1 shows the final m_{bb} invariant mass distribution after subtraction of the non-resonant backgrounds. The events are weighted according the expected S/B in each bin of the m_{bb} distribution. To interpret the result, the signal strength μ is defined as the ratio between the observed (fitted) cross section and the expectation from the SM. The ATLAS experiment quotes an observed (expected) exclusion limit at 95% confidence level on μ of 1.3 (1.4) for $m_H = 125$ GeV and an extracted signal strength of

$$\mu = 0.2 \pm 0.6. \tag{1}$$

The CMS experiment quotes an observed (expected) exclusion limit at 95% confidence level on μ of 1.7 (0.94) for $m_H = 125$ GeV, and an extracted signal strength of

$$\mu = 1.0 \pm 0.5. \tag{2}$$

The ATLAS result is compatible with both the signal and background only hypothesis, while the CMS experiment has an excess in the observed data with respect to the expected background of about two standard deviations.

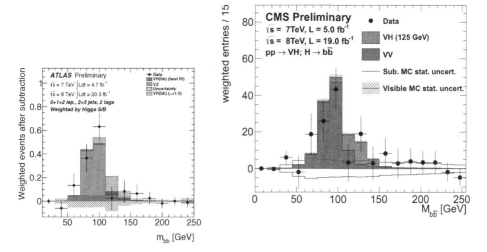

Fig. 1. Weighted m_{bb} distribution after subtraction of the non-resonant backgrounds for the ATLAS (left) and CMS (right) experiments.

3.2. *Optimized search for the vector boson fusion production*

The search for $qqH(\to b\bar{b})$ is particularly challenging because of the important multijet background. The CMS experiment has designed a dedicated four-jet final state trigger for this search, taking advantage of the two VBF forward jets, namely with a high rapidity gap and a high invariant mass. The other key signature is the presence of two b-jets inside this rapidity gap. Finally, to further reduce the multijet background where the two forward jets are due to gluon radiation, a q/g-jet discriminant is built based on the expected differences between quark and gluon jet shapes. In order to efficiently exploit the full information, the b-jet identification variable, the VBF tagging jet variables and the q/g-jet discriminant are combined in a neural network (NN). Four analysis categories are defined based on the NN output. For each category, an unbinned fit of the m_{bb} distribution is performed to test the presence of a signal. The final result, given by the combination of these categories, reaches an observed (expected) limit on μ at the 95% confidence level of 3.6 (3.0) for $m_H = 125$ GeV. Figure 2 shows the distribution of the NN output and the expected and observed limit versus m_H.

4. Search for $H \to \tau\tau$

The $\tau\tau$ final state exploits all production modes. This final state is sensitive to the GGF process mainly for high transverse momentum of the $\tau\tau$ system, called boosted topology. However, also this search is most sensitive to the VBF production, using the tagging jets to significantly reduce the $Z \to \tau\tau$ background. The associated production provides an appreciable sensitivity in spite of its lower rate. Finally, the search for the $ttH(\to \tau\tau)$ process is described Section 6.

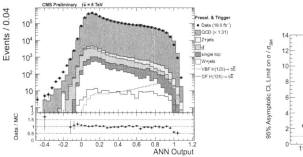 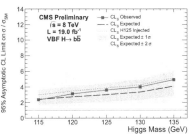

Fig. 2. NN distribution used to define the five categories:$NN < 0.52$, $0.52 < NN < 0.76$, $0.76 < NN < 0.90$, $0.90 < NN < 0.96$ and $NN > 0.96$ (left). Expected and observed exclusions limit on μ at the 95% confidence level (right). The expected limits are shown for the background only hypothesis as well as for the signal plus background hypothesis (signal of $m_H = 125$ GeV injected).

On the other hand, analysing di-τ final states is experimentally challenging for several reasons. First, each τ-lepton decays into stable particles before reaching the detector. This leads to three different final states with different background compositions: $\tau_{\text{had}}\tau_{\text{had}}$ (41.9%), $\tau_\ell\tau_\ell$ (12.4%) and $\tau_\ell\tau_{\text{had}}$ (45.8%), where τ_ℓ denotes the lepton ℓ of the $\tau \to \ell\nu_\tau\nu_\ell$ decay and τ_{had} generically denotes the hadronic decay products of the $\tau \to$ hadrons ν_τ decay. Each of these final states needs a dedicated analysis in order to maximize the overall sensitivity. Secondly, the measurement of the di-τ invariant mass ($m_{\tau\tau}$) is particularly challenging due to the escaping neutrinos. In both the ATLAS and CMS experiments, the kinematic of the neutrinos is assessed given the observed missing transverse energy and the matrix element of the τ-lepton decay. This kinematic fit, applied in a slightly different way in the two experiments, leads to an improvement of the order of 25% in the sensitivity due to a better $m_{\tau\tau}$ resolution. Finally, it is important to keep the energy response of hadronically decaying τ lepton under control, which is achieved within a precision of the order of $\pm 5\%$.[14]

4.1. *Optimized search for gluon fusion and vector boson fusion*

The common strategy is to search for an excess of events localized around $m_{\tau\tau} = 125$ GeV on top of the $Z \to \tau\tau$ background. The estimation of this most important background is based on $Z \to \mu\mu$ data events for both experiments. After selection of a pure data sample of $Z \to \mu\mu$, muon energy deposits are removed and are replaced by those of simulated τ with the kinematics as the initial muons. In this way, the so-called embedding technique, the jet kinematics, the detector noise and all the soft interactions are directly taken from data. Furthermore, the hadronic τ final states suffer from background due to jets wrongly identified as τ_{had}, such as $W(\to \ell\nu)$+jets where one jet is misidentified as a τ_{had}. In order to better control the fake backgrounds, data-driven estimates are used. In the ATLAS $\tau_{\text{had}}\tau_{\text{had}}$ analysis, the multijet normalisation is obtained by performing a two-dimensional template fit to the track multiplicity distribution of the two τ_{had} candidates, where the multijet template is modelled from data using same sign electric charge candidates. In the CMS analysis, the multijet background is estimated using a non isolated τ_{had} control sample in data.

To optimize the sensitivity, the events are categorized following the two main topologies in a similar way for all three sub-channels:high momentum of the (τ, τ) pair (sensitive to GGF) and high invariant mass between two forward jets (sensitive to VBF). This last topology heavily relies on jet kinematic, properties which are modelled using data ensuring the background estimation robustness in this most sensitive phase space region. To further suppress the backgrounds, different properties can be exploited depending on the process to reject. For example, the missing transverse energy points in the direction between the two τ-lepton candidates for $Z/H \to \tau\tau$, but not for non-resonant backgrounds like W+jets or $t\bar{t}$ production. The ATLAS and CMS experiments exploit this property using the transverse mass of the lepton candidate and the missing energy $(m_T(\ell, E_T^{\text{miss}}))$ and the E_T^{miss} projection on the direction of the $\tau\tau$-axis, respectively.

By combining the three $\tau\tau$ final states, the ATLAS analysis reaches an observed (expected) limit on μ at 95% confidence level of 1.9 (1.2) for $m_H = 125$ GeV. The CMS analysis achieves an observed (expected) limit of 0.77 (1.8) at the same mass. The ATLAS and CMS experiments measures a signal strength of $\mu = 0.7 \pm 0.7$ and $\mu = 1.1 \pm 0.4$, respectively. For CMS, this corresponds to a 2.8 standard deviation excess of observed events compatible with a 125 GeV Higgs boson. The ATLAS result is compatible with both the signal and the background-only hypothesis. The ATLAS collaboration is currently working on an update of this search to inlude the full available dataset. Figure 3 shows the invariant mass distributions of the di-τ system for both analyses.

4.2. *Optimized search for the associated production*

Even if vector boson (semi)-leptonic decays only is considered, the combinatorics due to the three $\tau\tau$ pair decays lead to several possible final states. The analysed channels are then chosen taking into account hadronic τ identification efficiency, branching ratios and background rates. The CMS experiment analyses the following channels: $\ell\ell\tau_{\text{had}}$ enriched in $W(\to \ell\nu)H(\to \tau_\ell\tau_{\text{had}})$, $\ell\ell\ell\ell$ enriched in $Z(\to \ell\ell)H(\to \tau_\ell\tau_\ell)$ and $\ell\tau_{\text{had}}\tau_{\text{had}}$ mostly enriched in $W(\to \ell\nu)H(\to \tau_{\text{had}}\tau_{\text{had}})$. The charge and/or the flavour correlation between the different particles represents a valuable information to reduce the background. The most important SM backgrounds are the WZ and ZZ production for $\ell\ell\tau_{\text{had}}$ and $\ell\ell\ell\ell$ channels, respectively, while the fake background is dominant in the $\ell\tau_{\text{had}}\tau_{\text{had}}$ final state. The signal extraction is based on the invariant mass spectrum of all the visible decay products of the reconstructed Higgs boson candidate. By combining the different channels, this search reaches an interesting

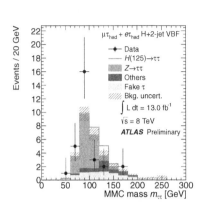
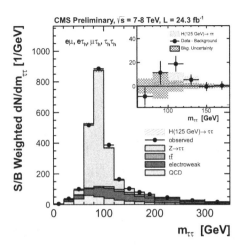

Fig. 3. Invariant mass distribution of the $\tau\tau$ system for the VBF category in the $\tau_\ell\tau_{\mathrm{had}}$ final state at the ATLAS experiment (left). S/B weighted invariant mass distribution of the $\tau\tau$ system for four channels at the CMS experiment (right). The upper box represents the background-subtracted weighted data overlaid with a 125 GeV signal.

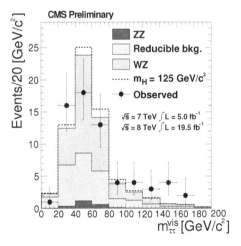
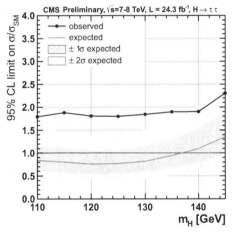

Fig. 4. Invariant mass of the $\tau_\ell\tau_{\mathrm{had}}$ system, for the $\ell\ell\tau_{\mathrm{had}}$ final state as measured in the CMS experiment (left). Expected and observed exclusion limits on the signal strength for different m_H (right).

sensitivity:the observed (expected) limit on μ at 95% confidence level is 3.1 (3.9) for $m_H = 125$ GeV, as illustrated Fig. 4.

5. Search for $H \to \mu\mu$

The search for $H \to \mu\mu$ is extremely challenging because of the low branching ratio of this decay mode and the large cross section of the $Z \to \mu\mu$ background.

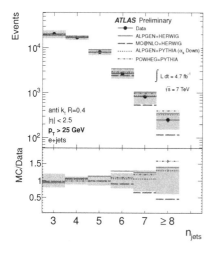 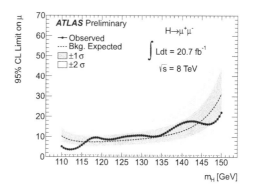

Fig. 5.   Invariant mass of the di-muon system as measured in the ATLAS experiment (left). Expected and observed exclusion limits on the signal strength for different m_H (right).

However, the good experimental resolution of the di-muon mass ($m_{\mu\mu}$) leads to a promising sensitivity ($S/\sqrt{B} \sim 0.25$ for 20 fb^{-1}). In addition, this channel offers a unique way to directly probe the coupling between the Higgs boson and second generation fermions. Finally, the $H \rightarrow \mu\mu$ branching ratio is sensitive to several SM extensions,[13] such that this channel becomes valuable also for new physics searches.

The analysis strategy is to search for an excess of events at $m_{\mu\mu} \sim m_H$ GeV. The events are split into two categories based on the $m_{\mu\mu}$ resolution. The background and signal modelling relies on analytical $m_{\mu\mu}$ shapes, determined using simulated distributions. The signal hypothesis is tested via an unbinned fit of $m_{\mu\mu}$. The observed (expected) sensitivity in terms of signal strength is 9.8 (8.2) for $m_H = 125$ GeV, as illustrated Fig. 5.

6. Dedicated Searches for $t\bar{t}H$ Production

Even though the ttH production represents only 0.4% of the total production cross section, the search for this process has several motivations. First, it provides direct access to the couplings between the Higgs boson and the top quark. This direct measurement, together with the $gg \rightarrow H$ cross-section measurement, could allow to constrain new physics involved in the loops of this process. Finally, it allows to test Higgs field properties at a higher mass scale where new physics is more likely to appear. However, this channel is experimentally challenging mainly because of a complex final state containing electrons, muons or τ-leptons, four b-jets and missing transverse energy. Some of the SM backgrounds, like $t\bar{t}+b$-jets and $t\bar{t}+W/Z$, are not well known and the associated systematic uncertainties can significantly degrade the search sensitivity.

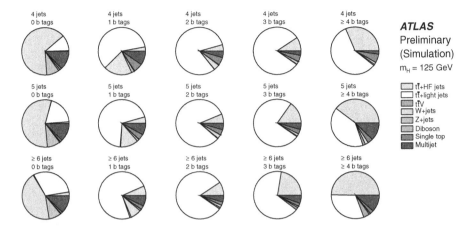

Fig. 6.   Expected background composition in the ATLAS $ttH(\to b\bar{b})$ analysis, for each defined category based on the number of reconstructed jets and b-jets.

In order to cope with these background uncertainties, both the ATLAS and CMS collaborations follow the same general strategy defining different phase space regions enriched in different backgrounds and performing a global fit to data in order to constrain each single background. Figure 6 illustrates how these samples are defined and what the expected background compositions for the ATLAS $ttH(\to b\bar{b})$ analysis are.

6.1. *ATLAS searches*

The ATLAS search focuses on $H \to b\bar{b}$ decay and the semi-leptonic decay of the $t\bar{t}$ pair, analysing the $\ell\nu + 4$ b-jets+2 jets final state. A search for $ttH(\to \gamma\gamma)$[15] is also performed but is not presented in this document. One of the specific challenge of this final state is the event kinematic reconstruction. Indeed, among the four reconstructed b-jets, it is *a priori* impossible to identify the pair coming from the Higgs boson decay. The ATLAS experiment developed a method based on an event-by-event kinematic fit in order to determine the b-jet pair most likely coming from the Higgs boson decay. As shown in Fig. 6, the background modelling is done by defining categories based on the number of reconstructed jets and b-jets. The global fit allows to significantly reduce the $t\bar{t} + b$-jets and the $t\bar{t} + W/Z$ background uncertainties. For each of these categories, the invariant mass of the b-quark pair, identified as coming from $H \to b\bar{b}$, is used to test the presence of a signal. After combining all categories, this analysis achieves an observed (expected) limit on μ of 13.1 (10.5) for $m_H = 125$ GeV as illustrated in Fig. 7.

6.2. *CMS searches*

The CMS searches focus on $H \to b\bar{b}$ and $H \to \tau\tau$ decays together with semi-leptonic and di-leptonic decays of the $t\bar{t}$ pair. The $H \to \gamma\gamma$ decay is also analysed and a

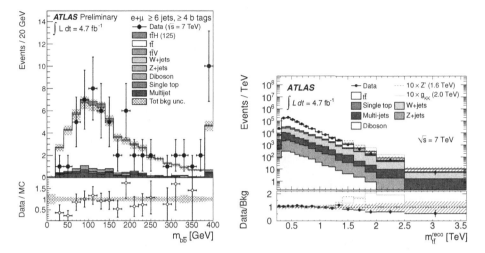

Fig. 7. Invariant mass of the b-quark pair identified as coming from $H \to b\bar{b}$ as measured in the ATLAS experiment(left). Expected and observed exclusion limit on the signal strength for different m_H (right).

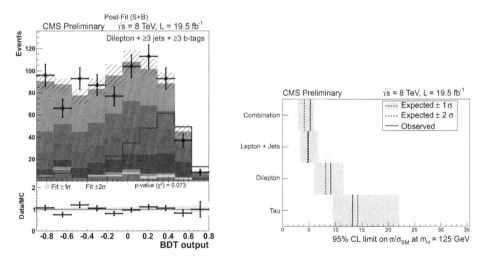

Fig. 8. BDT output distribution as measured in the CMS experiment (left). Expected and observed exclusion limit on the signal strength for different m_H (right).

$H \to bb/\tau\tau/\gamma\gamma$ combination is available as an inclusive search for ttH, but is not presented in this document. The two $t\bar{t}$ decays are treated as different categories for the $H \to b\bar{b}$ search while they are treated inclusively for the $H \to \tau\tau$ search. The signal is extracted using various discriminant variables combined in a BDT, like b-quark and τ-lepton identification variables as well as b-jet pairing related variables. By combining the three channels, this analysis yields an observed (expected) limit of 5.2 (4.1) for $m_H = 125$ GeV. The improvement provided by the $\tau\tau$ channel is illustrated Fig. 8.

7. Conclusions

Higgs boson fermionic decay modes are experimentally challenging to extract in the presence of large backgrounds. However, they are crucial to fully understand the electroweak symmetry breaking sector and how the Higgs boson couples to elementary fermions. The results of these searches constitute an essential input to identify the nature of the newly discovered Higgs boson. A wide LHC program to address this fundamental question was designed and first exciting results are obtained by the ATLAS and CMS experiments in the $H \to b\bar{b}$ and $H \to \tau\tau$ decay modes. A 3.4 standard deviation excess in the combined $H \to \tau\tau$ and $H \to bb$ searches from the CMS collaboration is observed. The result obtained by the ATLAS collaboration is compatible with the signal hypothesis as well as the background-only hypothesis. An update of the ATLAS search inluding the full available dataset is under preparation. In addition, some valuable secondary channels reach very promising sensitivities and are expected to contribute to our understanding of the Higgs sector in the upcoming years.

References

1. ATLAS Collaboration, *JINST* **3**, S08003 (2008).
2. CMS Collaboration, *JINST* **3** S08004, (2008).
3. ATLAS Collaboration, *Phys. Lett. B* **716**, 1–29 (2012).
4. CMS Collaboration, *Phys. Lett. B* **716**, 30 (2012).
5. CMS Collaboration, CMS PAS HIG-13-011.
6. ATLAS Collaboration, ATLAS-CONF-2013-79.
7. CMS Collaboration, CMS PAS HIG-13-012.
8. ATLAS Collaboration, ATLAS-CONF-2012-135.
9. CMS Collaboration, CMS PAS HIG-13-019.
10. ATLAS Collaboration, ATLAS-CONF-2012-60.
11. CMS Collaboration, CMS PAS HIG-13-004.
12. CMS Collaboration, CMS PAS HIG-12-053.
13. ATLAS Collaboration, ATLAS-CONF-2013-10.
14. ATLAS Collaboration, ATLAS-CONF-2013-044.
15. ATLAS Collaboration, ATLAS-CONF-2013-080.

Physics in Collision (PIC 2013)
International Journal of Modern Physics: Conference Series
Vol. 31 (2014) 1460288 (12 pages)
© The Author
DOI: 10.1142/S2010194514602889

Higgs searches beyond the Standard Model

R. Mankel

(On behalf of the ATLAS, CMS, CDF and D0 collaborations)

Deutsches Elektronen-Synchrotron (DESY),
Notkestr 85, D-22603 Hamburg, Germany
Rainer.Mankel@desy.de

Published 15 May 2014

While the existence of a Higgs boson with a mass near 125 GeV has been clearly established, the detailed structure of the entire Higgs sector is yet unclear. Besides the Standard Model interpretation, various possibilities for extended Higgs sectors are being considered. The minimal supersymmetric extension (MSSM) features two Higgs doublets resulting in five physical Higgs bosons, which are subject to direct searches. Alternatively, more generic Two-Higgs Doublet models (2HDM) are used for the interpretation of results. The Next-to-Minimal Supersymmetric Model (NMSSM) has a more complex Higgs sector with seven physical states. Also exotic Higgs bosons decaying to invisible final states are considered. This article summarizes recent findings based on results from collider experiments.

Keywords: Higgs; boson; BSM; MSSM; NMSSM; supersymmetry; 2HDM.

1. Introduction

The electroweak symmetry breaking mechanism of the Standard Model (SM) predicts the Higgs particle as a scalar boson. The discovery of a Higgs boson with a mass near 125 GeV[1,2] by the experiments ATLAS[3] and CMS[4] at the CERN Large Hadron Collider (LHC) is an important milestone, and it is of fundamental interest to study the properties of this state, its quantum numbers and couplings. An equally important aim is the unravelling of the overall structure of the Higgs sector. While at the level of current measurements, the observed state is compatible with the Higgs boson as predicted by the SM, the mass of the Higgs boson is divergent at high energies.[5] For many other open questions, related e.g. to dark matter, naturalness and CP violation in the universe, solutions might evolve through more detailed studies of the Higgs sector. From investigation of the observed boson alone one obtains relatively weak contraints on Higgs decays beyond the Standard Model

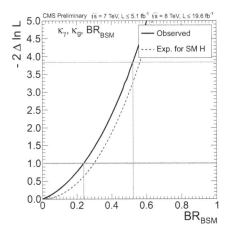

Fig. 1.   1D test statistics as a function of the Higgs branching fraction into non-SM decay modes.[6] Only the photon and gluon couplings within the loops are modified. The tree-level couplings are fixed at SM level.

(BSM); for example, a study of the CMS experiment (Fig. 1)[6] derives an upper limit of 52% for the branching fraction into BSM decay modes at 95% confidence level. Significant improvement of such limits can at least be expected to take a long time. For this reason, direct searches of BSM Higgs signatures are essential to clarify the structure of the Higgs sector.

2. The MSSM Higgs Sector

While the SM features a single complex Higgs doublet resulting in one physical scalar Higgs boson, the Minimal Supersymmetric Model (MSSM) is characterized by an extended Higgs sector with two complex Higgs doublets. This gives rise to five physical Higgs bosons, three of them neutral, labeled h, H (CP-even) and A (CP-odd) and jointly referred to as ϕ, and the other two charged, named $H^{\pm}$. At tree level, the MSSM Higgs sector is governed by two parameters: the mass m_A and $\tan \beta$, which is the ratio of the vacuum expectation values of the two Higgs doublets. Beyond tree level, additional parameters enter via radiative corrections, and benchmark scenarios fixing these parameters are used to compare different measurements. In many cases, the m_h^{max} benchmark scenario[7] is used.

The mass of the CP-odd Higgs boson is usually approximately degenerate with one of the CP-even bosons, either the H for large m_A, or the h if m_A is small. With the exception of the di-muon channel, this degeneracy cannot be resolved within the experimental mass resolution, and thus the visible cross section effectively doubles. The coupling to the b quark is proportional to $\tan \beta$. Thus production cross sections in association with b quarks are enhanced by a factor of $\approx 2 \tan^2 \beta$. It is important to note that the observation of a Higgs boson near 125 GeV with SM-like properties does not exclude additional heavy Higgs bosons at large $\tan \beta$, as discussed in detail in Ref. 8. If m_A is large compared to the Z mass, the light MSSM Higgs boson (h) becomes SM-like, a situation referred to as the decoupling limit. At the current level of measurements, both SM and MSSM are found to fit the measurements about equally well, see for example Ref. 9.

2.1. *MSSM searches in the $\tau\tau$ channel*

The $\phi \to \tau\tau$ channel can be seen as a good compromise between a relatively large branching fraction and manageable backgrounds. Searches in this channel have been performed by the Tevatron[10] and LHC experiments; the latter will be described in more detail. The main production mechanisms are associated production with b quarks and gluon-gluon fusion (GGF). In correspondence, the data are sub-divided in event categories with at least one, or no b-tagged jet. The decay modes of the τ are grouped into leptonic (e or μ) and hadronic decays ("had"). This results in six decay patterns, five of which are covered by the analyses of ATLAS[11] and CMS:[12] $e + \mu$, $e + had$, $\mu + had$, $had + had$ (only ATLAS), $\mu + \mu$ (only CMS). The invariant mass of the τ pair is reconstructed from the visible decay products and the missing transverse energy, using a likelihood technique (CMS) and a method named Missing Mass Calculator[13] (ATLAS).

A very important background arises from $Z \to \tau\tau$ decays. It is addressed with an embedding technique: events with reconstructed $Z \to \mu\mu$ decays are taken from the data, and the muon objects are replaced with simulated τ decays. Additional backgrounds arise from $Z \to ee(\mu\mu)$ decays, $t\bar{t}$ and di-boson production, QCD multijet and W+jets events. Their relevance differs across the various $\tau\tau$ decay patterns. The results shown here are based on the 7 TeV data sample in the case of ATLAS, while the CMS results include the first part of the 8 TeV data. Figure 2

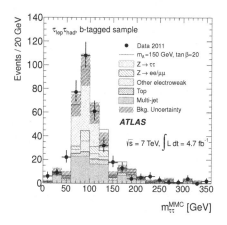 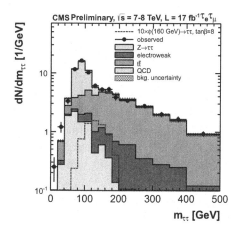

Fig. 2. Distributions of reconstructed invariant mass of the $\tau\tau$ system in the btag category; from ATLAS[11] in the combined $e + had$ and $\mu + had$ channels (left) and CMS[12] (right) in the $e + \mu$ channel. Estimated backgrounds are also shown.

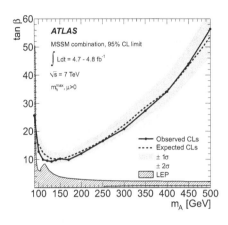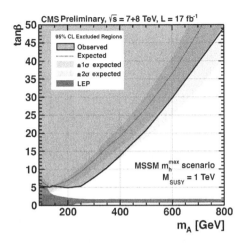

Fig. 3. Upper limits for $\tan\beta$ vs. m_A obtained in the $\tau\tau$ channel from the ATLAS[11] (left) and CMS experiments[12] (right). The ATLAS measurement also includes the $\phi \to \mu\mu$ channel.

shows examples of the reconstructed $m_{\tau\tau}$ spectra for selected decay patterns in the b-tag category for ATLAS and CMS together with the estimated backgrounds. The summed background estimates describe the data very well, and there is no indication of a signal.

The results from the various event categories and decay patterns are combined, and are used to compute upper limits on the MSSM parameter $\tan\beta$ at the 95% confidence level as a function of m_A. The ATLAS analysis also includes the $\phi \to \mu^+\mu^-$ decay channel, which is not discussed here in detail. The results are shown in Fig. 3 in the m_h^{max} scenario. At low masses (below $m_A \approx 250$ GeV), the upper limit on $\tan\beta$ reaches down to values of 5, and touches the lower limit obtained by the LEP experiments. At larger m_A, there is still a wide range of $\tan\beta$ allowed. Inclusion of the full set of 8 TeV data can be expected to give further improvements in the sensitivity of both experiments.

2.2. *MSSM searches in the $b\bar{b}$ channel*

In the MSSM, the Higgs decay into two b quarks is the dominating channel for $\tan\beta$ significantly larger than one. Furthermore, the associated Higgs production with b quarks is enhanced by a factor of $\tan^2\beta$ due to the modified coupling. On the other hand, there is a copious background from QCD multi-jet production, which makes this analysis very difficult and in particular challenging from the trigger aspect.

The analyses shown here search for $\phi \to b\bar{b}$ signatures with at least one additional b-tagged jet. This channel was first successfully analyzed by the Tevatron experiments CDF and D0.[14–16] The signal is searched in the invariant mass of the two leading jets, which must be b-tagged. CDF uses a second variable which functions as a global b-tag. The background estimation, which is a key component of the analyses, is performed with different methods. The CDF analysis derives

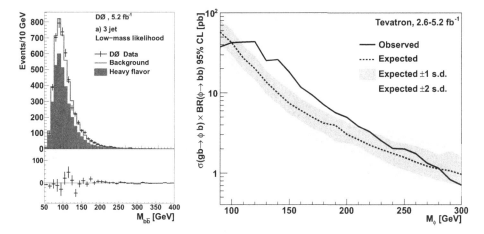

Fig. 4. Left: Invariant mass spectrum of the two leading b jets in triple-b-tag events from the D0 experiment, together with the estimated background.[15] A likelihood-based selection optimized for the low-mass region is shown. Right: Combined cross section times branching fraction upper limit from the CDF and D0 experiments.[16]

background templates from the double b-tag sample by applying b-tag efficiency weights. The combination of the various background templates, together with a signal template, is fitted to the data. In the D0 analysis, fractional contributions of the various multi-jet processes are determined by fitting p_T distributions from simulation to the data. The invariant mass spectrum from the D0 analysis is shown in Fig. 4 (left).[15] The data are shown along with the data-driven background estimate. The combined upper limits on the cross section times branching fraction is shown in Fig. 4 (right).[16] Neither experiment sees a signal above the background expectation, but there are modest excesses in the observed vs. the expected upper limits of $\approx 2.8\sigma$ in the CDF and of $\approx 2.5\sigma$ in the D0 case, which are both visible in the combination plot at relatively low masses.

The first analysis of this channel at the LHC has been carried out by CMS.[17] The search is performed both in the all-hadronic final state with three b-tagged jets, as well as in semi-leptonic signatures requiring in addition a non-isolated muon in one of the b-tagged jets. The all-hadronic analysis is inspired by the CDF method, using a background model of templates determined from the double-b-tag sample. The invariant mass spectrum of the two leading jets is shown in Fig. 5 (left). The observed spectrum can be fitted well with a combination of five background templates, and there is no indication of a signal or excess. Upper limits for the cross section times branching ratio are converted to limits in the MSSM parameter space in the m_h^{max} scenario. In Fig. 5 (right) these limits are shown as a function of $\tan\beta$ in comparison to the combined Tevatron results. The CMS numbers have been converted for a Higgsino mass parameter of $\mu = -200$ GeV to allow direct comparison with the Tevatron numbers. The CMS results shows a much higher sensitivity

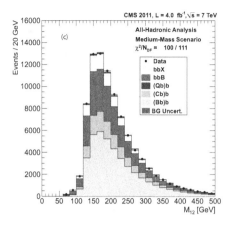 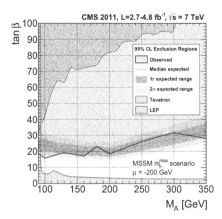

Fig. 5. Left: Invariant mass spectrum of the two leading b jets in triple-b-tag events from the all-hadronic analysis in CMS.[17] The shaded stacked histograms show the contribution of the different background templates resulting from the background-only fit. Right: Observed upper limits at 95% confidence level on $\tan \beta$ as a function of m_A for the combined all-hadronic and semi-leptonic analyses from CMS.[17] Exclusion regions from LEP and Tevatron are also shown.

already with the 7 TeV data, with limits ranging between 18 and 32 in the whole mass range up to 350 GeV.

2.3. Charged Higgs boson search

Discovery of a charged Higgs boson would be an immediate indication of physics beyond the SM. For $\tan \beta > 3$, the dominant decay mode for light charged Higgs bosons is $H^+ \to \tau \nu_\tau$, for heavy charged Higgs bosons the branching fraction to $\tau \nu_\tau$ can still be sizable. The main production modes depend on the mass of the charged Higgs. For $m(H^+) < m_t$, it can be produced in top quark decays, while for $m(H^+) > m_t$, associated production together with top quarks takes over. The ATLAS analysis shown here[18] uses $t\bar{t}$ events with a τ lepton decaying hadronically in the final state, with veto on any other leptons. The analysis requires at least three or four jets, with at least one of them b-tagged, and large missing E_T. The discriminating variable is the transverse invariant mass of the τ products combined with the missing E_T, defined as

$$m_T = \sqrt{2p_T^\tau E_T^{miss}(1 - \cos \Delta\phi_{\tau,miss})} \tag{1}$$

where $\Delta\phi_{\tau,miss}$ is the azimuthal angle between the hadronic decay products of the τ lepton and the direction of the missing transverse momentum. Main backgrounds are general $t\bar{t}$, single top, W/Z+jets and di-boson production, as well as QCD.

The full 2012 dataset is used for this measurement. The low and high mass selections cover H^+ mass ranges of 90–160 and 180–600 GeV, respectively. Figure 6 shows the m_T spectra as well as the model dependent limits for both selections. No evidence for a H^+ signal is found. At low masses, large parts of the MSSM

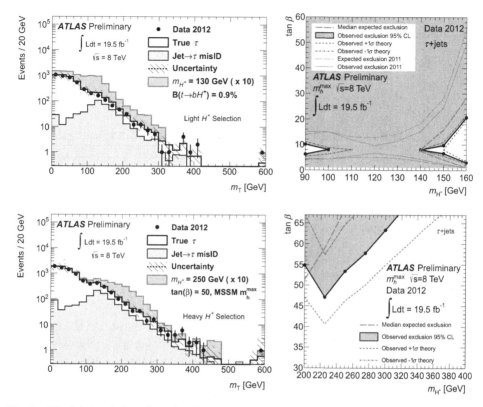

Fig. 6. The left-hand plots show the data and background predictions for the ATLAS H^+ boson search[18] as function of m_T , both for the light (top) and heavy (bottom) Higgs mass selections. Expected Higgs signals for Higgs masses of 130 and 250 GeV are also superimposed, scaled up by a factor of ten, assuming $BR(t \rightarrow bH^+) = 0.9\%$ and $\tan\beta = 50$, respectively. The right-hand plots show the resulting model-dependent limits in the m_{H^+} and $\tan\beta$ parameter plane, again for the light (top) and heavy (bottom) Higgs mass selections.[18]

parameter space are excluded, while at high masses relatively large values of $\tan\beta$ are still allowed.

3. Generic 2HDM Searches

While the MSSM is a very specific implementation of a Higgs sector motivated by supersymmetry, the Two-Higgs Doublet Model (2HDM) is a phenomenological approach, which allows data interpretations based on two Higgs doublets without committing to a particular theory. As opposed to MSSM at tree level, the 2HDM can accomodate CP violation, as well as flavor-changing couplings. Examples for categories of 2HDM models with natural flavor conservation are the Type-I, in which all quarks couple only to one Higgs doublet, and the Type-II, in which the up-type quarks couple to one and the down-type quarks couple to the other Higgs doublet. Key parameters are $\tan\beta$, which is the ratio of the vacuum expectation

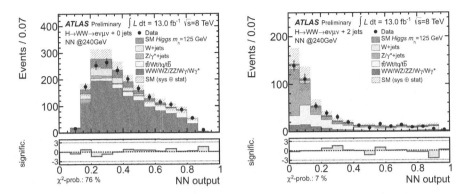

Fig. 7. Distributions of the neural network discriminant variable for the ATLAS $h/H \to WW^{(*)} \to e\nu\mu\nu$ analysis,[19] optimized for a Higgs mass of 240 GeV, for the GGF (left) and VBF selections (right).

values of the two Higgs doublets, and the scalar mixing angle α. The MSSM is a special case of a Type-II 2HDM. As the MSSM, the 2HDM feature three neutral and two charged Higgs bosons.

A recent analysis from ATLAS[19] searches for decays of the scalar Higgs bosons h and H in the channel $h/H \to WW^{(*)} \to e\nu\mu\nu$, assuming $m_h = 125$ GeV. The pseudo-scalar A does not decay into W pairs. The analysis requires exactly two leptons of opposite charge and missing energy. In order to cover the GGF and vector boson fusion (VBF) production mechanisms, either zero or two jets are required, respectively. A neural network combining several kinematic variables is trained for each mass point to enhance the signal over the background. The Higgs boson observed near 125 GeV is also treated as background. The resulting distributions of the neural network discriminant are shown in Fig. 7 for an assumed Higgs mass of 240 GeV. The data are compared to a model of simulated events representing the various background processes. No indication of a signal is seen. The results are translated into exclusion contours within the 2HDM parameter space. Figure 8 shows examples for such contours for two values of $\tan\beta$, each for Type-I and Type-II 2HDM models. For low masses, significant parts of the $\cos\alpha$ range are excluded.

4. NMSSM Higgs Search

The next-to-minimal supersymmetric model (NMSSM) features two complex Higgs doublets and an additional scalar field. The physical states are mixtures: three CP-even (h_1, h_2, h_3), two CP-odd (a_1, a_2), and two charged bosons ($h^\pm$). The NMSSM requires less fine tuning for the Higgs mass, and solves the so-called "μ problem" of the MSSM.

A recent CMS analysis[20] searches for the decay of a non-standard Higgs into two very light bosons, resulting in two boosted pairs of muons. The corresponding NMSSM interpretation is a decay chain $h_{1,2} \to a_1 a_1 \to (\mu\mu)(\mu\mu)$, where either the

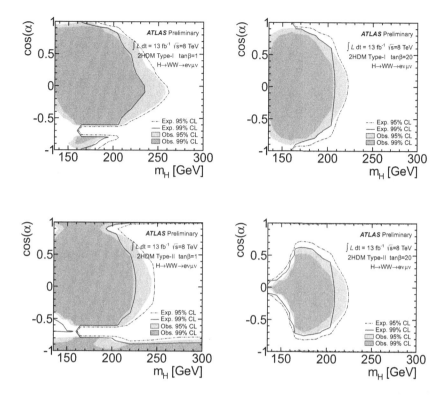

Fig. 8. Exclusion contours in the $\cos\alpha - m_H$ parameter space for the ATLAS $h/H \to WW^{(*)} \to e\nu\mu\nu$ analysis[19] . The top row shows Type-I and the bottom row Type-II 2HDM models. The left and right column plots assume $\tan\beta = 1$ and $\tan\beta = 20$, respectively.

h_1 or the h_2 could correspond to the boson observed near 125 GeV, and a_1 is a new CP-odd Higgs boson lighter than twice the τ mass. But also an interpretation within dark-SUSY models is possible, based on the decay chain $h \to 2n_1 \to 2n_D + 2\gamma_D \to 2n_D + (\mu\mu)(\mu\mu)$, where n_1 is the lightest visible neutralino, n_D is a light dark fermion and γ_D a light massive dark photon with weak couplings to SM particles.

The analysis selects events with two isolated, boosted muon pairs, considering the mass ranges $0.25 < m_a < 3.55$ GeV and $m_h > 86$ GeV. The signal region is defined by requiring the masses of the two muon pairs to be equal. Main backgrounds in the analysis are direct production of J/Ψ pairs, and $b\bar{b}$ production with subsequent di-muon decays, either semi-leptonic or via quarkonia resonances. The $b\bar{b}$ background is estimated from a bb-enriched control sample and the double-J/Ψ production from PYTHIA. Figure 9 (left) shows the distribution in the space of the two di-muon masses with the diagonal signal region still blinded. Eight events are observed in the off-diagonal sideband. After unblinding, only one event is observed in the diagonal signal region (Fig. 9 (right)), consistent with an expected background of 3.8 ± 2.1 events. The results are interpreted in the context of NMSSM

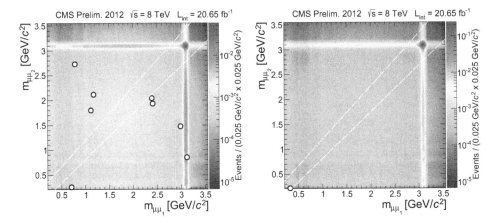

Fig. 9. Distribution in the invariant masses of the two muon pairs, in the off-diagonal control region (left) and the diagonal signal region (right) in the CMS NMSSM search.[20] The empty circles show the surviving events in the data. The histogram represented by shades shows the $b\bar{b}$ background template.

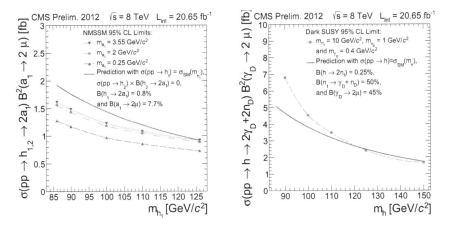

Fig. 10. Interpretation of the CMS results in the NMSSM (left) and the dark-SUSY (right) benchmark models. (See text for details).

and dark-SUSY models. The NMSSM interpretation (Fig. 10 (left)) shows 95% CL upper limits on the cross section times branching fraction vs. the Higgs mass m_{h_1} for different values of m_{a_1}. The dark-SUSY interpretation (Fig. 10 (right)) displays corresponding limits vs. m_h for a simplified scenario for $m_{n_1} = 10$ GeV and $m_{n_D} = 1$ GeV and $m_{\gamma_D} = 0.4$ GeV. The comparison shows that the experimental limits of this analysis are already able to exclude certain models.

5. Searches for Invisible Higgs Bosons

If a Higgs boson would decay with a significant probability into final states consisting entirely of invisible particles, this might still be measurable if the Higgs is produced

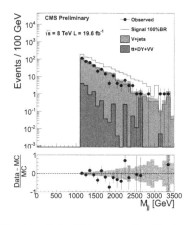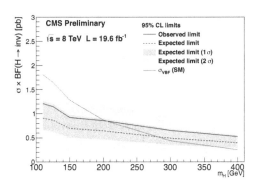

Fig. 11. Left: Di-jet mass distribution for VBF events together with the estimated backgrounds in the invisible Higgs search of CMS.[21] The signal expected for a 100% invisible branching fraction is also shown. Right: Upper limits for $\sigma \times BR$ into invisible final states. The full VBF Higgs production cross section is also shown.

in association with other, well detectable particles. Searches of invisible Higgs decays in association with Z bosons have lead to upper limits on the invisible branching fraction of 65% (ATLAS[22]) and 75% (CMS[23]) for a Higgs with SM cross section at a mass of 125 GeV at 95% confidence level, leaving still plenty of room for such decays.

Recently, invisible Higgs decays have also been searched for by CMS in VBF topology,[21] which has a higher cross section than associated Z production. The final state is characterized by two scattered jets with a large rapidity gap, very little other activity in the event and large missing E_T. Special triggers combining VBF and missing E_T signatures are crucial for this analysis, and large efforts have been undertaken to reduce the QCD background. The signal is searched for in the invariant mass of the two jets. The signature of an invisible Higgs should manifest as an excess which is growing with the dijet mass. Figure 11 (left) shows the observed distribution in the data with the background expectation and the expected signal in case the Higgs near 125 GeV would decay to 100% into invisible modes. No signal is observed. Figure 11 (right) shows the obtained upper limits on the cross section times branching fraction as a function of the Higgs mass, in comparison with the full VBF cross section. At $m_H = 125$ GeV, the upper limit on the invisible branching fraction is 69%.

6. Summary

The observation of SM-like properties of the established Higgs state near 125 GeV does not imply that the Higgs sector must have SM structure. The best way of clarification is the direct search for additional Higgs signatures.

A broad attack is launched to clarify whether the Higgs sector reaches beyond the SM. In the context of MSSM, at low masses m_A the limits from LEP and LHC start to close. Large values of m_A and $\tan\beta$ are still possible. The constraints from the H^+ searches have significantly improved. Recently, analyses interpret their findings also in the 2HDM approach. Concerning the NMSSM, only few channels have been targeted so far, and there is still a wide open range of possibilities. The search for invisible Higgs particles and decay modes yields first results in vector boson associated production and VBF signatures, but the limits on the invisible branching fraction are still large.

In summary, the Higgs searches beyond the SM have just scratched the surface. Many LHC analyses are being updated with the full 8 TeV statistics, and the Run-II of the LHC at $\approx$13 TeV will further extend the reach towards higher masses.

References

1. ATLAS Collab., *Phys. Lett.* **B716** (2012) 1.
2. CMS Collab., *Phys. Lett.* **B716** (2012) 30.
3. ATLAS Collab., *J. Inst.* **3** (2008) S08003.
4. CMS Collab., *J. Inst.* **3** (2008) S08004.
5. E. Witten, *Phys. Lett* **B105** (1981) 267.
6. CMS Collab., CMS PAS HIG-13-005 (2013).
7. M. Carena *et al.*, *Eur. Phys. J.* **C26** (2003) 601; M. Carena *et al.*, *Eur. Phys. J.* **C45** (2006) 797.
8. M. Carena *et al.*, arXiv:1302.7033 (2013).
9. P. Bechtle *et al.*, arXiv:1211.1955 (2012).
10. D0 Collab., *Phys. Lett.* **B710** (2012) 569; CDF Collab., *Phys. Rev. Lett.* **103** (2009) 201801.
11. ATLAS Collab., *JHEP* **02** (2013) 095.
12. CMS Collab., CMS PAS HIG-12-050 (2012).
13. A. Elagin *et al.*, *Nucl. Instr. Meth.* **A654** (2011) 481.
14. CDF Collab., *Phys. Rev.* **D85** (2012) 032005.
15. D0 collab., arXiv:1011.1931, *Phys. Lett.* **B698** (2011) 97.
16. CDF and D0 Collab., arXiv:1207.2757, *Phys. Rev.* **D86** (2012) 091101.
17. CMS Collab., *Phys. Lett.* **B722** (2013) 207.
18. ATLAS Collab., ATLAS CONF-2013-090 (2013).
19. ATLAS Collab., ATLAS CONF-2013-027 (2013).
20. CMS Collab., CMS PAS HIG-13-010 (2013).
21. CMS Collab., CMS PAS HIG-13-013 (2013).
22. ATLAS Collab., ATLAS CONF-2013-011 (2013).
23. CMS Collab., CMS PAS HIG-13-018 (2013).

Physics in Collision (PIC 2013)
International Journal of Modern Physics: Conference Series
Vol. 31 (2014) 1460289 (13 pages)
© The Author
DOI: 10.1142/S2010194514602890

Prospects for Higgs boson scenarios
beyond the standard model

Oscar Stål

The Oskar Klein Centre, Department of Physics
Stockholm University, SE-106 91, Stockholm, Sweden
oscar.stal@fysik.su.se

Published 15 May 2014

The new particle recently discovered at the Large Hadron Collider has properties compatible with those expected for the Standard Model (SM) Higgs boson. However, this does not exclude the possibility that the discovered state is of non-standard origin, as part of an elementary Higgs sector in an extended model, or not at all a fundamental Higgs scalar. We review briefly the motivations for Higgs boson scenarios beyond the SM, discuss the phenomenology of several examples, and summarize the prospects and methods for studying interesting models with non-standard Higgs sectors using current and future data.

Keywords: Higgs physics; beyond the standard model; supersymmetry; LHC.

1. Introduction

After the 2012 discovery of a new particle in the $\gamma\gamma$ and ZZ final states by ATLAS[1] and CMS,[2] experimental Higgs physics has entered a new era. Using already up to the full dataset from the first LHC run, corresponding to an integrated luminosity (per experiment) of ~ 5 fb^{-1} at $\sqrt{s} = 7$ TeV and ~ 25 fb^{-1} at $\sqrt{s} = 8$ TeV, measurements of key properties for this new state have been reported by both collaborations.[3, 4] To set the stage for the ensuing discussion on Higgs sectors beyond the SM, we summarize the key experimental findings here. The new particle has a measured mass of[4, 5]

$$M_H = 125.5 \pm 0.2(\text{stat.})^{+0.5}_{-0.6}(\text{syst.}) \text{ GeV} \quad (\text{ATLAS}),$$

$$M_H = 125.7 \pm 0.3(\text{stat.}) \pm 0.3(\text{syst.}) \text{ GeV} \quad (\text{CMS}).$$

Although the SM Higgs mass is in principle a free parameter, the measured value is in good agreement with earlier predictions performed using electroweak precision

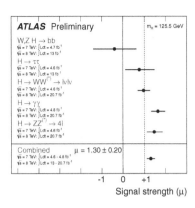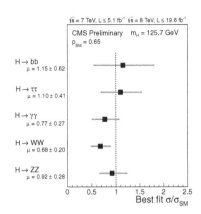

Fig. 1. Higgs signal strengths, relative to the SM value $\mu_i = 1$, measured in different final states by ATLAS[3] (left) and CMS[4] (right).

data. With the accurate Higgs mass determination, the fate of the SM vacuum has been analyzed to great precision,[6] with the outcome that the SM can be extrapolated to very high scales ($Q \sim 10^{11}$ GeV); the SM vacuum is likely metastable with a lifetime exceeding the age of the universe.

In the SM, the Higgs signal rates ($\sigma \times \mathrm{BR}$) into various channels are completely fixed by the Higgs mass. Precise predictions in various channels can then be combined with data to measurements of individual signal strenghts, μ_i, normalized such that the SM corresponds to $\mu_i = 1$. The current status of the Higgs signal strengths measurements is shown in Fig. 1. As can be seen from this figure, there is currently no statistically significant deviation from $\mu_i = 1$ observed in the data.

To establish that the observed particle is indeed a fundamental scalar, as predicted by the Higgs mechanism, it is also necessary to determine its spin-parity quantum numbers, J^P (the SM Higgs has $J^P = 0^+$). First attempts at this have rejected minimal alternative hypotheses like (pure) 0^- or 2^+ states at the 95% confidence level (CL),[7,8] which adds further credibility to the conclusion that the observed state is indeed a 0^+ scalar, exactly as predicted by the SM. However, it should be remembered that this is based on much less conclusive evidence than the discovery itself and that there is still room for, e.g., a sizeable $\mathcal{CP}$-odd admixture.

That said, we now turn to the question what alternative theories may be behind this discovery? After all, the SM is known to have several shortcomings, such as the gauge hierarchy problem, the absence of a suitable dark matter candidate and, ultimately, the problem of how to incorporate gravity into the theory. Even if the Higgs data currently tell us that the discovered scalar boson fits nicely the minimal Higgs structure of the SM, it is well-known that several extensions incorporate the SM Higgs sector in a particular (decoupling) limit, and that the observations therefore make perfect sense also in these theories. There are two principal ways in which a non-minimal Higgs sector would manifest itself experimentally: either directly through the discovery of additional scalar states, which would prove immediately that the Higgs sector is non-minimal, or through precise measurements of the Higgs properties, that could eventually reveal deviations from the SM predictions for the discovered 125 GeV state. Below we shall describe a few examples of physics beyond the SM where both strategies could be relevant.

2. Supersymmetry

Low-energy supersymmetry is widely recognized as an attractive extension of the SM, since the symmetry naturally protects the scalar Higgs mass against large quantum corrections. Other advantages of the minimal supersymmetric Standard Model (MSSM) is that it can offer unification of the gauge couplings and radiative generation of the electroweak scale. On the more phenomenological side, the MSSM with $\mathcal{R}$-parity conservation can provide a stable candidate particle for the cold dark matter in the lightest neutralino. Unfortunately, LHC searches for supersymmetric particles have so far turned up nothing,[9] which starts to put the naturalness arguments under question. Current limits, which imply for example that the gluino mass $m_{\tilde{g}} \gtrsim 1.2$ TeV, introduce already a certain degree of unavoidable fine-tuning.[10] This adds to the potential fine-tuning arising from the Higgs mass measurement, which we shall discuss in more detail below.

2.1. *A 125* GeV *Higgs boson in minimal SUSY*

The MSSM, which in many cases also serve as a template for weakly-coupled extended Higgs sectors, contains two complex Higgs doublets. Following electroweak symmetry breaking, there are five physical Higgs bosons in the spectrum. When $\mathcal{CP}$ is conserved, they are classified as two $\mathcal{CP}$-even scalars, h and H ($M_h < M_H$ and the two states mix with an angle α), one $\mathcal{CP}$-odd scalar, A, and a charged Higgs pair, $H^\pm$. At tree level, the Higgs sector can be specified using only two parameters, conveniently chosen as either M_A or $M_{H^\pm}$ and $\tan\beta$, the ratio of vacuum expectation values of the two doublets. The remaining Higgs masses and the mixing angle α become predictions of the theory; in particular one finds that $M_h^{\text{tree}} \leq M_Z$, with equality in the decoupling limit ($M_A \gg M_Z$, $\tan\beta \gg 1$) where the couplings of h approach their SM values.

Beyond leading order, large radiative corrections to the Higgs masses arise, in particular from the coupling to the top quark. The dominant corrections to the lightest Higgs mass at one loop can be written as

$$M_h^2 = M_Z^2 + \frac{3m_t^4}{2\pi^2 v^2}\left[\log\frac{M_S^2}{m_t^2} + \frac{X_t^2}{M_S^2}\left(1 - \frac{12X_t^2}{M_S^2}\right)\right], \qquad (1)$$

where $M_S = \sqrt{m_{\tilde{t}_1} m_{\tilde{t}_2}}$ and $X_t = A_t - \mu\cot\beta$ is the mixing parameter appearing in the off-diagonal entries of the scalar top (stop) mass matrix. According to Eq. 1, the lightest Higgs mass can be made heavier than M_Z (and even reach $M_h \sim 125$ GeV), either by increasing the average stop mass, M_S, or by having a large mixing in the stop sector; the maximum attained for $X_t = \pm\sqrt{6}\,M_S$. In theories like the MSSM, where the Higgs masses are predictions, constraints on the model parameters can be derived from the Higgs mass value measured at the LHC.[11–14] The resulting constraints are particularly severe for high-scale models with mediation mechanisms that do not generate large A-terms.[14,15]

In a bottom-up approach, the low-energy MSSM parameters can be treated as input directly without reference to a "higher" model with e.g. grand unification. This fixes the radiative corrections to Higgs masses and mixing, and allows higher-order SQCD corrections to Higgs production and decay to be calculated. With a complete spectrum, all aspects of the model phenomenology can be studied. This is the basis for MSSM Higgs sector benchmark scenarios,[16] which were successfully used at LEP[17] and the Tevatron. With the Higgs discovery, many of these original scenarios have become obsolete, and updated scenarios for LHC Higgs phenomenology were recently proposed.[18] We are going to use some of these scenarios here to discuss the prospects for current and future MSSM Higgs searches.

Numerical tools are available to calculate the MSSM Higgs spectrum to the two-loop level, which is necessary for precision phenomenology. As an example, we use numerical predictions from `FeynHiggs`[19] (version 2.9.4) for two benchmark scenarios: $M_h^{\max}$, which maximizes M_h for given values of M_S, M_A and $\tan\beta$, and M_h^{mod}, in which the stop mixing is selected to maximize the region in $\tan\beta$ that is compatible with the observed Higgs mass value. The current experimental status for these scenarios is shown in Fig. 2. In this figure, direct search limits from LEP, Tevatron and the LHC are evaluated using `HiggsBounds`[20] (version 4). `HiggsBounds` is a convenient program to take into account limits from direct Higgs searches in arbitrary models. Comparing the exclusion limits for the two scenarios in Fig. 2, several features are common. The bulk exclusion at high $\tan\beta$ comes from combined MSSM $h/H/A \to \tau\tau$ searches,[21] a channel which receives a strong $\tan\beta$ enhancement.

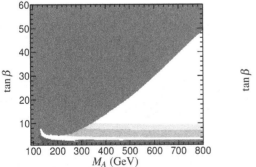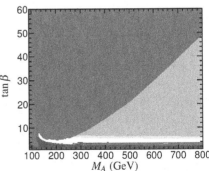

Fig. 2. Exclusion limits from direct Higgs searches at LEP (blue) and the LHC (red) in the MSSM benchmark scenarios $M_h^{\max}$ (left) and M_h^{mod} (right). The dark (bright) green regions are compatible with the direct limits and has the lightest MSSM Higgs boson in the mass range $M_h = 125.5 \pm 2(3)$ GeV. The results have been produced using `FeynHiggs` and `HiggsBounds`.

The same parameter region at low M_A, high $\tan\beta$ is also the one most strongly constrained by indirect measurements, such as e.g. $B_s \to \mu^+\mu^-$,[22] and this region therefore appears strongly disfavoured in the MSSM. This conclusion is corroborated by global analyses of the (low-energy) "pMSSM" parameter space taking into account the measured Higgs rates,[23] which now favour the region $M_A \gtrsim 300$ GeV.[24] Finally, we note from Fig. 2 that for SUSY-breaking masses M_S around the TeV-scale (here: $M_S = 1$ TeV), the low $\tan\beta$ region is not accessible for any value of M_A, since M_h becomes too low (even below the LEP limit).

The global MSSM analyses are useful to identify interesting phenomenology which can still be viable in the allowed parameter space regions. For example, while the jury is still out on the possibility of an enhancement in the two-photon signal in the data (as weakly indicated by ATLAS, but not supported by CMS, see Fig. 1), it is useful to assess under which circumstances such an enhancement can be reproduced. In the MSSM, $h \to \gamma\gamma$ can be enhanced primarily in two ways: either by suppressing the main decay mode, $h \to b\bar{b}$, through mixing between the two Higgs doublets, or by direct contributions to the partial width $\Gamma(h \to \gamma\gamma)$ from new particles. The largest possible new physics contribution comes from the scalar taus (staus),[25] which unlike a contribution from e.g. stops would not suppress Higgs production through gluon fusion since the staus do not carry colour charge. It is found that an enhancement of $R_{\gamma\gamma} = (\sigma \times \text{BR})/\text{SM} \gtrsim 1.5$ is possible if $m_{\tilde{\tau}_1} \sim 100$ GeV, close to the current limit.

Due to the high degree of compatibility between measurements and the SM, the current Higgs data favours the decoupling limit ($M_A \gg M_Z$, $\tan\beta \gg 1$), in which the couplings of h become SM-like. Nevertheless, MSSM Higgs phenomenology could be strikingly different from that of the SM, in particular due to the presence of additional, heavier, Higgs states, which can still be as light as ~ 300 GeV. This may also lead to interesting possibilities for interactions with the light Higgs through, for example, the cascade decays $H \to hh$ and $A \to hZ$ which could be significant. The MSSM Higgs bosons could also be produced in SUSY cascades, such as e.g. $\tilde{\chi}_i^0 \to \tilde{\chi}_j^0 h$. Finally, if there are light EW SUSY states, the heavy MSSM Higgs bosons may predominantly decay into light -ino pairs, $H/A \to \tilde{\chi}_i^0 \tilde{\chi}_j^0$, leading to quite unusual signatures. These options should be considered for future search strategies.

2.2. Heavy Higgs interpretation

Another, quite unusual, scenario which can arise is that the observed state is not the lightest state of an extended Higgs sector. An interpretation of the LHC data in terms of the second, heavier, $\mathcal{CP}$-even Higgs has been proposed in the MSSM,[12] as well as in the next-to-minimal model (NMSSM).[26] Perhaps surprisingly, the main constraints on these scenarios do typically *not* come from the presence of a lighter Higgs state (with a mass that is often below the SM LEP limit $M_h \gtrsim 114$ GeV), but rather from the presence of additional doublet-like states around 125 GeV. In the MSSM these are the usual A and $H^\pm$, where in particular the searches for

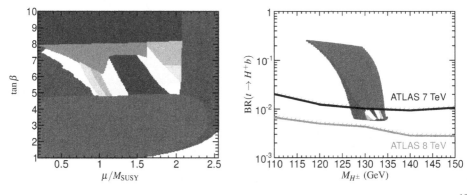

Fig. 3. Left: exclusion limits from different direct Higgs searches in the MSSM *low-M_H* scenario.[18] The experimentally allowed regions are shown in green ($M_H = 125.5 \pm 3$ GeV) and yellow ($M_H = 125.5 \pm 3$ GeV and SM-like rates). Right: predictions for $\text{BR}(t \rightarrow bH^+) \times \text{BR}(H^+ \rightarrow \tau^+ \nu_\tau)$ in this scenario, with the 7 TeV ATLAS upper limit[28] (black) and updated 8 TeV results[27] (red).

the charged Higgs in top decays poses unavoidable constraints. This is illustrated in Fig. 3, which shows as an example the parameter space of the so-called *low-M_H* scenario[18] (left), and how the prediction for $\text{BR}(t \rightarrow bH^+)$ compares to the latest ATLAS 8 TeV limit[27] (right). As the figure shows, there is not much room anymore for this interpretation in the MSSM. In non-minimal models, where the mass relations between $H^\pm$ and the other Higgs bosons are relaxed, the existence of Higgs bosons lighter than 125 GeV remains a viable possibility.

2.3. *Beyond minimal SUSY*

Besides the prospects to continue MSSM heavy Higgs searches with more data in the "standard" channels with high sensitivity — mainly for high $\tan \beta$ — it has recently been proposed[29] to extend searches into a regime at low $\tan \beta$ which is usually not accessible in the MSSM. This idea takes seriously the possibility that the stop mass scale is not only unnatural, but really multi-TeV, something which is certainly compatible with all (non-)observations. In this case, the lightest Higgs mass can be made sufficiently heavy also for $\tan \beta \sim 1$–2, where the heavy $\mathcal{CP}$-even scalar, H, has dominant decays into ZZ, W^+W^-, hh or $t\bar{t}$. While this comes at a high price in the MSSM, it should be noted that the same interesting parameter regime could also be accessible in more natural models with non-minimal supersymmetry, such as the NMSSM. In the NMSSM, a singlet superfield $\hat{S}$ is added to the MSSM, with a superpotential $W_{\text{NMSSM}}^{(3)} = \lambda \hat{S} \hat{H}_u \cdot \hat{H}_d + (\kappa/3)\hat{S}^3$. The resulting spectrum contains seven physical Higgs states and the lightest NMSSM Higgs mass receives an additional tree-level contribution proportional to λ. Figure 4 shows the degree of fine-tuning resulting from M_h, calculated in[11] for the MSSM (left) and the NMSSM (right). As can be seen from this figure, the Higgs mass fine-tuning can be significantly reduced in the NMSSM. The NMSSM could also

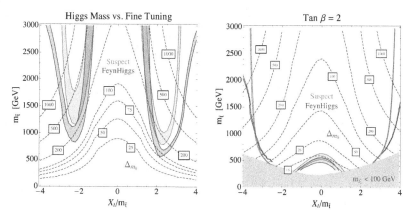

Fig. 4. Degree of Higgs mass fine-tuning Δm_h (contours) from.[11] Higgs mass predictions (coloured bands) are shown for $m_{\tilde{t}} \equiv M_S$ versus X_t in the MSSM (left), and the NMSSM (right) with $\tan\beta = 2$ and $\lambda = 0.7$ (the maximal value perturbative up to the GUT scale).

offer attractive features from both phenomenological and experimental points-of-view, such as singlet-doublet mixing as an alternative to enhance loop-mediated $h \to \gamma\gamma$ decays[30] and the existence of light Higgs bosons[31] that can be searched for in different ways.[32] A comprehensive review of phenomenological studies in the NMSSM Higgs sector can be found in Ref. 33.

3. General Two-Higgs-Doublet Models

An alternative framework to interpret Higgs boson searches, including searches sensitive to new Higgs sectors beyond the SM, is the general two-Higgs-doublet model (2HDM).[34] This model in itself does not address the SM hierarchy problem, but it is interesting as a possible low-energy limit of a UV-complete theory (for which the MSSM is one example). The 2HDM offers a large freedom, since there are enough free parameters to specify e.g. all four physical Higgs masses as input. The current data already leads to non-trivial constraints on the 2HDM parameter space.[35] Most analyses so far focus on the 2HDM types I/II with a (softly broken) Z_2-symmetry, but there are also studies with more general Yukawa couplings.[36] Due to the large freedom in choosing the parameters, the heavy Higgs bosons are less constrained by direct searches than in SUSY models. On the other hand, the absence of additional new states typically leads to strong constraints from flavour physics.[37] First experimental LHC results from a 2HDM Higgs search in the $H \to WW$ channel have been presented,[38] but there are many other channels that could be interesting for a 2HDM interpretation. In case future searches for additional Higgs bosons reveal nothing, this would indicate for the 2HDM — like in the case of the MSSM — that the most probable scenario is the decoupling limit. In this limit, the heavy Higgs states decouple, leaving a single light (SM-like) Higgs boson in the spectrum. To measure the small deviations in the Higgs couplings predicted in this case would likely require the precision of a linear e^+e^- collider.[39]

Another instance of the 2HDM leading to similar results is the Inert Doublet Model (IDM).[40] In the IDM, the Z_2-symmetry is made exact, rendering the lightest scalar particle stable. In this way the model can accommodate a dark matter (DM) candidate. Recent analyses[41] demonstrate that this is compatible with the Higgs discovery. However, taking into account data on invisible Higgs decays,[42] in most of the allowed parameter space the mass of the DM candidate is $M_{DM} > 500$ GeV, and it appears difficult to discover any of the additional Higgs bosons at the LHC. The only exception is when the scalar DM exhibits resonant annihilation on the Higgs pole, $M_{DM} \gtrsim M_h/2 \sim 60\text{--}80$ GeV, which could be an interesting mass region.

4. Fits of the Higgs Couplings

A generic way to test for (absence of) new physics is to parametrize coupling deviations from the SM, and then use the available data to determine constraints on these deviations. Here we focus on deviations in the signal *strengths*, for which a simple "interim" framework has been proposed.[43] In the most general case, each SM coupling is assigned a scale factor, κ_i, defined such that, for example,

$$\sigma(gg \to H) \times \mathrm{BR}(H \to WW) = \frac{\kappa_g^2 \kappa_W^2}{\kappa_H^2} \left[\sigma(gg \to H) \times \mathrm{BR}(H \to WW)\right]_{\mathrm{SM}}, \quad (2)$$

where κ_H is a scale factor for the Higgs total width (which is not measurable at the LHC). This definition allows the κ_i to be extracted from data using the most accurate predictions available for the SM production cross sections and decay branching ratios.[a] Using this framework, different fits to the (SM) measurements can be performed where some of the coupling scale factors are allowed to vary. This strategy has been pursued by several theory groups (there are too many results to list them here), and also directly by the experimental collaborations.[3,4]

A significant deviation from the SM point ($\forall \kappa_i : \kappa_i = 1$) in one of these coupling fits would signal a shortcoming of the SM to account for the measured data. Currently, there is no such indication, and the constraints on the allowed deviation in individual κ_i from highly constrained analyses are of the order $10 - 20\%$ (but larger if more scale factors are varied). In the absence of direct observations of new physics, it will be a very important goal of the LHC to improve the constraints on the SM Higgs couplings and search for deviations at a greater level of precision. From general arguments, new physics entering at a scale f modifying (some of) the Higgs couplings can be expected to contribute at the level

$$\kappa_i \simeq 1 \pm c \frac{v^2}{f^2}, \quad (3)$$

where c is a number of $\mathcal{O}(1\text{--}100\%)$.[45] This already tells us that percent-level precision is desirable to approach the TeV-scale. While an improvemed precision for

---

[a]To allow for deviations in the $\mathcal{CP}$, spin, or tensor structure of the SM Higgs couplings, more elaborate parametrizations are required.[44]

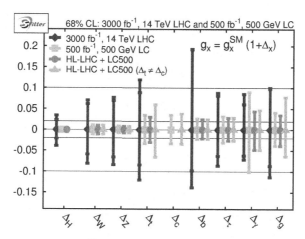

Fig. 5.   Projected sensitivity from[47] for Higgs coupling measurements (deviations from the SM) at the 14 TeV HL-LHC and a 500 GeV e^+e^- linear collider (LC).

the coupling measurements is certainly expected at LHC-300, and even more so at a high-luminosity (HL-LHC) with 3000 fb^{-1}, the best precision on the Higgs couplings would only be reached at a lepton collider, such as the ILC.[39,46] This statement can be made more quantitative by comparing the expected performance for Higgs coupling determination in different future collider scenarios. In Fig. 5 we show a projection by the SFitter collaboration.[47] It can be seen from this figure that a precision at the few percent level on nearly all the Higgs couplings can be achieved with a linear collider at 500 GeV alone, and that further improvement on several couplings, in particular Δ_γ, can be reached by combining LHC and ILC results.

4.1. *Composite Higgs models*

An alternative to a weakly coupled natural Higgs sector — as in supersymmetry — would be if the observed scalar is not fundamental, but results as a composite pseudo-Goldstone boson from some broken global symmetry of a new strong inter-action[48] (much like pions in QCD). These ideas are exemplified by the minimal composite Higgs models[49] (MCHM). In the minimal case, a global SO(5) symmetry is assumed to be broken dynamically down to SO(4) at a scale f, leading to a single doublet of SU(2)$_L$, precisely as in the SM. The compositeness leads to modifications of the SM Higgs couplings, which in the minimal case amounts to universal effects on the coupling scale factors to vector bosons, κ_V, and fermions, κ_F. In this scenario, we have explicitly[50]

$$\kappa_V = \sqrt{1 - \xi^2}, \quad \kappa_F = \frac{1 - (1 + n)\xi^2}{\sqrt{1 - \xi^2}},$$

where $n = 0, 1, \ldots$ is an integer that depends on the realization and $\xi = v/f$. The scale f plays a role similar to M_S is SUSY theories when it comes to naturalness. A

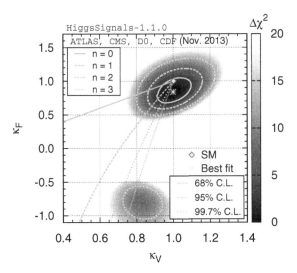

Fig. 6. Two-parameter fit of Higgs coupling scale factors (κ_V, κ_F) to all available rate measurements performed with HiggsSignals.[52] The colour coding shows $\Delta\chi^2 = \chi^2 - \chi^2_{\min}$, and the green contours the MCHM predictions for varying compositeness scale ξ and values of n.

higher symmetry-breaking scale f implies decoupling to the SM limit, but also that the compositeness has no role in a natural solution of the SM hierarchy problem. Conversely, for $f \simeq v$, the SM "Higgs" would be a fully composite state of the strong interaction (technicolor limit), an option which is already strongly disfavoured by the data. Current measurements constrain the MCHM, as can be seen in Fig. 6 (for an earlier analysis of this type, see[51]). To produce these constraints we have used the public code HiggsSignals,[52] which provides a generic test for compatibility between Higgs measurements and the predictions of an arbitrary model. Using all available Higgs channels from Tevatron and the LHC, the combined constraints corresponds to a 95% CL lower limit of $f \gtrsim 500$–800 GeV (depending on n), which starts to challenge the naturalness of the theory. As already mentioned, to probe deviations in the Higgs couplings up to much higher scales (TeV and beyond) is likely require the full precision of the (upgraded) LHC and/or a lepton collider. However, one particular aspect of the composite Higgs theories is that they quite generically predict the existence of exotic particles around the TeV-scale.[50,53] To discover (or rule out) the existence of these states at the LHC will therefore provide another important test of the Higgs compositeness idea.

5. Conclusions

The new particle discovered by ATLAS and CMS has properties compatible with the SM Higgs boson; no significant deviations from the SM are observed in the data. This has far-reaching implications also for scenarios beyond the SM, which must now face the additional constraint that there should be a SM-like Higgs boson in the spectrum. At the same time, many of these theories (like, for example, supersymmetry) face quite stringent constraints from the non-observation of new particles below the TeV-scale. Looking at these constraints from a bottom-up perspective on the prospects for extended Higgs scenarios, the picture is however quite different. For theories with a decoupling limit it is fully consistent to expect the lightest state to be a SM-like Higgs boson, and for this particle to become more SM-like the more additional (heavier) degrees of freedom decouple. This still motivates searches for additional Higgs states in e.g. the MSSM (where the decoupling limit is favoured by global analyses), as well as in the context of general 2HDM. Large effects on the individual rates of the 125 GeV state are not excluded, with modifications of up to 20–100% still allowed in some cases. To explicitly highlight these options and spur experimental activity, updated benchmark scenarios for MSSM Higgs searches have been presented, and new signatures are discussed for heavy Higgs searches.

Beyond the MSSM, extended supersymmetric scenarios offer a more natural explanation for the "high" Higgs mass. The phenomenology of these scenarios can differ substantially from that in the MSSM, and most limits obtained from searches for heavy MSSM Higgs bosons are not directly applicable in this context. The prospects to constrain these models in the near future therefore depend to a large degree on the feasibility to include additional decay modes into the LHC heavy Higgs searches. The situation is similar in the general 2HDM, which offers a parametrization with significant freedom remaining to choose the Higgs masses and couplings.

Precision measurements pertaining to the 125 GeV state, first at the LHC and later at a future lepton collider, will eventually be sensitive to deviations in individual Higgs rates associated with scales that are not accessible to direct Higgs searches. This could offer the intriguing, but also somewhat frustrating, scenario that the SM is rejected without a direct answer to the question what lies behind. The best prospects for the LHC to prove that the Higgs boson is non-standard is through new discoveries, which would give immediate proof for the existence of physics beyond the SM. Continued Higgs searches in motivated SM extensions should therefore remain a top priority at the LHC.

Acknowledgments

I would like to thank the organizers of PIC2013 for the invitation and their kind hospitality. I am also grateful to my collaborators on the various projects from which material has been borrowed for this review; in particular I would like to thank Tim Stefaniak for useful assistance with `HiggsSignals` in preparing Fig. 6.

References

1. ATLAS Collaboration, G. Aad *et al. Phys. Lett. B* **716** (2012) 1, [arXiv:1207.7214].
2. CMS Collaboration, S. Chatrchyan *et al. Phys. Lett. B* **716** (2012) 30, [arXiv:1207.7235].
3. ATLAS Collaboration. ATLAS-CONF-2013-034.

4. CMS Collaboration. CMS-PAS-HIG-13-005.
5. ATLAS Collaboration. ATLAS-CONF-2013-014.
6. G. Degrassi *et al. JHEP* **1208** (2012) 098, [arXiv:1205.6497].
7. ATLAS Collaboration. ATLAS-CONF-2013-040.
8. CMS Collaboration, S. Chatrchyan *et al. Phys. Rev. Lett.* **110** (2013) 081803, [arXiv:1212.6639].
9. https://twiki.cern.ch/twiki/bin/view/AtlasPublic/SupersymmetryPublicResults; https://twiki.cern.ch/twiki/bin/view/CMSPublic/PhysicsResultsSUS.
10. C. Brust, A. Katz, S. Lawrence, and R. Sundrum *JHEP* **1203** (2012) 103, [arXiv:1110.6670]; M. Papucci, J. T. Ruderman, and A. Weiler *JHEP* **1209** (2012) 035, [arXiv:1110.6926].
11. L. J. Hall, D. Pinner, and J. T. Ruderman *JHEP* **1204** (2012) 131, [arXiv:1112.2703].
12. S. Heinemeyer, O. Stål, and G. Weiglein *Phys. Lett. B* **710** (2012) 201, [arXiv:1112.3026].
13. H. Baer, V. Barger, and A. Mustafayev *Phys. Rev. D* **85** (2012) 075010, [arXiv:1112.3017].
14. A. Arbey, M. Battaglia, A. Djouadi, F. Mahmoudi, and J. Quevillon *Phys. Lett. B* **708** (2012) 162, [arXiv:1112.3028]; P. Draper, P. Meade, M. Reece, and D. Shih *Phys. Rev. D* **85** (2012) 095007, [arXiv:1112.3068].
15. F. Brümmer, S. Kraml, and S. Kulkarni *JHEP* **1208** (2012) 089, [arXiv:1204.5977].
16. M. S. Carena, S. Heinemeyer, C. E. M. Wagner, and G. Weiglein hep-ph/9912223.
17. LEP Higgs WG, S. Schael *et al. Eur. Phys. J. C* **47** (2006) 547, [hep-ex/0602042].
18. M. Carena, S. Heinemeyer, O. Stål, C. Wagner, and G. Weiglein *Eur. Phys. J. C* **73** (2013) 2552, [arXiv:1302.7033].
19. S. Heinemeyer, W. Hollik, and G. Weiglein *Comput. Phys. Commun.* **124** (2000) 76, [hep-ph/9812320]; *Eur. Phys. J. C* **9** (1999) 343, [hep-ph/9812472]; G. Degrassi, S. Heinemeyer, W. Hollik, P. Slavich, and G. Weiglein *Eur. Phys. J. C* **28** (2003) 133, [hep-ph/0212020]; M. Frank, T. Hahn, S. Heinemeyer, *et al. JHEP* **0702** (2007) 047, [hep-ph/0611326].
20. P. Bechtle, O. Brein, S. Heinemeyer, G. Weiglein, and K. E. Williams *Comput. Phys. Commun.* **181** (2010) 138, [arXiv:0811.4169]; ibid. **182** (2011) 2605, [arXiv:1102.1898]; P. Bechtle, O. Brein, S. Heinemeyer, O. Stål, T. Stefaniak, and G. Weiglein arXiv:1311.0055.
21. CMS Collaboration. CMS-HIG-12-050.
22. LHCb collaboration, R. Aaij *et al. Phys. Rev. Lett.* **111** (2013) 101805, [arXiv:1307.5024].
23. J. Cao, Z. Heng, J. M. Yang, and J. Zhu *JHEP* **1210** (2012) 079, [arXiv:1207.3698]; P. Bechtle, S. Heinemeyer, O. Stål, T. Stefaniak, G. Weiglein, and L. Zeune *Eur. Phys. J. C* **73** (2013) 2354, [arXiv:1211.1955]; A. Arbey, M. Battaglia, A. Djouadi, and F. Mahmoudi *Phys. Lett. B* **720** (2013) 153, [arXiv:1211.4004].
24. A. Djouadi arXiv:1311.0720.
25. M. Carena, S. Gori, N. R. Shah, and C. E. Wagner *JHEP* **1203** (2012) 014, [arXiv:1112.3336]; M. Carena, S. Gori, N. R. Shah, C. E. Wagner, and L.-T. Wang *JHEP* **1207** (2012) 175, [arXiv:1205.5842].
26. J. F. Gunion, Y. Jiang, and S. Kraml *Phys. Rev. D* **86** (2012) 071702, [arXiv:1207.1545]; G. Belanger, U. Ellwanger, J. F. Gunion, Y. Jiang, S. Kraml, *et al. JHEP* **1301** (2013) 069, [arXiv:1210.1976].
27. ATLAS Collaboration. ATLAS-CONF-2013-090.
28. ATLAS Collaboration, G. Aad *et al. JHEP* **1206** (2012) 039, [arXiv:1204.2760].
29. A. Djouadi and J. Quevillon arXiv:1304.1787.

30. U. Ellwanger *Phys. Lett. B* **698** (2011) 293, [arXiv:1012.1201]; J. Cao, Z. Heng, T. Liu, and J. M. Yang *Phys. Lett. B* **703** (2011) 462, [arXiv:1103.0631]; U. Ellwanger *JHEP* **1203** (2012) 044, [arXiv:1112.3548]; J.-J. Cao, Z.-X. Heng, J. M. Yang, Y.-M. Zhang, and J.-Y. Zhu *JHEP* **1203** (2012) 086, [arXiv:1202.5821]; R. Benbrik, M. Gomez Bock, S. Heinemeyer, *et al. Eur. Phys. J. C* **72** (2012) [arXiv:1207.1096].

31. R. Dermisek and J. F. Gunion *Phys. Rev. D* **79** (2009) 055014, [arXiv:0811.3537].

32. O. Stål and G. Weiglein *JHEP* **1201** (2012) 071, [arXiv:1108.0595]; U. Ellwanger *JHEP* **1308** (2013) 077, [arXiv:1306.5541]; J. Cao, F. Ding, C. Han, J. M. Yang, and J. Zhu *JHEP* **1311** (2013) 018, [arXiv:1309.4939].

33. U. Ellwanger *Eur. Phys. J. C* **71** (2011) 1782, [arXiv:1108.0157].

34. G. Branco, P. Ferreira, Lavoura, *et al. Phys. Rept.* **516** (2012) 1, [arXiv:1106.0034].

35. P. Ferreira, R. Santos, M. Sher, and J. P. Silva arXiv:1305.4587; G. Belanger, B. Dumont, U. Ellwanger, J. Gunion, and S. Kraml *Phys. Rev. D* **88** (2013) 075008, [arXiv:1306.2941].

36. A. Celis, V. Ilisie, and A. Pich arXiv:1310.7941.

37. F. Mahmoudi and O. Stål *Phys. Rev. D* **81** (2010) 035016, [arXiv:0907.1791].

38. ATLAS Collaboration. ATLAS-CONF-2013-027.

39. D. Asner, T. Barklow, C. Calancha, K. Fujii, N. Graf, *et al.* arXiv:1310.0763.

40. N. G. Deshpande and E. Ma *Phys. Rev. D* **18** (1978) 2574; R. Barbieri, L. J. Hall, and V. S. Rychkov *Phys. Rev. D* **74** (2006) 015007, [hep-ph/0603188]; E. Ma *Phys. Rev. D* **73** (2006) 077301, [hep-ph/0601225].

41. A. Goudelis, B. Herrmann, and O. Stål *JHEP* **1309** (2013) 106, [arXiv:1303.3010]; M. Krawczyk, D. Sokolowska, P. Swaczyna, and B. Swiezewska *JHEP* **1309** (2013) 055, [arXiv:1305.6266].

42. ATLAS Collaboration. ATLAS-CONF-2013-011.

43. LHC Higgs Cross Section Working Group arXiv:1209.0040; arXiv:1307.1347.

44. R. Contino, M. Ghezzi, C. Grojean, M. Mühlleitner, and M. Spira *JHEP* **1307** (2013) 035, [arXiv:1303.3876].

45. H. Baer, T. Barklow, K. Fujii, Y. Gao, A. Hoang, *et al.* arXiv:1306.6352.

46. M. E. Peskin arXiv:1207.2516.

47. M. Klute, R. Lafaye, T. Plehn, M. Rauch, and D. Zerwas *Europhys. Lett.* **101** (2013) 51001, [arXiv:1301.1322].

48. G. Giudice, C. Grojean, A. Pomarol, and R. Rattazzi *JHEP* **0706** (2007) 045, [hep-ph/0703164].

49. K. Agashe, R. Contino, and A. Pomarol *Nucl. Phys. B* **719** (2005) 165, [hep-ph/0412089]; R. Contino, L. Da Rold, and A. Pomarol *Phys. Rev. D* **75** (2007) 055014, [hep-ph/0612048].

50. A. Pomarol and F. Riva *JHEP* **1208** (2012) 135, [arXiv:1205.6434].

51. A. Falkowski, F. Riva, and A. Urbano *JHEP* **1311** (2013) 111, [arXiv:1303.1812].

52. P. Bechtle, S. Heinemeyer, O. Stål, T. Stefaniak, and G. Weiglein arXiv:1305.1933.

53. D. Marzocca, M. Serone, and J. Shu *JHEP* **1208** (2012) 013, [arXiv:1205.0770].

Physics in Collision (PIC 2013)
International Journal of Modern Physics: Conference Series
Vol. 31 (2014) 1460290 (9 pages)
© The Author
DOI: 10.1142/S2010194514602907

World Scientific
www.worldscientific.com

Review of B and B_s decays

Concezio Bozzi

INFN Sezione di Ferrara
Via G. Saragat 1, I-44121 Ferrara, Italy
bozzi@fe.infn.it

Published 15 May 2014

A review of B and B_s decays is presented. Emphasis is given to processes most sensitive to physics beyond the Standard Model, such as radiative, electroweak and "Higgs" penguin decays, and tree-level decays involving tau leptons in the final state. An outlook on future perspectives is also given.

1. Introduction

The Standard Model of Electroweak Interactions gives stringent predictions in the flavor sector. The decays of B and B_s mesons are among the most useful probes to test the presence of a single source of CP violation in charged weak currents, the suppressions due to the hierarchy of CKM matrix elements, and the suppression of flavor-changing neutral currents. Currently, the CKM matrix unitarity has been tested at a few percent level.[a] Flavor physics is thus sensitive to physics processes beyond the standard model, which might not respect the many suppressions of the SM. Complementary to a *relativistic way*, based on increasing the available energy and directly producing new particles, a *quantum way* can be exploited to search for new physics, where collider luminosity is increased and indirect effects due to virtual particles in higher-order diagrams result in experimental observables being different from SM predictions. In many cases, this indirect way probe higher mass scales than direct searches. In the following, recent results on the most promising two classes of B and B_s decays, i.e. penguin decays and leptonic/semileptonic decays with tau leptons in the final state, are reviewed.

[a]A comprehensive review of CP Violation has been given in this Conference by S. Monteil.

2. The Experimental Facilities

B meson physics is extensively studied in two experimental environments: e^+e^- B Factories, running on the $\Upsilon(4S)$ mass, and proton colliders. The former offer the advantages of an initial state with well define energy-momentum and quantum numbers, and low multiplicity events. The two B Factory experiment Babar and Belle showed that full event reconstruction is achievable, the decay products of both B mesons from the $\Upsilon(4S)$ decay can be disentangled, particle identification is excellent, reconstruction of neutral hadrons is efficient, and neutrinos can be indirectly detected by missing mass. Hadron colliders offer much higher production rates and the possibility to produce all flavored mesons and baryons. However, high background rates require selective triggers and the acceptance is restricted. The LHCb experiment at the LHC exploits long decay paths of B hadrons produced at the LHC and, due to very precise charged particle tracking and identification, is able to discriminate B decays from background and search for the rare decays most sensitive to new physics.

3. Penguin Decays

In penguin decays, non-SM particles might give their contribution in loop diagrams. These decays are conventionally split in three classes:

(i) radiative penguins, with a single photon accompanying the hadronic system,
(ii) electroweak (EW) penguins, where two leptons are emitted instead of a photon, and
(iii) *Higgs penguins, which are the s-channel version of the previous ones.*

The branching ratios for radiative penguin decays are typically 10^{-4} or less. One might expect EW penguins to be suppressed in the SM about a factor $\alpha_{em} \approx 1/100$ with respect to radiative ones, resulting in typical BRs of 10^{-6}. Higgs penguins are further helicity suppressed, with predicted BRs at the 10^{-9} level. The effective hamiltonian describing these processes can be written by means of the Operator Product Expansion technique, with Wilson coefficients calculable from perturbation theory and matrix elements of operators which need to be computed non perturbatively. A parametrization in terms of the Lorentz structure of the operators can be written as

$$H_{eff} = -\frac{4G_F}{\sqrt{2}} V_{tb} V_{ts}^* \sum_i [C_i(\mu)O_i(\mu) + C_i'(\mu)O_i'(\mu)] \tag{1}$$

where C_i are Wilson coefficients and O_i Lorentz-Invariant operators. Primed and unprimed quantities refer to right- and left-handed couplings, the former being suppressed in the SM. The relevant operator for radiative penguins is $O_7 \sim m_b \bar{s}_L \sigma_{\mu\nu} b_R F^{\mu\nu}$. The operators $O_9 \sim \bar{s}_L \gamma_\mu b_L \bar{\ell} \gamma^\mu \ell$ and $O_{10} \sim \bar{s}_L \gamma_\mu b_L \bar{\ell} \gamma^\mu \gamma_5 \ell$ dominate EW penguis, while the scalar and pseudoscalar $O_S \sim \bar{s}_L b_R \bar{\ell}\ell$, $O_P \sim \bar{s}_L b_R \bar{\ell} \gamma_5 \ell$ contribute to Higgs penguins.

Starting from this effective hamiltonian, it is possible to obtain SM predictions with reduced theoretical uncertainties for sets of observables in several decays. Accurate measurements can then be used to check the agreement with the SM. By combining several observables, it is possible to check each building block of the theoretical calculation (e.g. Wilson coefficients).

3.1. *Radiative penguin decays*

The branching fraction of the inclusive $b \to s\gamma$ decay has been precisely measured at the B Factories in the last ten years. The experimental average,[1] $Br(b \to s\gamma) = (3.43 \pm 0.22) \times 10^{-4}$ is in good agreement with the SM prediction $Br_{SM}(b \to s\gamma) = (3.15 \pm 0.23) \times 10^{-4}$ and, together with the measured Higgs boson mass, gives one of the strongest constraints in the parameter space of Minimal Supersymmetric models. Many exclusive decays have been studied as well, in particular at hadron colliders where backgrounds to inclusive decays are unmanageable. LHCb performed a first measurement in the B_s system,[2] on a sample of about 700 $B_s \to \phi\gamma$ decays, giving $Br(B_s \to \phi\gamma) = (3.5 \pm 0.4) \times 10^{-5}$. Given the size of the samples collected at LHCb, it is now possible to perform the measurement of the photon polarization in radiative decays, which is predominantly left-handed in the SM. The photon polarization is inferred by measuring the up-down asymmetry with respect to the hadronic decay plane in $B^+ \to K_{res}\gamma$, where K_{res} is an intermediate resonance decaying in $K^+\pi^+\pi^-$. LHCb obtained[3] an up-down asymmetry, on a sample of 8000 decays into a mixture of K resonances, $\mathcal{A}_{ud} = -0.085 \pm 0.019(stat) \pm 0.004(syst)$, giving evidence for photon polarization in $b \to s\gamma$ decays at 4.6 σ. However, due to multiple overapping resonances, it is difficult to measure the photon polarization for each single decay. More theoretical studies are needed in order to do so.

3.2. *EW penguin decays*

The most widely studied[1] EW penguin decay is $B^0 \to K^{*0}\mu^+\mu^-$. The accumulated data samples are now adequate to perform an angular analysis. The differential distribution can be parametrized in terms of three angles and observables which depend on Wilson coefficients and form factors, which in turn depend on q^2, the di-muon squared invariant mass. A widely used set of observables include A_{FB}, the forward-backward asymmetry of the positive lepton with respect to the B flight path, F_L, the fraction of longitudinally polarized K^{*0}, S_3, the asymmetry in the K^{*0} transverse polarization, and A_9, a T-odd CP asymmetry. A zero crossing point is precisely predicted in the SM for A_{FB}: $q_0^2 = (4.36^{+0.33}_{-0.31})GeV^2/c^4$. The results obtained by LHCb[4] on a sample of about 900 events reconstructed on data corresponding to an integrated luminosity of 1 fb^{-1} are shown in Fig. 1. There is good agreement with SM predictions within uncertainties for all observables. LHCb has also obtained a first measurement of the A_{FB} zero crossing point, $q_0^2 = (4.9 \pm 0.9)GeV^2/c^4$, which is consistent with the SM. In addition to the observables described before and the other introduced in Fig. 1, new ones were proposed,[5] which have limited dependence

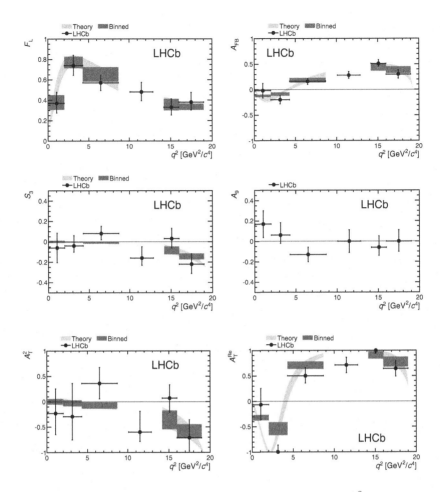

Fig. 1. Top four plots: fraction of longitudinal polarisation of the K^{*0}, F_L, dimuon system forward-backward asymmetry, A_{FB}, and the angular observables S_3 and A_9 from the $B^0 \to K^{*0}\mu^+\mu^-$ decay as a function of the dimuon invariant mass squared, q^2. The two bottom plot report the theoretically cleaner transverse asymmetries $A_T^2 = 2S_3/(1 - F_L)$ and $A_T^{Re} = 4/3 A_{FB}/(1 - F_L)$.

on the uncertainties on form-factors at low q^2. LHCb has analyzed their data in terms of these new observables,[6] and found good agreement for some of them and local disagreements up to 3.7 σ. These disagreements could be interpreted in terms of large new physics contributions to the Wilson coefficient of the O_9 operator.[7] However, other theory groups[8] quote more conservative uncertainties on their calculated observables, thereby reducing the disagreement with the SM. Clearly, no definitive conclusions can be reached at this point. More data need to be analyzed and more theoretical studies are needed.

Many other modes have been studied at B Factories and hadron colliders, such as $B_d^0 \to K^{*0}e^+e^-$, $B^+ \to K^+\mu^+\mu^-$, $B_s^0 \to \phi\mu^+\mu^-$, $\Lambda_b \to \Lambda\mu^+\mu^-$.[1]

Cabibbo-suppressed $B \to h_d \ell^+ \ell^-$ decays have also been studied by Babar[9] and LHCb,[10] and found in agreement with SM rates within current experimental uncertainties.

3.3. *Higgs penguin decays*

This class of decays is rare in the SM. Besides proceedings through flavour-changing neutral currents, they are also helicity suppressed. For this reason, they are sensitive to scalar and pseudoscalar contribution due to non-SM particles. The most promising of such decays is $B^0_{(s)} \to \mu^+ \mu^-$, whose BR is precisely predicted in the SM.[11] By taking into account the finite decay width difference of the B_s meson mass eigenstates, one can compute the time-integrated branching fractions[12] reported in the last column of Table 1. The experimental measurement of $B^0_{(s)} \to \mu^+ \mu^-$ is conceptually simple. Muons are triggered and reconstructed with high efficiency at hadron colliders and the current luminosities give enough sensitivity to the predicted SM BRs. However, a huge combinatorial background needs to be taken into account and properly subtracted. Since most of this background is due to muons not originating from the same B meson, it can be discriminated from signal by using topological and kinematical criteria. Backgrounds with muon candidates originating from the same B meson are due to partially reconstructed decays where two, one or zero particles are misidentified as muons, such as $B^0_{(s)} \to h^+ h^-$, $B^0_s \to K^- \mu^+ \nu$, $\Lambda_b \to p \mu^- \nu$, $B^0 \to \pi^- \mu^+ \nu$, $B^{0,+} \to \pi^{0,+} \mu^+ \mu^-$, $B^+_c \to J/\psi(\to \mu\mu)\mu^+\nu$. All of them can be parametrized and subtracted by using data-driven methods. The most sensitive analyses of $B^0_{(s)} \to \mu^+ \mu^-$ decays were obtained by LHCb[13] and CMS.[14] Both measurements classify their events in the plane $m_{\mu\mu} - BDT$, where the former is the dimuon invariant mass and the latter is the output of a Boosted Decision Tree (BDT) aimed at discriminating combinatorial background. The $B^0_{(s)} \to \mu^+ \mu^-$ branching fraction is determined by a simultaneous fit to $m_{\mu\mu}$ in BDT bins (LHCb) or categories (CMS). Upper limits are given following the CLs technique. Fit results are shown in Fig. 2 and summarized in Table 1.

Evidence at 4.0 σ or above is found for the $B^0_s \to \mu^+ \mu^-$ decay, while upper limits can be set for $B^0 \to \mu^+ \mu^-$. A weighted average of the results from LHCb and CMS gives 5σ and 3σ significances, respectively. These results are compatible with the SM prediction and gives stringent limits on the parameter space of theories beyond SM.[15]

Table 1. Measurements of $B_{(s)} \to \mu^+ \mu^-$ decays and comparison with SM prediction.

Decay	LHCb	CMS	weighted average	SM
$Br(B^0_s \to \mu^+\mu) \times 10^{-9}$	$2.9^{+1.1}_{-1.0}\,^{+0.3}_{-0.1}$	$3.0^{+1.0}_{-0.9}$	2.9 ± 0.7	3.56 ± 0.30
	4.0σ	4.3σ	$> 5\sigma$	
$Br(B^0_s \to \mu^+\mu) \times 10^{-10}$	$3.7^{+2.4}_{-2.1}\,^{+0.6}_{-0.4}$	$3.5^{+2.1}_{-1.8}$	$3.6^{+1.6}_{-1.4}$	1.07 ± 0.10
	< 7.4 at 95% CL	< 11 at 95% CL	$> 3\sigma$	

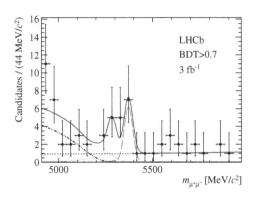
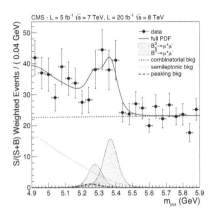

Fig. 2.   Left: Invariant mass distribution of the selected LHCb $B^0 \to \mu^+\mu^-$ candidates (black dots) with BDT > 0.7. Right: Invariant mass distribution of the CMS $B^0 \to \mu^+\mu^-$ candidates (black dots) for the combination of all BDT categories. In both plots the result of the fit is overlaid (blue solid line) and the different components detailed.

4. Tree-Level Decays with Tau Leptons in the Final State

Leptonic and semileptonic decays with a τ lepton in the final state are sensitive to processes involving non-Standard Model particles, and particularly to charged Higgs bosons which would appear as propagator in the Feynman diagram. The SM expectation for the leptonic decay is $Br(B^+ \to \tau^+\nu) = (1.01 \pm 0.29) \times 10^{-4}$. This 30% uncertainty is dominated by the knowledge of the CKM matrix element V_{ub} and, to a lesser extent, the B meson pseudoscalar constant f_B. In models with an enlarged Higgs sector, such as the two Higgs doublet model (2HDM) of MSSM, this rate is multiplied by a factor $(1 - \tan^2\beta m_{B^+}^2/m_{H^+}^2)$, where $\tan\beta$ is the ratio of the vacuum expectation values of the two doublets and m_{H^+} is the charged Higgs mass. The ratios $R(D^{(*)})$ of decays with taus and lighter leptons in the final state is predicted at the 5-10% level in the Standard model :[16, 17]

$$R(D) = \frac{\mathcal{B}(B \to D\tau\bar{\nu})}{\mathcal{B}(\overline{B} \to D\ell\bar{\nu})} = 0.297 \pm 0.017$$

$$R(D^*) = \frac{\mathcal{B}(B \to D^*\tau\bar{\nu})}{\mathcal{B}(\overline{B} \to D^*\ell\bar{\nu})} = 0.252 \pm 0.003.$$

These ratios are modified in models with a charged Higgs, such as in the 2HDM of type-II, by adding a scalar helicity amplitude to the SM ones. This might enhance or suppress the SM ratio, depending on the decay as well as the phase space region under study.

Experimentally, due to the presence of at least two neutrinos in the final state, the B Factories offer the most favorable environment to measure these decays. B mesons from the $\Upsilon(4S)$ decay can be both fully reconstructed, one ("tag")in an hadronic or semileptonic decay, the other in the detectable particles of the decay of interest. The presence of missing neutrinos is then inferred in leptonic decays by looking at E_{extra}, the sum of calorimeter energy not associated with the signal or

Table 2. Measurements of $B \to \tau\nu$ decays and comparison with SM prediction.

Experimenrt	Tag	Branching Ratio ($\times 10^{-4}$)
Babar	hadronic	$1.83^{+0.53}_{-0.49} \pm 0.24$
Babar	semileptonic	$1.7 \pm 0.8 \pm 0.2$
Belle	hadronic	$0.72^{+0.27}_{-0.25} \pm 0.11$
Belle	semileptonic	$1.54^{+0.38}_{-0.37} {}^{+0.29}_{-0.31}$
	Average	1.15 ± 0.23
	SM prediction	1.01 ± 0.29

the other B. This variable peaks at zero for signal events, while it has a smooth increasing distribution to positive values for background. For semileptonic decays, the squared missing mass m^2_{miss} is used. For the most dominant background due to semileptonic decays into lighter leptons, this variable peaks at zero. Additional missing particles, e.g. neutrinos from signal decays or hadrons in semileptonic decays into higher D resonances, shift this variable to positive, smoothly varying values.

Several measurements of the branching ratio of $B \to \tau\nu$ decays were performed by Babar and Belle, as reported in Table 2. The average of the experimental measurements is in agreement with SM within uncertainties.

The most recent analysis of $B \to D\tau\bar{\nu}$ and $B \to D^*\tau\bar{\nu}$ with *BABAR* data[18] will be now described. Four samples are defined, according to the reconstructed D meson $(D^0, D^+, D^{*0}, D^{*+})$, and the tau meson is reconstructed in leptonic final states. In this way, signal samples have the same detectable particles as *normalization samples*, i.e. decays into lighter leptons. A poorly known remaining background is due decays into P-wave charmed mesons (D^{**}). These are studied with four control samples (one for each signal sample), selected by adding a neutral pion to the decay chain. A simultaneous fit is performed to the distribution of the lepton momentum in the B rest frame, p^*_ℓ, and the square of the missing mass. The results of the fit are shown in Fig. 3. The signal yields, signal significance and measured values of $R(D)$ and $R(D^*)$ are reported in table 3. Systematic uncertainties are dominated by the assumptions made on the D^{**} background, the statistical accuracy of the simulation and the knowledge of continuum and $B\bar{B}$ backgrounds. The correlation between $R(D)$ and $R(D^*)$ is -27%.

The two measurements of $R(D)$ and $R(D^*)$ exceed the Standard Model prediction by 2.0σ and 2.7σ, respectively. Their combination, including their correlation, gives a $p-$value of 6.9×10^{-4}. Therefore, the possibility of both measurements to agree with the Standard Model is excluded at the 3.4σ level.

The effect that a charged Higgs boson, in a two-Higgs-doublet model of type II,[19,20] would have on the measurements is evaluated by reweighting the relevant matrix element in the simulation according to 20 different values of $\tan\beta/m_{H^+}$, where $\tan\beta = v_2/v_1$ is the ratio of the expectation values, and repeating the analysis. The regions allowed by the $R(D)$ and $R(D^*)$ measurements are $\tan\beta/m_{H^+} = 0.44 \pm 0.02$ and $\tan\beta/m_{H^+} = 0.75 \pm 0.04$, respectively. Their combination excludes

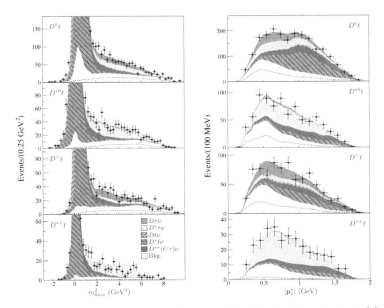

Fig. 3. Data and the fit projections in the m^2_{miss} variable (left plots) for the four $D^{(*)}\ell$ samples. The right plot show the p^*_ℓ projections for $m^2_{miss} > 1\,\mathrm{GeV}^2$, which excludes most of the normalization modes. In the background component, the region above the dashed line corresponds to charge cross-feed, and the region below corresponds to continuum and $B\bar{B}$.

Table 3. Results of the isospin-unconstrained (top four rows) and isospin-constrained fits (last two rows). The columns show the signal and normalization yields, $R^{(*)}$, branching fractions, and Σ_{stat} and Σ_{tot}, the statistical and total significances.

Decay	N_{sig}	N_{norm}	$R^{(*)}$	$\mathcal{B}(B \to D^{(*)}\tau\nu)\,(\%)$	Σ_{stat}	Σ_{tot}
$B^- \to D^0\tau^-\bar{\nu}_\tau$	314 ± 60	1995 ± 55	$0.429 \pm 0.082 \pm 0.052$	$0.99 \pm 0.19 \pm 0.13$	5.5	4.7
$B^- \to D^{*0}\tau^-\bar{\nu}_\tau$	639 ± 62	8766 ± 104	$0.322 \pm 0.032 \pm 0.022$	$1.71 \pm 0.17 \pm 0.13$	11.3	9.4
$\bar{B}^0 \to D^+\tau^-\bar{\nu}_\tau$	177 ± 31	986 ± 35	$0.469 \pm 0.084 \pm 0.053$	$1.01 \pm 0.18 \pm 0.12$	6.1	5.2
$\bar{B}^0 \to D^{*+}\tau^-\bar{\nu}_\tau$	245 ± 27	3186 ± 61	$0.355 \pm 0.039 \pm 0.021$	$1.74 \pm 0.19 \pm 0.12$	11.6	10.4
$\bar{B} \to D\tau^-\bar{\nu}_\tau$	489 ± 63	2981 ± 65	$0.440 \pm 0.058 \pm 0.042$	$1.02 \pm 0.13 \pm 0.11$	8.4	6.8
$\bar{B} \to D^*\tau^-\bar{\nu}_\tau$	888 ± 63	11953 ± 122	$0.332 \pm 0.024 \pm 0.018$	$1.76 \pm 0.13 \pm 0.12$	16.4	13.2

the full parameter space of this model with 99.8% probability, in the region $m_{H+} > 10\,GeV$. An interpretation of these results in the framework of 2HDM of type III has recently been presented.[21]

5. Conclusion and Outlook

The decays of B and B_s mesons are a very useful probe to search for new physics effects, induced by non-SM virtual particles in tree and loop diagrams. New physics has not been discovered yet, however there are intriguing *tensions* in the angular analysis of $B \to K^*\mu + \mu^-$ decays, in the isospin analysis of $B \to K\mu + \mu^-$ decays (not discussed here) and in $B \to D^{(*)}\tau + \nu$ decays. These tensions should be followed up by collecting more data, analyzing other decays modes and performing more accurate theoretical studies. B Factories and hadron colliders complement each other in several ways. Experiments at the B Factories have nearly completed to analyze their final samples, LHCb is now the major driver in many modes. ATLAS and CMS can nevertheless give substantial contribution in very rare decays with muons. Belle-II will extend the dataset from B Factories by almost two orders of magnitudes, thereby allowing studies in previosly statistically limited decay channels.

References

1. Y. Amhis *et al.*, Averages of b-hadron, c-hadron, and tau-lepton properties as of early 2012, *arXiv:1207.1158* and online update at *http://www.slac.stanford.edu/xorg/hfag*
2. LHCb Collab. (R. Aaji *et al.*), *Nucl. Phys.* B**867**, 1 (2013).
3. LHCb Collab. (R. Aaji *et al.*), *CP* and up-down asymmetries in $B^{\pm} \to K^{\pm}\pi^{\mp}\pi^{\pm}\gamma$ decays), *LHCB-CONF-2013-009*.
4. LHCb Collab. (R. Aaji *et al.*), *JHEP* **08** (2013) 131.
5. C. Bobeth, G. Hiller, and D. van Dyk, *JHEP* **07** (2011) 067.
   J. Matias, F. Mescia, M. Ramon, and J. Virto, *JHEP* **04** (2012) 104.
6. LHCb Collab. (R. Aaji *et al.*), *Phys. Rev. Lett.* **111** (2013) 191801.
7. S. Descotes-Genon, J. Matias, and J. Virto *Phys. Rev. D* **88** (2013) 074002.
8. S. Jager and J.M. Camalich, *JHEP* **05** (2013) 043.
9. Babar Collab. (J.P. Lees *et al.*), *Phys. Rev. D* **88** (2013) 032012.
10. LHCb Collab. (R. Aaji *et al.*), *JHEP* **12** (2012) 125.
11. A. J. Buras, J. Girrbach, D. Guadagnoli, and G. Isidori, *Eur. Phys. J. C* **72** (2012) 2172.
12. K. de Bruyn *et al.*, *Phys. Rev. Lett.* **109** (2012) 041801.
13. LHCb Collab. (R. Aaji *et al.*), *Phys. Rev. Lett.* **111** (2013) 101805.
14. CMS Collab. (S. Chatrchyan *et al.*), *Phys. Rev. Lett.* **111** (2013) 101804.
15. D. Straub, Overview of constraints on new physics in rare B decays, *arxiv:1205.6094*, in *Proceedings of the 47th Rencontres de Moriond on Electroweak Interactions and Unified Theories, 2012*
16. J. Kamenik and F. Mescia, *Phys. Rev. D* **78** (2008) 014003.
17. S. Fajfer, J. Kamenik, and I. Nisandzic, *Phys. Rev. D* **85** (2012) 094025.
18. Babar Collab. (J.P. Lees *et al.*), *Phys. Rev. Lett.* **109** (2012) 101802.
19. M. Tanaka and R. Watanabe, *Phys. Rev. D* **82** (2010) 034027.
20. V. Barger, J. Hewett, and R. Phillips, *Phys. Rev. D* **41** (1990) 3421.
21. Babar Collab. (J.P. Lees *et al.*), *Phys. Rev. D* **88** (2013) 072012.

Physics in Collision (PIC 2013)
International Journal of Modern Physics: Conference Series
Vol. 31 (2014) 1460291 (11 pages)
© The Author
DOI: 10.1142/S2010194514602919

World Scientific
www.worldscientific.com

$D\bar{D}$ mixing and CP-symmetry violation
in the charm sector, and rare D-meson decays

Sergey Barsuk

(For the LHCB Collaboration)

Laboratoire de l'Accelerateur Lineaire
Centre Scientifique d'Orsay, Bat. 200, 91898 Orsay, France
sergey.barsuk@lal.in2p3.fr

Published 15 May 2014

An overview of selected recent results on $D\bar{D}$ mixing and CP-symmetry violation in the charm sector, and D meson rare decays, was presented at the Physics In Collision conference 2013 in Beijing.

Keywords: Charm mixing; CP-violation; rare decays.

PACS Numbers: 14.40.Lb, 11.30.Er

Flavour mixing and CP-symmetry violation are well established effects in the K and B systems, while charm is the unique up-type system where these effects can occur. In the Standard Model (SM) the neutral charm meson mass eigenstates are related to flavour eigenstates: $|D_{1,2}> = p|D^0> \pm q|\bar{D}^0>$. In the limit of CP-symmetry conservation, $p = q$ (which holds to $O(10^{-4})$ in SM), decays become modulated by mixing, occuring via box-diagrams, and the decay time distribution follows $\frac{e^{-\Gamma t}}{4}(cosh(\Delta\Gamma t/2) \pm 2\cos\Delta mt)$, where Γ is the average decay width, and Δm and $\Delta\Gamma$ are the mass and decay width splitting between the two mass eigenstates. Denoting by *mixed* transitions $\bar{D}^0 \to D^0$ and $D^0 \to \bar{D}^0$ between initial and final D flavours, and by *unmixed* the $\bar{D}^0 \to \bar{D}^0$ and $D^0 \to D^0$ transitions, the time-dependent mixing amplitude $A_{mix}(t)$ is written as

$$A_{mix}(t) \equiv \frac{N_{unmixed} - N_{mixed}}{N_{unmixed} + N_{mixed}}(t) = \frac{cos(\Delta mt)}{cosh(\Delta\Gamma t/2)}.$$

Mixing parameters $x = \Delta m/\Gamma$ and $y = \Delta\Gamma/\Gamma$ can be evaluated by considering the box diagram for mass and decay width splitting, and generic tree diagram to estimate the decay width. Fast or slow oscillations occur for $x \gg 1$ and $x \ll 1$ respectively. According to the above naive considerations, mixing probability is high, about 50% for the $K^0 \bar{K}^0$ and $B^0_s \bar{B}^0_s$ systems, well established at the level of 18% for the $B^0 \bar{B}^0$ system, and small at the level of 10^{-6} for the $D^0 \bar{D}^0$ system. The corresponding mixing parameters are $x, y \sim O(1)$ for the $K^0 \bar{K}^0$ system, $x \sim 1$, $y \ll 1$ for the $B^0 \bar{B}^0$ system, $x \gg 1$, $y \sim O(1)$ for the $B^0_s \bar{B}^0_s$ system and $x, y \ll 1$ for the $D^0 \bar{D}^0$ system. Experimentally, most challenging are the opposite cases of rapid $B^0_s \bar{B}^0_s$ oscillations, requiring high decay time resolution to resolve them, and of small $D^0 \bar{D}^0$ oscillations.

The small $D^0 \bar{D}^0$ flavour mixing can be explained by the fact, that unlike $B^0_{(s)} \bar{B}^0_{(s)}$ systems, there is no contribution from the heavy t quark to the box diagram, and the b quark contribution is suppressed by λ^{10} due to the Cabibbo-Kobayashi-Maskawa (CKM) matrix elements. Non-perturbative long-range effects are more difficult to calculate with typically 10^{-3} value with large uncertainty, see e.g. Ref. 1. The $D^0 \bar{D}^0$ flavour mixing is well-established, however until 2013 no single 5σ measured was available.[2]

Three types of CP-symmetry violation (CPV) are distinguished in SM. *Direct CPV* can occur when the final state is reached via tree and penguin diagrams, and thus Cabibbo-suppressed decays (CSD) of both charged and neutral D mesons are involved. If $A_f = <f|H|D>$ and $\bar{A}_{\bar{f}} = <\bar{f}|H|\bar{D}>$, then $|\bar{A}_{\bar{f}}/A_f| \neq 1$ quantifies the CPV effect. Direct CPV effect is expected to be up to few $\times 10^{-3}$ in SM, and is searched for by studying asymmetries in two-body CSD and the asymmetries across the phase space in multi-body decays. The *CPV in meson mixing* arises when $|p| \neq |q|$, and is expected at the level $O(10^{-4})$ in SM. The condition for *CPV in the interference between the decay with and without mixing* to the common final state is $arg(q/p \times \bar{A}_f/A_f) \neq 0$.

Charm sector CSD and flavour mixing involve primarily the 2×2 Cabibbo matrix, where CP-violating phase enters the V_{cd} CKM element with a factor λ^5. The V_{ub} element, where CPV phase enters with a factor λ^3, contributes via loops. In the framework of SM, all CPV effects are described by a single phase, providing a strong predictive power for CP asymmetries. Direct CPV of $O(1\%)$ in charm decays could signal physics beyond SM.

A set of decisive SM tests has been proposed in theory papers. Taking into account the fact, that if CPV is small, $\sin \phi \ll 1$, which is valid for D system, the authors of Ref. 3 obtain $y/x = \frac{1 - |q/p|}{\tan \phi}$. If the above relation is violated, physics beyond the SM contributes to direct CPV.[3] Other examples are proposed in Ref. 4. In the limit of SM direct CPV, the CP-asymmetries in CSD into final CP eigenstates are universal, and $(a_{KK} - a_{\pi\pi})/(a_{KK} + a_{\pi\pi}) \sim 10^{-3}$, which provides another important test. The authors also suggest to separate direct against indirect CPV in CSD using time-integrated CP asymmetry measurements. Assuming negligible new CPV in Cabibbo-allowed decays (CAD) and doubly Cabibbo-suppressed decays (DCSD), time-integrated CP asymmetry in CAD to CP eigenstate arises from a universal indirect asymmetry. It can be subtracted from the time-integrated CP asymmetry in CSD to CP eigenstate, to obtain the non-universal direct CP asymmetry. Finally, the authors reiterate that any observed non-vanishing CP asymmetry in charged D decays is necessarily direct CPV.

Experimental results in the charm sector come from both e^+e^- and hadron machines. The B-factories, BaBar and Belle operated at the center-of-mass energy corresponding to the $\Upsilon(4S)$ resonance. They accumulated large samples of D mesons from B meson decays, with more than 10^9 $B\bar{B}$ pairs produced, and large D samples from continuum production, with more than 10^9 charm hadron pairs.

The $D\bar{D}$-factories, CLEO-c and now BES, use quantum correlations at $D\bar{D}$ threshold. They collect data at energies around the charmonium production region, and use beam energy constraint to search for a simple signature for CPV, $\psi(3770) \to D^0\bar{D}^0 \to CP(\pm)CP(\pm)$. Clean data samples are available, corresponding to few $\times10^7$ $\psi(3770)$ at CLEO-c, and more than 2×10^6 D-mesons in the 2010-2011 data sample of BES III.[5]

The pp or $p\bar{p}$ colliders operate in a more challenging environment. The results are delivered by the CDF and D0 experiments at TEVATRON, and the LHCb experiment at LHC. High production cross-section at small rapidities at CDF, and at forward LHCb rapidities, $\sigma(c\bar{c})_{LHCb,7TeV} \sim 1400$ μb,[6] provide large samples of even rare charm decays. The LHCb reconstructed about 3.5×10^8 charm decays per 1 fb^{-1} of integrated luminosity, the number being limited by the trigger bandwidth. The D mesons have a large boost and fly about 3 mm before decaying. Precision track and vertex reconstruction, as well as powerful particle identification at LHCb, favour precision studies in the charm sector by these experiments.

In the following, selected recent results on flavour mixing and CPV in the charm sector, as well as charm rare decays will be discussed.

Among new results on the $D^0\bar{D}^0$ **mixing**, the BaBar experiment measured[7] $y_{CP} = (\Gamma^{CP=+} + \bar{\Gamma}^{CP=+})/2\Gamma - 1$ via D^0 lifetime difference measurement, where $\Gamma^{CP=+}$ ($\bar{\Gamma}^{CP=+}$) is the partial decay rate of D^0 ($\bar{D}^0$) to the CP-even final state. Assuming no CPV, y_{CP} measures directly y mixing parameter, and not the combination of x and y parameters, as accessible in other methods. BaBar measured the effective D^0 lifetime in three different two-body final states: CSD to CP-even K^-K^+ and $\pi^-\pi^+$ final states, and the $K^\mp\pi^\pm$ final state without a defined CP-parity, or in other words a mixture of $CP = +$ and $CP = -$. The BaBar result, $y_{CP} = (0.72 \pm 0.18 \pm 0.12)\%$, is the most precise single measurement to date, with no-mixing hypothesis excluded at 3.3σ level.

Using the D meson sample from the $D^{*+} \to D^0\pi^{+a}$ decay mode, the initial D flavour can be tagged by the sign of bachelor pion. Final D flavour is tagged using the final state of the D decay. For the $D^0 \to K\pi$ transition, the "right-sign" (RS) $K^-\pi^+$ final state can be reached dominantly via CAD, or alternatively via $D^0 \to \bar{D}^0$ transition, followed by DCSD $D^0 \to K^+\pi^-$. The D^0 "wrong-sign" (WS) $K^+\pi^-$ final state is reached by competing DCSD, and $D^0 \to \bar{D}^0$ transition, followed

---

[a]Throughout the paper charge-conjugated states are equally assumed unless stated otherwise.

by CAD $D^0 \rightarrow K^-\pi^+$. Assuming $|x|, |y| \ll 1$ and CP symmetry conservation, the time-dependent WS rate can be written as the sum of the "mixing" term, DCSD term and the interference term,

$$dN_{WS}/dt \approx e^{-\Gamma t} \times ((x'^2 + y'^2)/2 \cdot \Gamma^2 t^2/2 + D^2 + D \cdot y' \cdot \Gamma t),$$

where x' and y' are the x and y mixing parameters rotated by the strong phase δ:

$$y' = y \cos \delta - x \sin \delta, x' = x \cos \delta + y \sin \delta.$$

And the time-dependent ratio is written as:

$$N_{WS}/N_{RS}(t) \approx (x'^2 + y'^2)/2 \cdot \Gamma^2 t^2/2 + R_D + \sqrt{R_D} \cdot y' \cdot \Gamma t,$$

where $\bar{A}(K^-\pi^+)/A(K^-\pi^+) = \sqrt{R_D}e^{-i\delta}$. The effect of mixing is distinguished from the DCSD by measuring time dependence, and x'^2 and y' are determined. For semileptonic decays only the first term in the above expressions is non-zero, and the value $x'^2 + y'^2 = x^2 + y^2$ is accessed.

The LHCb experiment studied[8] WS and RS decays in D^0 decay time bins, with 3.6×10^4 and 8.4×10^6 candidates in the D^{*+} peak for WS and RS respectively. Background of WS in the analysis was dominated by combinations of real D^0 and a random pion. Contamination from the D^0 from inclusive b-hadron decays was reduced by the impact parameter cuts on D^0 and bachelor pion, with residual contamination of $(2.7 \pm 0.2)\%$. Doubly misidentified RS events contributed to the WS signal at the level of $(0.4 \pm 0.2)\%$. Finally, the mixing parameters were found to be $x'^2 = (-0.09 \pm 0.13) \times 10^{-3}$ and $y' = (7.2 \pm 2.4) \times 10^{-3}$. Note that there is a strong correlation between x'^2 and y'. The R_D parameter was measured to be $R_D = (3.52 \pm 0.15) \times 10^{-3}$ and $(4.25 \pm 0.04) \times 10^{-3}$ for mixing and no-mixing hypothesis respectively. The no-mixing hypothesis was excluded at more than 9σ level.

The same technique was used[9] by the CDF experiment. The mixing parameters were found to be $x'^2 = (0.08 \pm 0.18) \times 10^{-3}$ and $y' = (4.3 \pm 4.3) \times 10^{-3}$, and the R_D parameter $R_D = (3.51 \pm 0.35) \times 10^{-3}$. The no-mixing hypothesis was excluded at more than 6σ level.

The HFAG[10] averages of mixing parameters, $\bar{x} = (0.53^{+0.16}_{-0.17})\%$ and $\bar{y} = (0.67 \pm 0.09)\%$ for no-CPV hypothesis, and $\bar{x} = (0.39^{+0.16}_{-0.17})\%$ and $\bar{y} = (0.67^{+0.07}_{-0.08})\%$ with CPV allowed, exclude no-mixing hypothesis at more than 12σ level.

Among new results on **CPV in charm sector**, the BaBar experiment searched[11] for direct CPV in $D^+ \rightarrow K^-K^+\pi^+$ decay, using 2.2×10^{-5} candidates in the D^+ signal. The asymmetry in the total rate

$$A_{CP} = \frac{\Gamma(D^0 \rightarrow f) - \Gamma(\bar{D}^0 \rightarrow \bar{f})}{\Gamma(D^0 \rightarrow f) + \Gamma(\bar{D}^0 \rightarrow \bar{f})}$$

was studied. A model independent search over the Dalitz plot yielded the p-value of 72% for no CPV, and a model-dependent search using simultaneous fit to D^+ and D^- Dalitz plots showed no indication of CPV in relative phases and magnitudes. Finally, the integrated asymmetry of $A_{CP} = (0.37 \pm 0.30 \pm 0.15)\%$ was obtained.

Several new results on the search for CPV in the $D_{(s)}^+ \to K_S^0 h^+$ decays have emerged. The $D^+ \to K_S^0 K^+$ decay proceeds via Feynman diagrams similar to those of the $D^0 \to K^- K^+$ mode, where the penguin contribution may be enhanced by physics beyond the SM. For the $D^+ \to K_S^0 \pi^+$ decay, direct CPV from the interference between the CAD and DCSD amplitudes is negligible in SM, but CPV from the kaon mixing may be detectable. Finally, $D_s^+ \to K_S^0 \pi^+$ decay proceeds via Feynman diagrams similar to those of the $D^+ \to K_S^0 K^+$ mode.

The CP asymmetry in the $D_{(s)}^+ \to K_S^0 h^+$ modes accounts also for the $K^0 \bar{K}^0$ mixing:

$$A_{CP}^{D_{(s)}^+ \to K_S^0 h^+} = \frac{\Gamma(D_{(s)}^+ \to \bar{K}^0 h^+)\Gamma(\bar{K}^0 \to \pi^+\pi^-) - \Gamma(D_{(s)}^- \to K^0 h^-)\Gamma(K^0 \to \pi^+\pi^-)}{\Gamma(D_{(s)}^+ \to \bar{K}^0 h^+)\Gamma(\bar{K}^0 \to \pi^+\pi^-) + \Gamma(D_{(s)}^- \to K^0 h^-)\Gamma(K^0 \to \pi^+\pi^-)}$$

$$= \frac{A_{CP}^{D_{(s)}^+ \to \bar{K}^0 h^+} + A_{CP}^{\bar{K}^0}}{1 + A_{CP}^{D_{(s)}^+ \to \bar{K}^0 h^+} \times A_{CP}^{\bar{K}^0}} \approx A_{CP}^{D_{(s)}^+ \to \bar{K}^0 h^+} + A_{CP}^{\bar{K}^0}.$$

The A_{CP} value determined at $e^+ e^-$ colliders is obtained from the asymmetry in the signal yields:

$$A_{raw}^{D_{(s)}^+ \to K_S^0 h^+} = \frac{N(D_{(s)}^+) - N(D_{(s)}^-)}{N(D_{(s)}^+) + N(D_{(s)}^-)} = A_{CP}^{D_{(s)}^+ \to K_S^0 h^+} + A_{FB}^{D_{(s)}^+} + A_{Det}^{h^+} + A_{Det}^{K^0},$$

where the forward-backward asymmetry A_{FB} is an odd function of $\cos\theta_D^*$:

$$A_{FB}^{D_{(s)}^+} = \frac{1}{2}\left(A_{raw}^{K_S^0 h^+}(+\cos\theta_D^*) - A_{raw}^{K_S^0 h^+}(-\cos\theta_D^*)\right),$$

the detector asymmetry $A_{Det}^{h^+}$ is determined from the CAD $D^0 \to K^-\pi^+$ and $D_s^+ \to \phi\pi^+$, where no CPV is expected, assuming $A_{FB}^{D^0} = A_{FB}^{D_s^+}$, and the detector asymmetry $A_{Det}^{K^0}$ is due to different interaction with matter for K^0 and $\bar{K}^0$ and is estimated numerically to be e.g. about 0.1% in Belle.[12]

The Belle experiment searched for[13,14] CP asymmetry in the $D^+ \to \bar{K}^0 K^+$ and $D^+ \to \bar{K}^0 \pi^+$ decay modes, in bins of $\cos\theta_D^*$. Having 2.8×10^5 and 17.4×10^5 candidates, the CP asymmetry was found to be $A_{CP}(D^+ \to \bar{K}^0 K^+) = (0.08\pm0.28\pm 0.14)\%$ and $A_{CP}(D^+ \to \bar{K}^0 \pi^+) = (-0.024 \pm 0.094 \pm 0.067)\%$ in the $D^+ \to \bar{K}^0 K^+$ and $D^+ \to \bar{K}^0 \pi^+$ decay modes respectively. The BaBar experiment searched for[15] CP asymmetry in the $D^+ \to \bar{K}^0 K^+$ and $D_s^+ \to \bar{K}^0 \pi^+$ decay modes, also in bins of $\cos\theta_D^*$. With 1.6×10^5 and 0.14×10^5 candidates, the CP asymmetry was found to be $A_{CP}(D^+ \to \bar{K}^0 K^+) = (0.46 \pm 0.36 \pm 0.25)\%$ and $A_{CP}(D_s^+ \to \bar{K}^0 \pi^+) = (0.3\pm2.0\pm0.3)\%$ in the $D^+ \to \bar{K}^0 K^+$ and $D_s^+ \to \bar{K}^0 \pi^+$ decay modes respectively. In all the three decay modes, the results were found to be compatible with no charm sector driven CPV.

The LHCb experiment searched for[16] direct CPV in the $D^+ \to \phi\pi^+$ and $D_s^+ \to K_S^0\pi^+$ decay modes. The CPV asymmetry in the ϕ region of the $D^+ \to K^-K^+\pi^+$ Dalitz plot was found as:

$$A_{CP}^{D^+\to\phi\pi^+} = A_{raw}^{D^+\to\phi\pi^+} - A_{raw}^{D^+\to K_S\pi^+} + A_{CP}^{K^0-\bar{K}^0}.$$

In addition, a complementary observable $A_{CP|S}$, sensitive to asymmetries that change sign, according to the strong phase variation along the ϕ band, $1.00~GeV/c^2 < M(K^+K^-) < 1.04~GeV/c^2$, was constructed. A concurrent measurement

$$A_{CP}^{D_s^+\to K_S^0\pi^+} = A_{raw}^{D_s^+\to K_S^0\pi^+} - A_{raw}^{D_s^+\to\phi\pi^+} + A_{CP}^{K^0-\bar{K}^0}$$

was performed. The $A_{CP}^{K^0-\bar{K}^0}$, correction for CPV in neutral kaon system, was found to be[17] $(-0.028 \pm 0.028)\%$. The analysis was performed in 12 bins of transverse momentum p_T and rapidity η. Totals of 1.6×10^6 (3.0×10^6) and 1.1×10^6 (2.4×10^4) candidates were found in the D^+ and D_s^+ signal peaks in the $D_{(s)}^+ \to \phi\pi^+$ and $D_{(s)}^+ \to K_S^0\pi^+$ decay modes respectively. The corresponding CP asymmetries were found to be $A_{CP}(D^+ \to \phi\pi^+) = (-0.04 \pm 0.14 \pm 0.13)\%$, $A_{CP|S}(D^+ \to \phi\pi^+) = (-0.18 \pm 0.17 \pm 0.18)\%$ and $A_{CP}(D_s^+ \to K_S^0\pi^+) = (0.61 \pm 0.83 \pm 0.13)\%$, providing no evidence of CPV.

The time-dependent CP asymmetry in D^0 decays to a CP eigenstate f is written as:

$$A_{CP}(f,t) = \frac{\Gamma(D^0 \to f) - \Gamma(\bar{D}^0 \to f)}{\Gamma(D^0 \to f) + \Gamma(\bar{D}^0 \to f)} \approx a_{CP}^{dir}(f) + (t/\tau)a_{CP}^{ind},$$

where the indirect asymmetry a_{CP}^{ind} is universal to a good approximation. Assuming the a_{CP}^{ind} universality, the difference of the CP asymmetries for the K^-K^+ and $\pi^-\pi^+$ final states, depends on the difference of the corresponding direct CP asymmetries, and suppressed a_{CP}^{ind} term present due to the difference in the average decay time:

$$\Delta A_{CP} = A_{CP}(K^-K^+) - A_{CP}(\pi^-\pi^+) = a_{CP}^{dir}(K^-K^+) - a_{CP}^{dir}(\pi^-\pi^+) + \Delta\bar{t}/\tau a_{CP}^{ind}.$$

Tagging the D^0 flavour using the sign of the pion from $D^{*+} \to D^0\pi^+$, the time-integrated raw asymmetry is written as:

$$A_{raw}(f) = \frac{N(D^{*+} \to D^0\pi^+) - N(D^{*-} \to \bar{D}^0\pi^-)}{N(D^{*+} \to D^0\pi^+) + N(D^{*-} \to \bar{D}^0\pi^-)},$$

and, to the first order,

$$A_{raw}(f) = A_{CP}(f) + A_D(f) + A_D(\pi^+) + A_P(D^{*+}),$$

where $A_D(K^-K^+) = A_D(\pi^-\pi^+) = 0$, and $A_D(\pi^+)$ and $A_P(D^{*+})$ are independent of the final state. Thus, we conclude, that $\Delta A_{CP} = A_{raw}(K^-K^+) - A_{raw}(\pi^-\pi^+)$.

By the end of 2012, the measurements[18–21] of ΔA_{CP} by the LHCb, CDF, Belle and BaBar experiments suggested a negative value of ΔA_{CP}, and poor agreement with the no-CPV hypothesis, with confidence level 2.0×10^{-5}. The BaBar, Belle and CDF experiments also quoted individual asymmetries for the the $D^0 \to K^- K^+$ and $D^0 \to \pi^- \pi^+$ channels with partially cancelled systematic uncertainty in the difference.

The LHCb experiment published an update[22] of the analysis using the integrated luminosity of 1 fb^{-1}, and 2.2×10^6 and 0.7×10^6 candidates in the $D^0 \to K^- K^+$ and $D^0 \to \pi^- \pi^+$ signal peaks respectively. The difference in the decay time was measured to be $\Delta \bar{t}/\tau = (11.19 \pm 0.13 \pm 0.17)\%$. In the analysis, fiducial cuts were used to exclude a kinematic region with large π^+ detection asymmetry. Impact parameter cuts were used to reduce contamination from D^0 originating from b hadron decays. The resolution on $\Delta M = M(h^- h^+ \pi^+) - M(h^- h^+) - M(\pi^+)$ was improved by constraining the D^{*+} vertex to match the primary vertex. In addition, weighting was applied to account for differences in kinematics of the two final states, in order to properly cancel the production and detection asymmetries. Finally, ΔA_{CP} was measured to be $\Delta A_{CP} = (-0.34 \pm 0.15 \pm 0.10)\%$. This result superseded the previous LHCb measurement,[18] and better agrees with the no-CPV hypothesis.

Additionally, LHCb measured[23] ΔA_{CP} by tagging the D^0 flavour using the muon charge in semileptonic $b \to D^0 \mu X$ decays. This analysis employs a sample which is statistically independent from the analysis discussed above, that was using D^{*+} to tag the D^0 flavour, and has different trigger composition and systematic uncertainty. The raw asymmetry is written as

$$A_{raw} = \frac{\Gamma(D^0 \to f)\varepsilon(\mu^-)P(D^0) - \Gamma(\bar{D}^0 \to f)\varepsilon(\mu^+)P(\bar{D}^0)}{\Gamma(D^0 \to f)\varepsilon(\mu^-)P(D^0) + \Gamma(\bar{D}^0 \to f)\varepsilon(\mu^+)P(\bar{D}^0)} \approx A_{CP}^f + A_D^\mu + A_P^B,$$

and ΔA_{CP} as

$$\Delta A_{CP} = A_{raw}(K^- K^+) - A_{raw}(\pi^- \pi^+) \approx A_{CP}(K^- K^+) - A_{CP}(\pi^- \pi^+).$$

The difference in the decay time was measured to be $\Delta \bar{t}/\tau = (1.8 \pm 0.7)\%$, so that $\Delta A_{CP} \approx \Delta a_{CP}^{dir}$. The raw asymmetries are determined from samples with 5.6×10^5 and 2.2×10^5 candidates in the signal peaks for the $D^0 \to K^- K^+$ and $D^0 \to \pi^- \pi^+$ decay modes respectively. Weighting was also applied to account for differences in kinematics of the two final states. The value of ΔA_{CP} was measured to be $\Delta A_{CP} = (0.49 \pm 0.30 \pm 0.14)\%$. The two measurements[22, 23] can be averaged, $< \Delta A_{CP} >_{LHCb} = (-0.15 \pm 0.16)\%$. This result shows no evidence of CPV.

Multi-body D^0 decays provide another promising tool to search for CPV, such decays proceed via resonances, and Dalitz-plot like techniques allow to sum up several resonances, providing increased statistics and information about the strong phase. The decays, leading to the $K^- K^+ \pi^- \pi^+$ and $\pi^- \pi^+ \pi^- \pi^+$ final states, comprise contributions from ρ, ϕ, $a_1(1260)^+$, $K_1(1270)^\pm$, $K^*(1410)^\pm$ and other resonances, with Cabibbo-suppressed tree and penguin amplitudes. Direct CPV can be searched for in regions of four-body phase space, providing five combinations of two- and three-body invariant mass squared. No direct CPV is expected in the $D^0 \to K^- \pi^+ \pi^- \pi^+$ decay mode, which can be used as a control channel. The LHCb experiment used[24] the charge of π^+ from the $D^{*+} \to D^0 \pi^+$ mode, to tag the D^0

flavour. The signal was extracted via *sPlot* technique, with 57k, 330k and 2900k candidates in the $D^0 \to K^- K^+ \pi^- \pi^+$, $D^0 \to \pi^- \pi^+ \pi^- \pi^+$ and $D^0 \to K^- \pi^+ \pi^- \pi^+$ signal peaks. Significance was determined in equally populated bins of the 5D phase space. The $D^0 \to K^- \pi^+ \pi^- \pi^+$ significance distribution showed no evidence for local asymmetries, and the $D^0 \to K^- K^+ \pi^- \pi^+$ and $D^0 \to \pi^- \pi^+ \pi^- \pi^+$ distributions were found to be consistent with no CPV.

The HFAG[10] average shows present data to be consistent with no CPV at $CL = 2.1\%$, with $\Delta a_{CP}^{dir} = (-0.329 \pm 0.121)\%$ and $\Delta a_{CP}^{ind} = (-0.010 \pm 0.162)\%$.

New important results on **rare charm decays** have been reported. The Belle experiment performed[25] precision study of the D_s^+ leptonic decays. The D_s^+ mesons were produced in the reaction $e^+ e^- \to c\bar{c} \to D_{tag} K_{frag} X_{frag} D_s^{*-}$, $D_s^{*-} \to D_s^- \gamma$, where a tagging charm hadron D_{tag} can be $D^{(*)}$ or Λ_c^+, K_{frag} is the additional kaon that conserves the strangeness in the event, and X_{frag} denotes additional pions produced in the course of hadronization. The signal was searched by reconstructing the missing mass in the event, and looking at the energy deposit in the electromagnetic calorimeter. Most stringent to date upper limit was obtained on the suppressed $D_s^+ \to e^+ \nu_e$ channel, $BR(D_s^+ \to e^+ \nu_e) < 0.83 \times 10^{-4}$@90%CL. This limit still exceeds the SM expectation of 10^{-7} by three orders of magnitude. Measurement of the branching fractions for the $D_s^+ \to \mu^+ \nu_\mu$ and $D_s^+ \to \tau^+ \nu_\tau$ decay modes, provides lepton flavour universality check, and allows determination of the D_s^+ decay constant. In the analysis τ^+ leptons were reconstructed using the $\tau^+ \to e^+ \nu_e \bar{\nu}_\tau$, $\tau^+ \to \mu^+ \nu_\mu \bar{\nu}_\tau$ and $\tau^+ \to \pi^+ \bar{\nu}_\tau$ decay modes. The Belle experiment obtained:

$$BR(D_s^+ \to \mu^+ \nu_\mu) = (5.31 \pm 0.28 \pm 0.20) \times 10^{-3},$$
$$BR(D_s^+ \to \tau^+ \nu_\tau) = (5.70 \pm 0.21^{+0.31}_{-0.30}) \times 10^{-2},$$

and their ratio,

$$R_{\tau/\mu}^{D_s} = (10.73 \pm 0.69^{+0.56}_{-0.53})$$

in agreement with the SM expectations,[2]

$$R_{\tau/\mu}^{D_s} = m_\tau^2/m_\mu^2 \cdot (1 - m_\tau^2/m_{D_s}^2)^2/(1 - m_\mu^2/m_{D_s}^2)^2 = (9.762 \pm 0.031).$$

The results are thus consistent with lepton flavour universality, though the precision of the measurement is still significantly larger than that of the expectation. The Belle result provides also most precise single measurement of the D_s^+ decay constant,

$$f_{D_s^+} = (255.5 \pm 4.2 \pm 4.8 \pm 1.8)\ MeV,$$

where the first error is statistical, the second error is the systematic uncertainty and the third one is due to the D_s^+ lifetime precision. This result removes the tension between the experimental data and unquenched Lattice QCD expectation,[26] $f_{D_s^+} = (248 \pm 2.5)\ MeV$.

The Belle experiment also measured[27] the branching fraction of the wrong sign $D^0 \to K^+\pi^-\pi^+\pi^-$ decay, important for $D^0\bar{D}^0$ mixing studies. The result

$$R_{WS} = (0.324 \pm 0.008 \pm 0.007)\%$$

is the most precise measurement to date. By combining this result with x and y mixing parameters from Ref. 10 and α and δ values from Ref. 28, the R_D parameter is obtained as $R_D = (0.327^{+0.019}_{-0.016})\%$. Using in addition the CAD branching fraction from Ref. 2, the absolute branching fraction for the wrong sign $D^0 \to K^+\pi^-\pi^+\pi^-$ decay,

$$BR(D^0 \to K^+\pi^-\pi^+\pi^-) = (2.61 \pm 0.06^{+0.09}_{-0.08}) \times 10^{-4},$$

significantly improves the PDG value precision,[2] $BR(D^0 \to K^+\pi^-\pi^+\pi^-) = (2.61^{+0.21}_{-0.19}) \times 10^{-4}$.

The $D^0 \to \mu^+\mu^-$ rare decay proceeds via loop diagrams and is strongly suppressed in the SM. Using a sample of D^0 mesons from the $D^{*+} \to D^0\pi^+$ decay, the LHCb experiment obtained[29] the most stringent upper limit to date on the branching fraction of the $D^0 \to \mu^+\mu^-$ decay, $BR(D^0 \to \mu^+\mu^-) < 7.6 \times 10^{-9}@95\%CL$. This result improves the previously available sensitivity[2] by a factor of 20, however the upper limit value is still three orders of magnitude higher than the SM expectations between 10^{-13} and 10^{-11}.

The $D^+_{(s)} \to \pi^+\mu^+\mu^-$ rare decay proceeds via box, penguin or annihilation diagram, with a branching fraction as large as 10^{-9} in SM, while physics beyond the SM may increase[30] its branching fraction to 10^{-8}. Physics beyond the SM can be searched for in the low and high di-muon invariant mass regions not clouded by resonances, while positive signals in the η, ρ/ω and ϕ regions can be used for normalization. The LHCb experiment set[31] upper limits on the $D^+_{(s)} \to \pi^+\mu^+\mu^-$ branching fractions,

$$BR(D^+ \to \pi^+\mu^+\mu^-) < 7.3(8.3) \times 10^{-8}@90(95)\%CL,$$
$$BR(D^+_s \to \pi^+\mu^+\mu^-) < 41(48) \times 10^{-8}@90(95)\%CL.$$

The LHCb experiment also searched for the $D^+_{(s)} \to \pi^-\mu^+\mu^+$ decay with two same sign muons, that measures lepton number violation via non-SM particles, e.g. Majorano neutrino. The upper limits on the $D^+_{(s)} \to \pi^-\mu^+\mu^+$ branching fractions,

$$BR(D^+ \to \pi^-\mu^+\mu^+) < 2.2(2.5) \times 10^{-8}@90(95)\%CL,$$
$$BR(D^+_s \to \pi^-\mu^+\mu^+) < 12(14) \times 10^{-8}@90(95)\%CL,$$

are the new world best limits, presenting a factor 50 improvement with respect to previous results.

LHCb searched[32] also for another flavour changing neutral current decay, $D^0 \to \pi^+\pi^-\mu^+\mu^-$, that can proceed via box and penguin diagrams. A sample of D^0 mesons from $D^{*+} \to D^0\pi^+$ decays was used. In each di-muon $m(\mu^+\mu^-)$ invariant mass bin, the 2D fit, $m(\pi^+\pi^-\mu^+\mu^-)$ and $\Delta m = m(\pi^+\pi^-\mu^+\mu^-\pi^+) -$

$m(\pi^+\pi^-\mu^+\mu^-)$, was performed. The ϕ resonance region, i.e. the decay $D^0 \rightarrow \pi^+\pi^-\phi \rightarrow \pi^+\pi^-\mu^+\mu^-$, was used for normalization. Both low and high di-muon invariant mass $m(\mu^+\mu^-)$ regions showed no significant $D^0 \rightarrow \pi^+\pi^-\mu^+\mu^-$ signal, and the upper limit on the branching fraction was set, $BR(D^0 \rightarrow \pi^+\pi^-\mu^+\mu^-) < 7.4 \times 10^{-7}@90\%CL$. This result presents a factor 70 improvement in sensitivity compared to previous results.

In **summary**, important progress has been achieved over the last years in understanding and improving precision on the mixing and CP-violation phenomena in the charm quark sector. Large samples of high quality data from both e^+e^- and hadronic machines are available. Charm mixing is well established with the x and y mixing parameters at the level of few $\times 10^{-3}$, with now precise single measurements available. Sensitivity of the CP-symmetry tests in the charm sector will soon approach Standard Model tolerated values. Many charm rare decay searches show low combinatorial background level, even at hadronic machines, suggesting further sensitivity improvement with increased statistics.

Acknowledgments

It is my pleasure to thank F. Anulli, I. Belyaev, M. Calvi, L. Dong, T. Gershon, R. Harr, H. Muramatsu, M. Williams and A. Zupanc for their help in preparing the talk.

References

1. A. F. Falk, Y. Grossman, Z. Ligeti, Y. Nir, A. A. Petrov, *Phys. Rev.* **D69**, 114021 (2004).
2. J. Beringer *et al.*), *Phys. Rev.* **D86**, 010001 (2012).
3. Y. Grossman, Y. Nir, G. Perez, *Phys. Rev. Lett.* **103**, 071602 (2009).
4. Y. Grossman, A. L. Kagan, Y. Nir, *Phys. Rev.* **D75**, 036008 (2007).
5. BES III Collab. (M. Ablikim *et al.*), *arXiv* 1307.2022 (2013).
6. LHCb Collab. (R. Aaij *et al.*), *Nucl. Phys.* **B871**, 1 (2013).
7. BABAR Collab. (J. P. Lees *et al.*), *Phys. Rev.* **D87**, 012004 (2013).
8. LHCb Collab. (R. Aaij *et al.*), *Phys. Rev. Lett.* **110**, 101802 (2013).
9. CDF Collab., *CDF NOTE* **10990** (2013).
10. HFAG group, http://www.slac.stanford.edu/xorg/hfag/charm/index.html
11. BABAR Collab. (J. P. Lees *et al.*), *Phys. Rev.* **D87**, 052010 (2013).
12. B. R. Ko, E. Won, B. Golob and P. Pakhlov, *Phys. Rev.* **D84**, 111501 (2011).
13. Belle Collab. (B. R. Ko *et al.*), *JHEP* **1302**, 098 (2013).
14. Belle Collab. (B. R. Ko *et al.*), *Phys. Rev. Lett.* **109**, 021601 (2012).
15. BABAR Collab. (J. P. Lees *et al.*), *Phys. Rev.* **D87**, 052012 (2013).
16. LHCb Collab. (R. Aaij *et al.*), *JHEP* **1306**, 112 (2013).
17. LHCb Collab. (R. Aaij *et al.*), *Phys. Lett.* **B718**, 902 (2013).
18. LHCb Collab. (R. Aaij *et al.*), *Phys. Rev. Lett.* **108**, 111602 (2012).
19. CDF Collab. (T. Aaltonen *et al.*), *Phys. Rev. Lett.* **109**, 111801 (2012).
20. Belle Collab. (B. R. Ko *et al.*), *arXiv* 1212.1975 (2012).
21. BABAR Collab. (B. Aubert *et al.*), *Phys. Rev. Lett.* **100**, 061803 (2008).
22. LHCb Collab. (R. Aaij *et al.*), *LHCb-CONF-2013-003* (2013).

23. LHCb Collab. (R. Aaij *et al.*), *Phys. Lett.* **B723**, 33 (2013).
24. LHCb Collab. (R. Aaij *et al.*), *Phys. Lett.* **B726**, 623 (2013).
25. Belle Collab. (A. Zupanc *et al.*), *JHEP* **1309** 139 (2013).
26. C. T. H. Davies, C. McNeile, E. Follana, G. P. Lepage, H. Na and J. Shigemitsu, *Phys. Rev.* **D82**, 114504 (2010).
27. Belle Collab. (E. White *et al.*), *Phys. Rev.* **D88**, 051101 (2013).
28. CLEO Collab. (N. Lowrey *et al.*), *Phys. Rev.* **D80**, 031105 (2009).
29. LHCb Collab. (R. Aaij *et al.*), *Phys. Lett.* **B725**, 16 (2013).
30. A. Paul, I. I. Bigi and S. Recksiegel, *Phys. Rev.* **D83**, 114006 (2011).
31. LHCb Collab. (R. Aaij *et al.*), *Phys. Lett.* **B724**, 203 (2013).
32. LHCb Collab. (R. Aaij *et al.*), *Phys. Lett.* **xB728**, 234 (2014).

Physics in Collision (PIC 2013)
International Journal of Modern Physics: Conference Series
Vol. 31 (2014) 1460292 (12 pages)
© The Author
DOI: 10.1142/S2010194514602920

Charmonium and light hadron spectroscopy

Chengping Shen

School of Physics and Nuclear Energy Engineering
Beihang University, Beijing, 100191, China
shencp@ihep.ac.cn

Published 15 May 2014

In this report I review some results on the charmonium and light hadron spectroscopy mainly from BESIII and Belle experiments. For the charmonium, the contents include the observation of $\psi(4040)/\psi(4160) \to \eta J/\psi$, the measurements of the $\eta_c/\eta_c(2S)$ resonance parameters and their decays, the evidence of the $\psi_2(1^3D_2)$ state in the $\chi_{c1}\gamma$ mass spectrum. For the light hadron spectroscopy, the contents include the $X(1835)$ research in $e^+e^- \to J/\psi + X(1835)$ and $\gamma\gamma \to \eta'\pi^+\pi^-$ processes, and the analysis of the $\eta\eta$, $\omega\phi$, $\phi\phi$ and $\omega\omega$ mass spectra in low mass region.

Keywords: Charmonium decays; light hadron spectroscopy.

PACS Numbers: 14.40.Pq, 13.25.-k, 13.25.Gv

1. $\psi(4040)/\psi(4160) \to \eta J/\psi$

Experimentally well established structures $\psi(4040)$, $\psi(4160)$, and $\psi(4415)$ resonances above the $D\bar{D}$ production threshold are of great interest but not well understood, even decades after their first observation.

BESIII accumulated a 478 pb^{-1} data sample at a center-of-mass (CMS) energy of $\sqrt{s} = 4.009$ GeV. Using this data sample, the processes $e^+e^- \to \eta J/\psi$ and $\pi^0 J/\psi$ cross section are measured.[1] In this analysis, the J/ψ is reconstructed through its decays into lepton pairs while η/π^0 is reconstructed in the $\gamma\gamma$ final state. After imposing all of some selection criteria, a clear J/ψ signal is observed in the $\mu^+\mu^-$ mode while indications of a peak around 3.1 GeV/c^2 also exist in the e^+e^- mode.

A significant η signal is observed in $M(\gamma\gamma)$ in both $J/\psi \to \mu^+\mu^-$ and $J/\psi \to e^+e^-$, as shown in Fig. 1. No significant π^0 signal is observed. The $M(\gamma\gamma)$ invariant mass distributions are fitted using an unbinned maximum likelihood method. For the η signal, the statistical significance is larger than 10σ while that for the π^0 signal is only 1.1σ. The Born cross section for $e^+e^- \to \eta J/\psi$ is measured to be

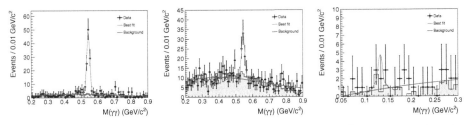

Fig. 1. Distributions of $M(\gamma\gamma)$ between 0.2 GeV/c^2 and 0.9 GeV/c^2 for $J/\psi \to \mu^+\mu^-$ (left panel) and for $J/\psi \to e^+e^-$ (middle panel) and distribution of $M(\gamma\gamma)$ below 0.3 GeV/c^2 for $J/\psi \to \mu^+\mu^-$ (right panel). Dots with error bars are data in J/ψ mass signal region, and the green shaded histograms are from normalized J/ψ mass sidebands. The curves show the total fit and the background term.

$(32.1 \pm 2.8 \pm 1.3)$ pb, and the Born cross section is found to be less than 1.6 pb at the 90% confidence level (C.L.) for $e^+e^- \to \pi^0 J/\psi$.

Belle used 980 fb^{-1} data to study the process $e^+e^- \to \eta J/\psi$ via ISR.[2] η is reconstructed in the $\gamma\gamma$ and $\pi^+\pi^-\pi^0$ final states. Due to the high background level from Bhabha scattering, the $J/\psi \to e^+e^-$ mode is not used in conjunction with the decay mode $\eta \to \gamma\gamma$.

Clear η and J/ψ signals could be observed. A dilepton pair is considered as a J/ψ candidate if $M_{\ell^+\ell^-}$ is within ± 45 MeV/c^2 of the J/ψ nominal mass. The η signal region is defined as $M_{\pi^+\pi^-\pi^0} \in [0.5343, 0.5613]$ GeV/c^2 and $M_{\gamma\gamma} \in [0.5, 0.6]$ GeV/c^2. -1 (GeV/c^2)$^2 < M_{\text{rec}}^2 < 2.0$ (GeV/c^2)^2 is required to select ISR candidates, where M_{rec}^2 is the square of the mass recoiling against the $\eta J/\psi$ system. After event selections, an unbinned maximum likelihood fit is performed to the mass spectra $M_{\eta J/\psi} \in [3.8, 4.8]$ GeV/c^2 from the signal candidate events and η and J/ψ sideband events simultaneously, as shown in Fig. 2. The fit to the signal events includes two coherent P-wave Breit-Wigner functions, BW_1 for $\psi(4040)$ and BW_2 for $\psi(4160)$, and an incoherent second-order polynomial background. Statistical significance is 6.5σ for $\psi(4040)$ and 7.6σ for $\psi(4160)$. There are two solutions with equally good fit quality: $\mathcal{B}(\psi(4040) \to \eta J/\psi) \cdot \Gamma_{e^+e^-}^{\psi(4040)} = (4.8 \pm 0.9 \pm 1.4)$ eV and $\mathcal{B}(\psi(4160) \to \eta J/\psi) \cdot \Gamma_{e^+e^-}^{\psi(4160)} = (4.0 \pm 0.8 \pm 1.4)$ eV for one solution and $\mathcal{B}(\psi(4040) \to \eta J/\psi) \cdot \Gamma_{e^+e^-}^{\psi(4040)} = (11.2 \pm 1.3 \pm 1.9)$ eV and $\mathcal{B}(\psi(4160) \to \eta J/\psi) \cdot \Gamma_{e^+e^-}^{\psi(4160)} = (13.8 \pm 1.3 \pm 2.0)$ eV for the other solution, where the first errors are statistical and the second are systematic. The partial widths to $\eta J/\psi$ are found to be about 1 MeV.

2. Some Results on η_c and $\eta_c(2S)$

The η_c mass and width have large uncertainties. The measured results of the η_c mass and width from J/ψ radiative transitions and two-photon fusion and B decays have large inconsistence. The most recent study by the CLEO-c experiment, using

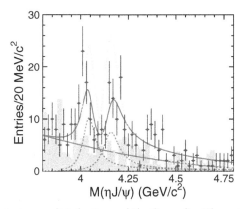

Fig. 2. The $\eta J/\psi$ invariant mass distribution and the fit results. The points with error bars show the data while the shaded histogram is the normalized η and J/ψ background from the sidebands. The curves show the best fit on signal candidate events and sideband events simultaneously and the contribution from each Breit-Wigner component. The dashed curves at each peak show the two solutions.

both $\psi(2S) \to \gamma\eta_c$ and $J/\psi \to \gamma\eta_c$, pointed out a distortion of the η_c line shape in $\psi(2S)$ decays.

With a $\psi(2S)$ data sample of 1.06×10^8 events, BESIII reported measurements of the η_c mass and width using the radiative transition $\psi(2S) \to \gamma\eta_c$.[3] Six modes are used to reconstruct the η_c: $K_S K^+\pi^-$, $K^+K^-\pi^0$, $\eta\pi^+\pi^-$, $K_S K^+\pi^+\pi^-\pi^-$, $K^+K^-\pi^+\pi^-\pi^0$, and $3(\pi^+\pi^-)$, where the K_S^0 is reconstructed in $\pi^+\pi^-$, and the η and π^0 in $\gamma\gamma$ decays.

Figure 3 shows the η_c invariant mass distributions for selected η_c candidates, together with the estimated backgrounds. A clear η_c signal is evident in every decay mode. Assuming 100% interference between the η_c and the non-resonant amplitude, an unbinned simultaneous maximum likelihood fit was performed. In the fit, the η_c mass, width, and relative phases are free parameters, and the mass and width are constrained to be the same for all decay modes. Two solutions of relative phase are found for every decay mode, one represents constructive interference, the other for destructive. The measured mass is $M = 2984.3 \pm 0.6(stat.) \pm 0.6(syst.)$ MeV$/c^2$ and width $\Gamma = 32.0 \pm 1.2(stat.) \pm 1.0(syst.)$ MeV. The interference is significant, which indicates previous measurements of the η_c mass and width via radiative transitions may need to be rechecked. The results are consistent with that from photon-photon fusion and B decays; this may partly clarify the discrepancy puzzle.

Similarly the properties of the $\eta_c(2S)$ are not well-established either. The $\eta_c(2S)$ was first observed by the Belle collaboration in the process $B^\pm \to K^\pm\eta_c(2S)$, $\eta_c(2S) \to K_S^0 K^\pm\pi^\mp$. It was confirmed in the two-photon production of $K_S^0 K^\pm\pi^\mp$, and in the double-charmonium production process $e^+e^- \to J/\psi c\bar{c}$. Combining the world-average values with the most recent results from Belle and BaBar on two-photon fusion into hadronic final states other than $K_S^0 K^\pm\pi^\mp$, one obtains updated averages of the $\eta_c(2S)$ mass and width of 3637.7 ± 1.3 MeV$/c^2$ and 10.4 ± 4.2 MeV,

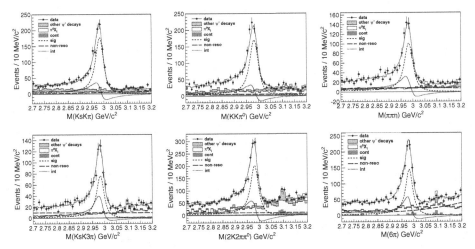

Fig. 3. The $M(X_i)$ invariant mass distributions for the decays $K_S K^+\pi^-$, $K^+K^-\pi^0$, $\eta\pi^+\pi^-$, $K_S K^+\pi^+\pi^-\pi^-$, $K^+K^-\pi^+\pi^-\pi^0$ and $3(\pi^+\pi^-)$, respectively, with the fit results (for the constructive solution) superimposed. Points are data and the various curves are the total fit results. Signals are shown as short-dashed lines; the non-resonant components as long-dashed lines; and the interference between them as dotted lines. Shaded histograms are (in red/yellow/green) for (continuum/$\pi^0 X_i$/other $\psi(2S)$ decays) backgrounds. The continuum backgrounds for $K_S K^+\pi^-$ and $\eta\pi^+\pi^-$ decays are negligible.

respectively. $\eta_c(2S)$ was also observed in six-prong final states in two-proton processes including $3(\pi^+\pi^-)$, $K^+K^-2(\pi^+\pi^-)$, $2(K^+K^-)\pi^+\pi^-$, $K_S^0 K^\pm\pi^\mp\pi^+\pi^-$ by Belle collaboration. The measured averaged mass and width of $\eta_c(2S)$ are $3636.9 \pm 1.1 \pm 2.5 \pm 5.0$ MeV/c^2 and $9.9 \pm 3.2 \pm 2.6 \pm 2.0$ MeV/c^2. The results were reported in ICHEP2010 meeting, but the results are still preliminary up to date.

Recently BESIII collaboration searched for the M1 radiative transition $\psi(2S) \rightarrow \gamma\eta_c(2S)$ by reconstructing the exclusive $\eta_c(2S) \rightarrow K_S^0 K^\pm\pi^\mp\pi^+\pi^-$ decay using 1.06×10^8 $\psi(2S)$ events.[4]

The final mass spectrum of $K_S^0 K^\pm\pi^\mp\pi^+\pi^-$ and the fitting results are shown in Fig. 4. The fitting function consists of the following components: $\eta_c(2S)$, $\chi_{cJ}(J = 0, 1,$ and $2)$ signals and $\psi(2S) \rightarrow K_S^0 K^\pm\pi^\mp\pi^+\pi^-$, $\psi(2S) \rightarrow \pi^0 K_S^0 K^\pm\pi^\mp\pi^+\pi^-$, ISR, and phase space backgrounds. The result for the yield of $\eta_c(2S)$ events is 57 ± 17 with a significance of 4.2σ. The measured mass of the $\eta_c(2S)$ is $3646.9 \pm 1.6(stat.) \pm 3.6(syst.)$ MeV/c^2, and the width is $9.9 \pm 4.8(stat.) \pm 2.9(syst.)$ MeV/c^2. The product branching fraction is measured to be $\mathcal{B}(\psi(2S) \rightarrow \gamma\eta_c(2S)) \times \mathcal{B}(\eta_c(2S) \rightarrow K_S^0 K^\pm\pi^\mp\pi^+\pi^-) = (7.03 \pm 2.10(stat.) \pm 0.70(syst.)) \times 10^{-6}$. This measurement complements a previous BESIII measurement of $\psi(2S) \rightarrow \gamma\eta_c(2S)$ with $\eta_c(2S) \rightarrow K_S K^+\pi^-$ and $K\bar{K}\pi$.

3. Evidence of the $1^3 D_2$ $c\bar{c}$ State (X(3823))

During the last decade, a number of new charmonium ($c\bar{c}$)-like states were observed, many of which are candidates for exotic states. The observation of a D-wave $c\bar{c}$

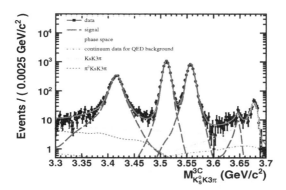

Fig. 4. The results of fitting the mass spectrum for χ_{cJ} and $\eta_c(2S)$. The black dots are the data, the blue long-dashed line shows the χ_{cJ} and $\eta_c(2S)$ signal shapes, the cyan dotted line represents the phase space contribution, the violet dash-dotted line shows the continuum data contribution, the green dash-double-dotted line shows the contribution of $\psi(2S) \to K_S^0 K^{\pm} \pi^{\mp} \pi^+ \pi^-$, and the red dashed line is the contribution of $\psi(2S) \to \pi^0 K_S^0 K^{\pm} \pi^{\mp} \pi^+ \pi^-$.

meson and its decay modes would test phenomenological models. The undiscovered 1^3D_2 $c\bar{c}$ (ψ_2) and 1^3D_3 $c\bar{c}$ (ψ_3) states are expected to have significant branching fractions to $\chi_{c1}\gamma$ and $\chi_{c2}\gamma$, respectively. So Belle used 772×10^6 $B\overline{B}$ events to search for the possible structures in $\chi_{c1}\gamma$ and $\chi_{c2}\gamma$ mass spectra in the processes $B \to \chi_{c1}\gamma K$ and $B \to \chi_{c2}\gamma K$ decays, where the χ_{c1} and χ_{c2} decay to $J/\psi\gamma$.[5] The J/ψ meson is reconstructed via its decays to $\ell^+\ell^-$ ($\ell = e$ or μ).

The $M_{\chi_{c1}\gamma}$ distribution from $B^{\pm} \to (\chi_{c1}\gamma)K^{\pm}$ and $B^0 \to (\chi_{c1}\gamma)K_S^0$ decays was shown in Fig. 5, where there is a significant narrow peak at 3823 MeV/c^2, denoted hereinafter as $X(3823)$. No signal of $X(3872) \to \chi_{c1}\gamma$ is seen. To extract the mass of the $X(3823)$, a simultaneous fit to $B^{\pm} \to (\chi_{c1}\gamma)K^{\pm}$ and $B^0 \to (\chi_{c1}\gamma)K_S^0$ is performed, assuming that $\mathcal{B}(B^{\pm} \to X(3823)K^{\pm})/\mathcal{B}(B^0 \to X(3823)K^0) = \mathcal{B}(B^{\pm} \to \psi'K^{\pm})/\mathcal{B}(B^0 \to \psi'K^0)$. The mass of the $X(3823)$ is measured to be $3823.1 \pm 1.8(stat.) \pm 0.7(syst.)$ MeV/c^2 and signal significance is estimated to be 3.8σ with systematic uncertainties included. The measured branching fraction product $\mathcal{B}(B^{\pm} \to X(3823)K^{\pm})\mathcal{B}(X(3823) \to \chi_{c1}\gamma)$ is $(9.7 \pm 2.8 \pm 1.1) \times 10^{-6}$. No evidence is found for $X(3823) \to \chi_{c2}\gamma$. The properties of the $X(3823)$ are consistent with those expected for the ψ_2 (1^3D_2 $c\bar{c}$) state.

4. Search for $X(1835)$

In the radiative decay $J/\psi \to \gamma\pi^+\pi^-\eta'$, the BESII Collaboration observed a resonance, the $X(1835)$, with a statistical significance of 7.7σ. Recently the structure has been confirmed by BESIII in the same process with 2.25×10^8 J/ψ events.

Many theoretical models have been proposed to interpret its underlying structure. Some interpret $X(1835)$ as radial excitation of η', a $p\bar{p}$ bound state, a glueball candidate, or a η_c-glueball mixture.

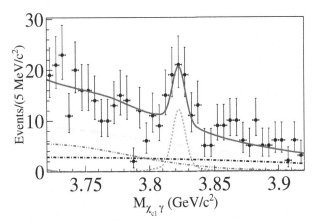

Fig. 5. Two-dimensional unbinned extended maximum likelihood fit projection of $M_{\chi_{c1}\gamma}$ distribution for the simultaneous fit of $B^{\pm} \to (\chi_{c1}\gamma)K^{\pm}$ and $B^0 \to (\chi_{c1}\gamma)K^0_S$ decays for $M_{\rm bc} > 5.27$ GeV/c^2.

Belle first tried to search for the $X(1835)$ in the two-photon process $\gamma\gamma \to \eta'\pi^+\pi^-$ using a 673 fb^{-1} data sample with $\eta' \to \eta\pi^+\pi^-$, and $\eta \to \gamma\gamma$.[6]

Significant background reduction is achieved by applying a $|\sum \vec{p}^*_t|$ requirement ($|\sum \vec{p}^*_t| < 0.09$ GeV/c), which is determined by taking the absolute value of the vector sum of the transverse momenta of η' and the $\pi^+\pi^-$ tracks in the e^+e^- center-of-mass system. The $|\sum \vec{p}^*_t|$ distribution for the signal peaks at small values, while that for both backgrounds decreases toward $|\sum \vec{p}^*_t| = 0$ due to vanishing phase space.

The resulting $\eta'\pi^+\pi^-$ invariant mass distribution was shown in Fig. 6. According to existing observations, two resonances, $X(1835)$ and $\eta(1760)$, have been reported in the lower mass region above the $\eta'\pi^+\pi^-$ threshold. A fit with the $X(1835)$ and $\eta(1760)$ signals plus their interference is performed to the lower-mass events. Here, the $X(1835)$ mass and width are fixed at the BES value. There are two solutions with equally good fit quality; the results are shown in Fig. 6. In either solution, the statistical significance is 2.9σ for the $X(1835)$ and 4.1σ for the $\eta(1760)$. Upper limits on the product $\Gamma_{\gamma\gamma}\mathcal{B}(\eta'\pi^+\pi^-)$ for the $X(1835)$ at the 90% C.L. are determined to be 35.6 eV/c^2 and 83 eV/c^2 for the constructive- and destructive-interference solutions, respectively.

C-even glueballs can be studied in the process $e^+e^- \to \gamma^* \to H + \mathcal{G}_J$, where H denotes a $c\bar{c}$ quark pair or charmonium state and $\mathcal{G}_J$ is a glueball. So if the $X(1835)$ was a candidate of glueball, it can also be searched for in the process $e^+e^- \to J/\psi X(1835)$ at $\sqrt{s} \approx 10.6$ GeV at Belle using a data sample of 672 fb^{-1}.

After all the event selections, the $M_{\rm recoil}$ distributions of the J/ψ are shown in Fig. 7. An unbinned simultaneous maximum likelihood fit to the $M_{\rm recoil}$ distributions was performed for the $\mu^+\mu^-$ and e^+e^- channels in the region of 0.8 GeV/$c^2 < M_{\rm recoil} < 2.8$ GeV/c^2, which constrains the expected signal from $J/\psi \to \mu^+\mu^-$ and

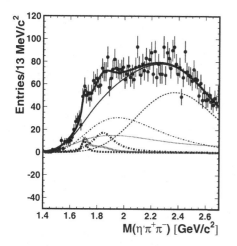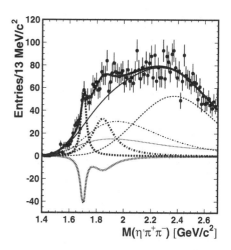

Fig. 6. Results of a combined fit for the $X(1835)$ and $\eta(1760)$ with interference between them. The points with error bars are data. The thick solid line is the fit; the thin solid line is the total background. The thick dashed (dot-dashed, dotted) line is the fitted signal for the $\eta(1760)$ ($X(1835)$, the interference term between them). The left (right) panel represents the solution with constructive (destructive) interference.

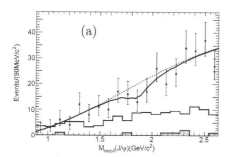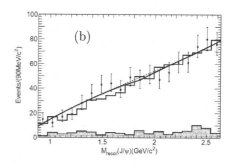

Fig. 7. The data points are for the distributions of the recoil mass against J/ψ reconstructed from (a) $\mu^+\mu^-$ and (b) e^+e^-. The histograms represent the backgrounds from the J/ψ sideband; the hatched histograms represent charmed- plus uds-quark backgrounds. The solid lines are results of the fits and the dashed lines are background shapes.

$J/\psi \rightarrow e^+e^-$ to be consistent with the ratio of ε_i and $\mathcal{B}_i$, where ε_i and $\mathcal{B}_i$ are the efficiency and branching fraction for the two channels, respectively. No significant evidence of $X(1835)$ is found, and an upper limit is set on its cross section times the branching fraction: $\sigma_{\mathrm{Born}}(e^+e^- \rightarrow J/\psi X(1835)) \cdot \mathcal{B}(X(1835) \rightarrow > 2$ charged tracks$) < 1.3$ fb at 90% C.L. This upper limit is three orders of magnitude smaller than the cross section of prompt production of J/ψ. According to this work, no evidence was found to support the hypothesis of $X(1835)$ to be a glueball produced with J/ψ in the Belle experiment.

5. $\eta\eta$ Mass Spectra

According to lattice QCD predictions, the lowest mass glueball with $J^{PC} = 0^{++}$ is in the mass region from 1.5 to 1.7 GeV/c^2. However, the mixing of the pure glueball with nearby $q\bar{q}$ nonet mesons makes the identification of the glueballs difficult in both experiment and theory. Radiative J/ψ decay is a gluon-rich process and has long been regarded as one of the most promising hunting grounds for glueballs. In particular, for a J/ψ radiative decay to two pseudoscalar mesons, it offers a very clean laboratory to search for scalar and tensor glueballs because only intermediate states with $J^{PC} = even^{++}$ are possible.

Recently the study of $J/\psi \to \gamma\eta\eta$ was made by BESIII using 2.25×10^8 J/ψ events,[7] where the η meson is detected in its $\gamma\gamma$ decay. There are six resonances, $f_0(1500)$, $f_0(1710)$, $f_0(2100)$, $f_2'(1525)$, $f_2(1810)$, $f_2(2340)$, as well as 0^{++} phase space and $J/\psi \to \phi\eta$ included in the basic solution. The masses and widths of the resonances, branching ratios of J/ψ radiative decaying to X and the statistical significances are summarized in Table 1. The comparisons of the $\eta\eta$ invariant mass spectrum, $\cos\theta_\eta$, $\cos\theta_\gamma$ and ϕ_η distributions between the data and the partial wave analysis (PWA) fit projections are displayed in Fig. 8. The results show that the dominant 0^{++} and 2^{++} components are from the $f_0(1710)$, $f_0(2100)$, $f_0(1500)$, $f_2'(1525)$, $f_2(1810)$ and $f_2(2340)$.

$\eta\eta$ mass spectrum was also ever studied by Belle in two-photon process $\gamma\gamma \to \eta\eta$ using 393 fb^{-1} data.[8] This pure neutral final states are selected with energy sum and cluster counting triggers, both of which information are provided by a CsI(Tl) electromagnetic calorimeter. The background was subtracted by studying sideband events in two-dimensional $M_1(\gamma\gamma)$ versus $M_2(\gamma\gamma)$ distributions. Further background effects are studied using $|\sum \vec{p}_t^{\,*}|$ distribution. Figure 9 shows the total cross sections. For the lower energy region 1.16 GeV $< W <$ 2.0 GeV, a PWA was performed to the differential cross section as shown in Fig. 10. In addition to the known $f_2(1270)$ and $f_2'(1525)$, a tensor meson $f_2(X)$ is needed to describe D_2 wave, which may correspond to $f_2(1810)$ state, and the mass, width and product of the two-photon

Table 1. Summary of the PWA results, including the masses and widths for resonances, branching ratios of $J/\psi \to \gamma X$, as well as the significance. The first errors are statistical and the second ones are systematic. The statistic significances here are obtained according to the changes of the log likelihood.

Resonance	Mass(MeV/c^2)	Width(MeV/c^2)	$\mathcal{B}(J/\psi \to \gamma X \to \gamma\eta\eta)$	Significance
$f_0(1500)$	1468^{+14+23}_{-15-74}	$136^{+41+28}_{-26-100}$	$(1.65^{+0.26+0.51}_{-0.31-1.40}) \times 10^{-5}$	8.2 σ
$f_0(1710)$	$1759\pm6^{+14}_{-25}$	$172\pm10^{+32}_{-16}$	$(2.35^{+0.13+1.24}_{-0.11-0.74}) \times 10^{-4}$	25.0 σ
$f_0(2100)$	$2081\pm13^{+24}_{-36}$	273^{+27+70}_{-24-23}	$(1.13^{+0.09+0.64}_{-0.10-0.28}) \times 10^{-4}$	13.9 σ
$f_2'(1525)$	$1513\pm5^{+4}_{-10}$	75^{+12+16}_{-10-8}	$(3.42^{+0.43+1.37}_{-0.51-1.30}) \times 10^{-5}$	11.0 σ
$f_2(1810)$	1822^{+29+66}_{-24-57}	$229^{+52+88}_{-42-155}$	$(5.40^{+0.60+3.42}_{-0.67-2.35}) \times 10^{-5}$	6.4 σ
$f_2(2340)$	$2362^{+31+140}_{-30-63}$	$334^{+62+165}_{-54-100}$	$(5.60^{+0.62+2.37}_{-0.65-2.07}) \times 10^{-5}$	7.6 σ

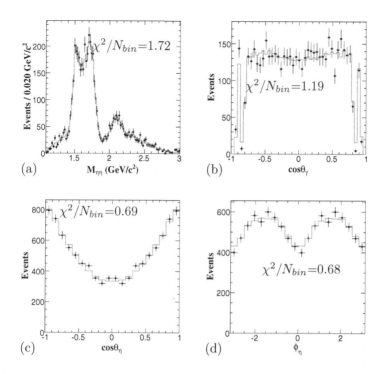

Fig. 8. Comparisons between data and PWA fit projections: (a) the invariant mass spectrum of $\eta\eta$, (b)-(c) the polar angle of the radiative photon in the J/ψ rest frame and η in the $\eta\eta$ helicity frame, and (d) the azimuthal angle of η in the $\eta\eta$ helicity frame. The black dots with error bars are data with background subtracted, and the solid histograms show the PWA projections.

decay width and branching fraction $\Gamma_{\gamma\gamma}B(\eta\eta)$ for $f_2(X)$ are obtained to be 1737 ± 9 MeV/c^2, 228^{+21}_{-20} MeV and $5.2^{+0.9}_{-0.8}$ eV, respectively.

6. $\omega\omega$, $\omega\phi$ and $\phi\phi$ Mass Spectra

An anomalous near-threshold enhancement, denoted as the $X(1810)$, in the $\omega\phi$ invariant-mass spectrum in the process $J/\psi \to \gamma\omega\phi$ was reported by the BESII experiment via PWA. The analysis indicated that the $X(1810)$ quantum number assignment favored $J^{PC} = 0^{++}$ over $J^{PC} = 0^{-+}$ or 2^{++} with a significance of more than 10σ. The mass and width are $M = 1812^{+19}_{-26}(stat.) \pm 18(syst.)$ MeV/c^2 and $\Gamma = 105 \pm 20(stat.) \pm 28(syst.)$ MeV/c^2, respectively, and the product branching fraction $\mathcal{B}(J/\psi \to \gamma\, X(1810))\, \mathcal{B}(X(1810) \to \omega\phi) = [2.61 \pm 0.27(stat.) \pm 0.65(syst.)] \times 10^{-4}$ was measured.

Possible interpretations for the $X(1810)$ include a tetraquark state, a hybrid, or a glueball state etc., a dynamical effect arising from intermediate meson rescattering, or a threshold cusp of an attracting resonance.

A PWA that uses a tensor covariant amplitude for the $J/\psi \to \gamma\omega\phi$ process was performed again in order to confirm the $X(1810)$ using $(225.3 \pm 2.8) \times 10^6 J/\psi$

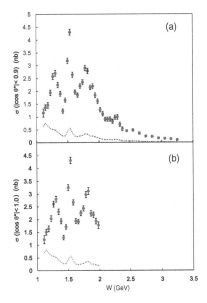

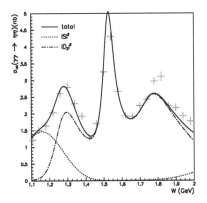

Fig. 9. (a) The cross section integrated over $|\cos\theta^*| < 0.9$ and (b) over $|\cos\theta^*| < 1.0$ for $W < 2.0$ GeV. Here θ^* is the angle of η in two-photon system. The dotted curve shows the size of the systematic uncertainty.

Fig. 10. Total cross sections and fitted curves for the nominal fit in the high mass region (solid curve). Dotted (dot-dashed) curves are $|S|^2$ ($|D_2|^2$) from the fit.

Table 2. Results from the best PWA fit solution.

Resonance	J^{PC}	M(MeV/c^2)	Γ(MeV/c^2)	Events	Significance
$X(1810)$	0^{++}	1795 ± 7	95 ± 10	1319 ± 52	$> 30\sigma$
$f_2(1950)$	2^{++}	1944	472	665 ± 40	20.4σ
$f_0(2020)$	0^{++}	1992	442	715 ± 45	13.9σ
$\eta(2225)$	0^{-+}	2226	185	70 ± 30	6.4σ
phase space	0^{-+}	—	—	319 ± 24	9.1σ

events.[9] A PWA was performed on the selected $J/\psi \to \gamma\omega\phi$ candidate events to study the properties of the $\omega\phi$ mass threshold enhancement. In the PWA, the enhancement is denoted as X, and the decay processes are described with sequential 2-body or 3-body decays: $J/\psi \to \gamma X, X \to \omega\phi, \omega \to \pi^+\pi^-\pi^0$ and $\phi \to K^+K^-$. The amplitudes of the 2-body or 3-body decays are constructed with a covariant tensor amplitude method. Finally, together with the contributions of the $X(1810)$ and phase-space, additional needed components are listed in Table 2 for the best solution of the PWA fit. The $J^{PC} = 0^{++}$ assignment for the $X(1810)$ has by far the highest log likelihood value among the different J^{PC} hypotheses, and the statistical significance of the $X(1810)$ is more than 30σ. The mass and width of the $X(1810)$ are determined to be $M = 1795 \pm 7(stat.)^{+13}_{-5}(syst.) \pm 19(mod.)$ MeV/c^2 and $\Gamma = 95 \pm 10(stat.)^{+21}_{-34}(syst.) \pm 75(mod.)$ MeV/c^2 and the product

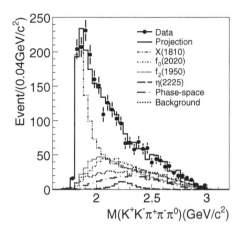

Fig. 11. The $K^+K^-\pi^+\pi^-\pi^0$ invariant-mass distribution between data and PWA fit projections.

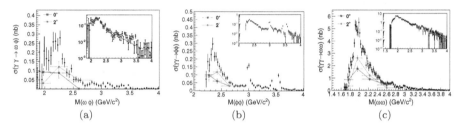

Fig. 12. The cross sections of $\gamma\gamma \to \omega\phi$ (a), $\phi\phi$ (b), and $\omega\omega$ (c) are shown as points with error bars. The fraction contributions for different J^P values as a function of $M(VV)$ are shown as the points and squares with error bars.

branching fraction is measured to be $\mathcal{B}(J/\psi \to \gamma X(1810)) \times \mathcal{B}(X(1810) \to \omega\phi) = (2.00 \pm 0.08(stat.)^{+0.45}_{-1.00}(syst.) \pm 1.30(mod.)) \times 10^{-4}$. The contributions of each component of the best solution of the PWA fit are shown in Fig. 11. The enhancement is not compatible with being due either to the $X(1835)$ or the $X(p\bar{p})$, due to the different mass and spin-parity. The search for other possible states decaying to $\omega\phi$ would be interesting.

In the two-photon processes $\gamma\gamma \to \omega J/\psi$ and $\phi J/\psi$, a state $X(3915)$ and an evidence for $X(4350)$ were observed. It is very natural to extend the above theoretical picture to similar states coupling to $\omega\phi$, since the only difference between such states and the $X(3915)$ or $X(4350)$ is the replacement of the $c\bar{c}$ pair with a pair of light quarks. States coupling to $\omega\omega$ or $\phi\phi$ could also provide information on the classification of the low-lying states coupled to pairs of light vector mesons.

The $\gamma\gamma \to VV$ cross sections are shown in Fig. 12.[10] The fraction of cross sections for different J^P values as a function of $M(VV)$ is also shown in Fig. 12. We conclude that there are at least two different J^P components ($J = 0$ and $J = 2$)

in each of the three final states. The inset also shows the distribution of the cross section on a semi-logarithmic scale, where, in the high energy region, we fit the $W_{\gamma\gamma}^{-n}$ dependence of the cross section.

We observe clear structures at $M(\omega\phi) \sim 2.2 \text{ GeV}/c^2$, $M(\phi\phi) \sim 2.35 \text{ GeV}/c^2$, and $M(\omega\omega) \sim 2.0 \text{ GeV}/c^2$. While there are substantial spin-zero components in all three modes, there are also spin-two components near threshold.

7. Conclusion

I have reviewed some results on the charmonium and light hadron spectroscopy mainly from BESIII and Belle experiments, including the observation of $\psi(4040)/\psi(4160) \to \eta J/\psi$, some measurements on the $\eta_c/\eta_c(2S)$ resonance parameters and their decays, the evidence of the $\psi_2(1^3D_2)$ state in the $\chi_{c1}\gamma$ mass spectrum, the X(1835) research in more processes, and the analysis of the $\eta\eta$, $\omega\phi$, $\phi\phi$ and $\omega\omega$ mass spectra.

Acknowledgments

This work is supported partly by the Fundamental Research Funds for the Central Universities of China (303236).

References

1. BESIII Collab. (M. Ablikim *et al.*), *Phys. Rev. D* **86**, 071101(R) (2012).
2. Belle Collab. (X. L. Wang *et al.*), *Phys. Rev. D* **87**, 051101(R) (2012).
3. BESIII Collab. (M. Ablikim *et al.*), *Phys. Rev. Lett.* **108**, 222002 (2012).
4. BESIII Collab. (M. Ablikim *et al.*), *Phys. Rev. D* **87**, 052005 (2013).
5. Belle Collab. (V. Bhardwaj *et al.*), *Phys. Rev. Lett.* **111**, 032001 (2013).
6. Belle Collab. (C. C. Zhang *et al.*), *Phys. Rev. D* **86**, 052002 (2012).
7. BESIII Collab. (M. Ablikim *et al.*), *Phys. Rev. D* **87**, 092009 (2013).
8. Belle Collab. (S. Uehara *et al.*), *Phys. Rev. D* **82**, 114031 (2010).
9. BESIII Collab. (M. Ablikim *et al.*), *Phys. Rev. D* **87**, 032008 (2013).
10. Belle Collab. (Z. Q. Liu *et al.*), *Phys. Rev. Lett.* **108**, 232001 (2012).

Physics in Collision (PIC 2013)
International Journal of Modern Physics: Conference Series
Vol. 31 (2014) 1460293 (12 pages)
© The Author
DOI: 10.1142/S2010194514602932

Hadron and quarkonium exotica

Sookyung Choi

Department of Physics, Gyeongsang National University
Jinju daero 501, Jinju, 660-773, Republic of Korea
schoi@gsnu.ac.kr

Published 15 May 2014

A number of charmonium-(bottomonium-)like states have been observed in B-factory experiments. Recently the BESIII experiment has joined this search with a unique data sample collected at the different center of mass energies ranging from 3.9 GeV to 4.42 GeV in which they found new charmonium-like states with non-zero electric charge. We review the status of experimental searchs for quarkonium-like states and also other types of non-$q\bar{q}$ meson or non-qqq baryons that are predicted by QCD-motivated models. We mainly focus on results from the B-factories and BESIII.

1. Introduction

Mesons and baryons are formed from color-singlet combinations of quarks. All non-$q\bar{q}$ mesons and non-qqq baryons predicted by QCD-motivated models are expected to be in color-singlet multi-quark combinations.

Color-singlet multiquark states can be formed from colored diquarks and diantiquarks: a pentaquark is predicted to contain two diquarks and one antiquark, the H-dibaryon contains three diquarks, and tetraquark mesons contain a diquark and diantiquark. Multiquark states can also be formed by molecules of two (or more) color-singlet $q\bar{q}$ mesons and/or qqq baryons that are bound together by the exchange of virtual $\pi\,\sigma$, ω, etc. mesons. In this way, Pentaquarks, the H-dibaryon, baryonium and tetraquark mesons can be formed. If these exist, where are they?

2. Pentaquark

A pentaquark is a predicted baryonic bound state of four quarks and one antiquark. In the 2003, two experiments, LEPS and CLAS, reported the existence of a positive strangeness baryon that they called the Θ^+. After that, a number of experiments reported the observation of confirming signals and even other new types of pentaquarks. But later, high-statistices experiments[1,2] with higher statistics could not reproduce these signals suggesting that the LEPS-CLAS results were due to statistical effects rather than a real resonance. Therefore the existence of the pentaquark is still an open question.

3. The H-Dibaryon

The possibility of a six-quark state was first proposed by R.L. Jaffe in 1977.[3] He dubbed it the H-dibaryon, and prdicted it to have a strangeness$=-2$ and baryon number$=+2$. Such a state could possibly be either a tightly bound diquark triplet or a hyperon-hyperon molecular type state. Jaffe's original prediction was 0^+ dihyperon with the mass that is about 80 MeV below the $2m_\Lambda$ threshold. His original specific prediction was ruled out by observation of a double-Λ hypernuclei events. Especially the "NAGARA" event[4] observed in double-Λ hypernuclei, $^6_{\Lambda\Lambda}$He, which greatly narrowed the mass window of the H to be $M_H > 2m_\Lambda$-$B_{\Lambda\Lambda}$ MeV, with binding energy $B_{\Lambda\Lambda}= 7.25 \pm 0.19^{+0.18}_{-0.11}$, corresponding to a lower limit of the H mass to be 2223.7MeV/c^2.

Although Jaffe's original prediction for $B_H \sim 80$ MeV has been ruled out, the theoretical case for an H-dibaryon with a mass near $2m_\Lambda$ remains concrete. Recent Lattice QCD results indicate the existence of an H-dibaryon with the mass close to the $2m_\Lambda$ threshold[5] and this motivated a Belle search for the H near the mass of the $2m_\Lambda$ threshold. For an H mass below $2m_\Lambda$ threshold, the H would decay via a $\Delta S = 1$ weak-interaction with a number of possible decay channels: $\Lambda p\pi$, λ n, Σ p, and Σn. For an H mass above the 2Λ threshold and below the mass of Ξp, the H would decay stongly into $\Lambda\Lambda$ almost 100% of the time.

Belle searched for the H-dibaryon production in inclusive $\Upsilon(1, 2S) \rightarrow HX$; $H \rightarrow \Lambda p\pi^-$ and $\Lambda\Lambda$ process using data sample containing 102 million events of $\Upsilon(1S)$ and 158 million events of $\Upsilon(2S)$ collected with Belle detector operating at the KEKB e^+e^- collider. The dominant decay mechanism for the narrow $\Upsilon(nS)$ resonances (n=1,2 and 3) is annihilation into three gluons. Each gluon materializes as quark-antiquark pairs and, since the Υ states are flavor SU(3) singlets, $u\bar{u}$, $d\bar{d}$ and $s\bar{s}$ pairs are produced in nearly equal numbers. This strangeness-rich high-density quark environment in a limited volume of phase space is ideal for producing multi-quark hadron states, especially $S = -2$, H-dibaryon.

These $\Upsilon(1, 2S)$ decays are a rich source of 6-quark antideuteron production, as demonstrated by large branching fraction of $\sim 3 \times 10^{-5}$ measured by the ARGUS and CLEO experiments.[6] The physics of near-threshold S-wave Feshbach resonances ensure that an H-dibaryon in this mass region should be very similar in character to a $\Lambda\Lambda$ version of the deuteron. Moreover, since the $\Upsilon(1S)$ and $\Upsilon(2S)$ are Flavor-SU(3) singlets, S$= -2$ counterparts of the antideuteron should be produced with sensitivities below or at the rate for antideuteron production.

The Fig. 1 shows results from the Belle search for H-dibaryon masses below and above the $2m_\Lambda$ threshold.[7] For masses below threshold (left), the H is searched for as a peak in the $\Lambda p\pi^-$ invariant mass distribution in inclusive $\Upsilon(1, 2S) \rightarrow \Lambda p\pi X$ and

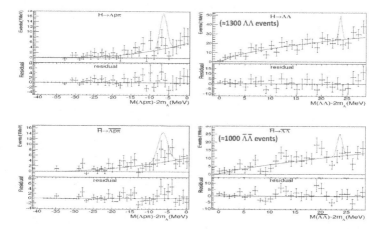

Fig. 1. The top and left (right) panel shows the continuum-subtracted $M(\Lambda p \pi^-)$ ($M(\Lambda\Lambda)$) distribution (upper) and fit residuals (lower) for the combined $\Upsilon(1S)$ and $\Upsilon(2S)$ data samples. The curve shows the results of the background-only fit using an ARGUS-like threshold function to model the background. Gaussian fit function for expected signals is superimposed assuming $\mathcal{B}(\Upsilon \to HX) = (1/20) \times \mathcal{B}(\Upsilon \to \bar{d}X)$. The corresponding $M(\bar{\Lambda}\bar{p}\pi^+)$ ($M(\bar{\Lambda}\bar{\Lambda})$) distributions in the bottom left (right) panel.

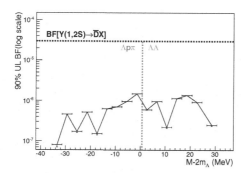

Fig. 2. Upper limits at 90% CL for $\mathcal{B}(\Upsilon(1,2S) \to HX)\mathcal{B}(H \to f)$ for a narrow ($\Gamma = 0$) H-dibaryon vs. $M_H - 2m_\Lambda$. The vertical dotted line indicate the $M_H = 2m_\Lambda$ threshold. The UL branching fraction below (above) the $2m_\Lambda$ threshold are for $f = \Lambda p \pi^-$ ($f = \Lambda\Lambda$). The horizontal dotted line indicates the average PDG value for $\mathcal{B}(\Upsilon(1,2S) \to \bar{D}X)$.

its charge conjugate decays. For masses above threshold (right), the H is searched for in the invariant mass distribution of $\Lambda\Lambda$ pairs from $\Upsilon(1,2S) \to \Lambda\Lambda X$. The continuum-subtracted $M(\Lambda p \pi^-)$ (left) and $M(\Lambda\Lambda)$ (right) distributions have no evident $H \to \Lambda p \pi^-$ and $H \to \Lambda\Lambda$ signals.

Figure 2 shows the corresponding $M_H - 2m_\Lambda$ dependent upper limits at 90% CL for $\mathcal{B}(\Upsilon(1,2S) \to HX)\mathcal{B}(H \to \Lambda p \pi^-$ and $\Lambda\Lambda)$. These are all more than an order of magnitude lower than the averaged PDG value of $\mathcal{B}(\Upsilon(1.2S) \to \bar{d}X)$, plotted as an horizontal dotted line. No evidence was found for a signal in any of these modes, and the most stringent branching-fraction upper limits on H-dibaryon production are determined for masses near the $2m_\Lambda$ threshold.

4. Candidate of Baryonium

Ten years ago, the BESII experiment reported the observation of a peculiar threshold enhancement in the $p\bar{p}$ invariant mass distribution in the radiative decay process $J/\psi \to \gamma p\bar{p}$ using 58M J/ψ events.[8] A fit to the enhancement in the $M(p\bar{p})$ distribution near the $M = 2m_p$ mass threshold with an S-wave Breit-Wigner function gave a below threshold peak value of $M = 1859 \, ^{+3}_{-10}(\text{stat}) \, ^{+5}_{-25}(\text{syst})$ MeV/c^2 and an upper limit on the full width of $\Gamma <30$ MeV at the 90% CL. The enhancement could not be explained by final state interactions (FSI) between the p and $\bar{p}$ and there was no corresponding any known resonance states listed in the standard table that could account for this state.

With $(225.2 \pm 2.8) \times 10^6$ J/ψ events sample collected with the BESIII detector, a partial wave analysis that included $I = 0$ FSI between p and $\bar{p}$ gave a mass and upper limit on the width of $M = 1832^{+19}_{-5} \, ^{+18}_{-17} \pm 19(\text{model})$ MeV/c^2 and $\Gamma <76$ MeV at the 90% CL.[9] The fitted mass peak is about 40 MeV below the $M = 2m_p$ mass threshold. Also the γ polar angle and $p\bar{p}$ decay angle distributions favor the $J^{PC} = 0^{-+}$ quantum number assignment over any other possibility with statistical significances that are larger than 6.8σ.

One proposed interpretation for this enhancement is that it is baryonium, *i.e.* a deuteron-like proton-antiproton bound state produced by standard nuclear forces mediated by π, σ and/or ω exchanges. For such a state with mass below the $2m_p$ mass threshold, the p and $\bar{p}$ would annihilate to mesons, while above threshold it would be expected to "fall apart" to $p\bar{p}$ final states almost 100% of the time. Since $p\bar{p}$ annihilation to $\pi^+\pi^-\eta'$ or $3(\pi^+\pi^-)$ are common and dominant channels for $I = 0$, $J^{PC} = 0^{-+}$ $p\bar{p}$ annihilations searches for the same state in the radiative processes $J/\psi \to \gamma\pi^+\pi^-\eta'$ and $J/\psi \to \gamma 3(\pi^+\pi^-)$ were pursued by BESIII with the same J/ψ data sample.

A dramatic peak was seen in the $M(\pi^+\pi^-\eta')$ spectrum[10] at $M = 1837 \pm 3^{+5}_{-3}$ MeV/c^2, very similar to the one measured from $p\bar{p}$ mass, also the cosine of the γ polar angle distribution is consistent with the form of 0^{-+}. Although the mass and J^{pc} are consistent with those in the $p\bar{p}$ channel, the broad width ($\Gamma = 190\pm9^{+38}_{-36}$ MeV) is not. Two more new states at higher masses, dubbed $X(2120)$ and $X(2370)$, are also observed in the same $M(\pi^+\pi^-\eta')$ mass distribution. Recently, BESIII reported the results of a study of the radiative process $J/\psi \to \gamma 3(\pi^+\pi^-)$, where a clear peak near $2m_p$ was seen with mass $M = 1842.2 \pm 4.2^{+7.1}_{-2.6}$ MeV/c^2 and width $\Gamma = 83 \pm 14 \pm 11$ MeV,[11] consistent with the the $p\bar{p}$ values. Are these peaks observed in the $p\bar{p}$, $\pi\pi\eta'$ and $3\pi^+\pi^-$ mass distribution all from the same state? The masses are consistent with each other, but the width measured from $p\bar{p}$ is significantly narrower than the width of the $\pi^+\pi^-\eta'$ enhancement. These resonances could come from different sources or there may be more than one resonance contributing to the $\pi\pi\eta'$ peak. To clarify all these, further study is needed including the measurement of J^{pc} for the $X(1842) \to 3(\pi^+\pi^-)$ signal.

5. Doubly Charged Partner States of $D_{s0}^+(2317)$

The $D_{s0}^+(2317)$, hereafter referred to as the D_{s0}^+ was first observed by BaBar[12] as a narrow peak in the $D_{s0}^+\pi^0$ mass spectrum produced in e^+e^- annihilation to $D_{s0}^+\pi^0+X$ process and subsequently confirmed by CLEO[13] and Belle. Its production in B meson decay were also established by both Belle[14] and BaBar.[16] It is generally considered to be the conventional $I(J^P) = 0(0^+)$ p-wave $c\bar{s}$ meson, but its mass $M_{D_{s0}^+} = 2317.8 \pm 0.6$ MeV$/c^2$[42] is very similar as the one $M_{D_0^*} = 2318 \pm 29$ MeV of the non-strange 0^+ P-wave D_0^* state, in spite of the mass difference between the s and u (or d) quarks which is known as $m_s - m_{u(d)} \sim 100$ MeV. Furthermore, Potential model and lattice QCD published prior to its discovery all predicted 0^+ P-wave $c\bar{s}$ meson mass to be well above the $m_{D^0} + m_{K^+} = 2358.6$ MeV threshold with a large partial decay width for $D_{s0}^+ \to DK$. This discrepancy has led to considerable theoretical speculation that the D_{s0}^+ is not a simple $c\bar{s}$ meson, but instead a four quarks state such as DK molecule or diquark-diantiquark state.

Among these four quarks models, Terasaki's[17] assignment to $I=1$ iso-triplet four-quark meson is favored by various existing experimental data. If this is the production process of D_{s0}^+, it should have doubly charged $I_Z=1$ (F_1^{++}) and neutral $I_Z = 0$ (F_1^0) partners.[18] The Ref. 18 also predicted that isospin invariance insures that the product branching fraction $\mathcal{B}(B^+ \to D^- F_1^{++,0})\mathcal{B}(F_1^{++,0} \to D_s^+\pi^{+,-})$ will be nearly equal to $\mathcal{B}(B^{+,0} \to \bar{D}D_{s0}^+)\mathcal{B}(D_{s0}^+ \to D_s^+\pi^0)$.

Preliminary results from a Belle search uncovered no evidence for the predicted state and established a stringent branching fraction limits to be $\mathcal{B}(B^+ \to D^- F^{++}(2317))\mathcal{B}(F^{++}(2317) \to D_s^+\pi^+) < 0.28 \times 10^{-4}$ at 90% CL that is a factor of about 30 below the predicted level. This makes possible eliminating the tetra-quark interpretation of the D_{sJ} meson family.

A by-product from this search are measurements of the branching fractions for the decay processes of $B^+ \to D_{s0}^+\bar{D}^0$ and $B^0 \to D_{s0}^+D^-$ with sub-decay $D_{s0}^+ \to D_s^+\pi^0$; $D_s^+ \to K^+K^-\pi^+$, which is significantly improved and used for the above upper limits calculations for the doubly charged partner state. Figure 3 shows the $M_{\rm bc}$ (left), $M(D_s^+\pi^0)$ (center) and ΔE (right) distribution for $B \to \bar{D}D_{s0}^+$ candidate events. Each distributions are projections of events that are in the signal regions of other two quantities for modes: a) $B^0 \to D^-D_{s0}^+$; $D^- \to K^+\pi^-\pi^-$, b) $B^\pm \to \bar{D}^0D_{s0}^\pm$; $\bar{D}^0 \to K^+\pi^-\pi^+\pi^-$ and c) $B^\pm \to \bar{D}^0D_{s0}^\pm$; $\bar{D}^0 \to K^+\pi^-$. The curves in each plot show the results of unbinned three-dimentional likelihood fit. From the fit result, the product branching fraction is determined to be $\mathcal{B}(B^0 \to D^-D_{s0}^+)\mathcal{B}(D_{s0}^+ \to D_s^+\pi^0) = 10.0 \pm 1.2 \pm 1.0 \pm 0.5) \times 10^{-4}$ and $\mathcal{B}(B^+ \to \bar{D}^0D_{s0}^+)\mathcal{B}(D_{s0}^+ \to D_s^+\pi^0) = (7.8^{+1.3}_{-1.2} \pm 1.0 \pm 0.5) \times 10^{-4}$, where the first error is statistical, the second is the systematic error, and the third one reflects the errors on the PDG world average D and D_s^+ branching fraction values.All these measurement used Belle full data

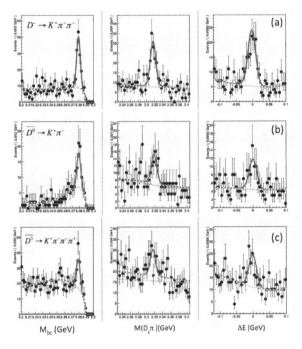

Fig. 3. The $M_{\rm bc}$ (left), $M(D_s^+\pi^0)$ (center) and ΔE (right) distribution for a) $B^0 \to D^- D_{s0}^+$ signal events for the $D^- \to K^+\pi^-\pi^-$ sub-decay modes, b) $B^\pm \to \bar{D}^0 D_{s0}^\pm; \bar{D}^0 \to K^+\pi^-\pi^+\pi^-$ and c) $B^\pm \to \bar{D}^0 D_{s0}^\pm; \bar{D}^0 \to K^+\pi^-$, with the results of the fit superimposed. The events in each distribution are in the signal regions of the two quantities not being plotted.

containing 772 million $B\bar{B}$ meson pairs collected at the $\Upsilon(4S)$ resonance. These results agree well with the average of the BaBar and previous Belle measurement with substantial improvement in precision.

6. The XYZ Quarkonium-Like Mesons

The XYZ states are known as states that decay into final states containing a $c\bar{c}$ (or $b\bar{b}$) quark pair but do not fit into the conventional charmonium (bottomonium) spectrum. The $X(3872)$, a key member of this family, was first discovered as a peak in the $\pi^+\pi^- J/\psi$ mass by Belle[19] in the $B \to \pi^+\pi^- J/\psi K$ decay process and confirmed by four different experiments and observed via five different decay channels. The quantum numbers were unambiguously determined to be $J^{PC} = 1^{++}$ by the LHCb experiment[20] by an analysis of the $B \to X(3872)K$; $X(3872) \to \pi^+\pi^- J/\psi$ decay chain. This result favors an exotic explanation of the $X(3872)$ state and disfavors a simple $c\bar{c}$ charmonium assignment. Recently, one more production channel was observed by BESIII,[21] where a 6.3σ significance $X(3872)$ signal is reported in the process $e^+e^- \to \gamma X(3872)$ using data collected at the four center of mass energies: $\sqrt{s} = 4.009, 4.229, 4.260$ and 4.360 GeV. Large cross sections of $\gamma X(3872)$ are seen at $\sqrt{s} = 4.229$ and 4.260, while cross section at $\sqrt{s} = 4.009$

Table 1. Fit results in the default model. Errors are statistical only.

J^P	0^-	1^-	1^+	2^-	2^+
Mass, MeV/c^2	4479±16	4477±4	4485±20	4478±22	4384±19
Width, MeV	110±50	22±14	200±40	83±25	52±28
Significance	4.5σ	3.6σ	6.4σ	2.2σ	1.8σ

and 4.360 GeV were consistent with zero. This suggests the $X(3872)$ might be produced from the radiative transition of the $Y(4260)$ rather than from the $\psi(4040)$ or $Y(4360)$, but, with the current limited statistics, continuum production cannot be ruled out.

The charged charmonium-like state $Z(4430)^+$, as a controversal member, was first observed by Belle[22, 23] as a peak in the $\pi^+\psi'$ mass in $B \to K\pi^+\psi'$ decays. However, the search by BaBar didn't confirm this signal.[24] Belle also observed two more charged states, Z_1^+ and Z_2^+, that are seen in the final states $\pi^+\chi_{c1}$ in the exclusive $B \to K\pi^+\chi_{c1}$ decays.[25] Recently Belle has reported the measurement of quantum numbers of the $Z(4430)^+$ by performing a full amplitude analysis of $B^0 \to \psi'K^+\pi^-$ in four dimensions. The Table 1 shows the results from 4-D fit, the preferred spin-parity hypotheses is 1^+, which is favored over 0^- by 2.9σ, but 0^- also cannot be ruled out with current statistics.[26]

The $Y(4260)$ is a $J^{PC}=1^{--}$ resonance peak that was discovered in the $e^+e^- \to \pi^+\pi^- J/\psi$ cross section by BaBar.[27] It was subsequently confirmed by CLEO,[29] Belle[30] and BESII.[31] A remarkable feature of the $Y(4260)$ is its large partial width of $\Gamma(Y(4260) \to \pi^+\pi^- J/\psi) >1$ MeV which is much larger than that for typical charmonium.[32] This motivated the Belle to investigate whether or not there is a corresponding structure in the bottomonium mass region. This study found that there is a large anomalous cross section for $e^+e^- \to \pi^+\pi^-\Upsilon(nS)$ (n=1,2,3) for e^+e^- cm energies around 10.9 GeV, near the $\Upsilon(5S)$ resonance.[33] Belle subsequently found that $\Upsilon(5S)$ decays to $\pi^+\pi^-\Upsilon(nS)$ final states are strong sources of charged, bottomonium-like states of $Z_b(10610)$ and $Z_b(10650)$.

6.1. *The bottomonium-like mesons*

Two charged bottomonium-like states, the $Z_b(10610)$ and $Z_b(10650)$, have been observed in the $\pi^\pm\Upsilon(nS)$ and $\pi^\pm h_b(mS)$ mass at Belle experiment in the decay process $\Upsilon(10860) \to \pi^+\pi^-\Upsilon(nS)$ (n = 1,2,3) and $\pi^+\pi^- h_b(mP)$ (m = 1,2).[34] Since they have non-zero electric charge, they must contain a minimum of four quarks.

The mass of the $Z_b(10610)$ is +(2.7 ± 2.1) MeV above the $M_B + M_B^*$ threshold mass, while the $Z_b(10650)$ is +(2.0 ± 1.8) MeV above the $2M_B^*$. The proximity to the $B\bar{B}^*$ and $B^*\bar{B}^*$ threshold mass suggest that these Z_b states may be a molecular type state formed by two mesons. This interpretaion can be checked by studying the $Z_b \to B\bar{B}^*$ and $Z_b \to B^*\bar{B}^*$ in the $\Upsilon \to \pi B(B^*)B^*$ three body decay.[35]

Figure 4 shows two distinct peaks at the masses of $Z_b(10610)$ and $Z_b(10650)$. In this fit the mass was fixed to the earlier measurements. This three body decay

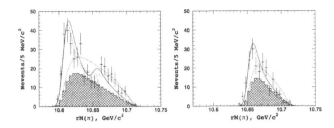

Fig. 4. $M_r(\pi)$ recoil mass distribution for $\Upsilon(10860) \to B\bar{B}^*\pi$ (left) and $\Upsilon(10860) \to B^*\bar{B}^*\pi$ (right) candidate events. Points with error bars are data, the solid line is the results of the fit with the nominal model the dashed line - fit to pure non-resonant amplitude, the dotted line - fit to a single Z_b state plus a non-resonant amplitude, and the dashed-dotted - two Z_b states and a non-resonant amplitude. The hatched histogram represents background component normalized to the estimated number of background events.

analysis is extended to $\Upsilon(10860) \to \Upsilon(nS)\pi^+\pi^-$ and $\Upsilon(10860) \to h_b\pi^+\pi^-$ to measure not only the newly observed Z_b states but also the fractions of individual contribution to the whole three-body signals. The relative fractions for Z_b are $\frac{\mathcal{B}(Z_b(10610)\to B\bar{B}^*)}{\mathcal{B}(Z_b(10610)\to \pi^+(b\bar{b}))} = 6.2 \pm 0.7$ and $\frac{\mathcal{B}(Z_b(10650)\to B^*\bar{B}^*)}{\mathcal{B}(Z_b(10650)\to \pi^+(b\bar{b}))} = 2.8 \pm 0.4$, here statistical errors added only.

We started these $\Upsilon(5S)$ mass region search to see $Y(4260)$ analogy in b quark sector. Then we found two charged bottomonium-like states which are the smoking guns for non $q\bar{q}$ mesons, then the next questions are "are there c-quark versions of Z_b s"? The answer is followed in the next section.

6.2. *The charmonium-like mesons*

Search for new states were mostly confined to analyses of B-factory data including Belle and BaBar, now it has been expanded to include investigations of $Y(4260)$ and $Y(4360)$ resonance decays using data from BESIII. BESIII/BEPCII(Beijing Electron Positron Collider) has been operating as a Y factory for the past six months, collecting large data samples of e^+e^- annihilation events at 13 different CM energies from 3.9 GeV to 4.42 GeV,[38] especially, at the peaks of the $Y(4230)$ (1090.0 pb^{-1}), $Y(4260)$ (826.8 pb^{-1}) and the $Y(4360)$ (5 pb^{-1}) resonances, allowing in-depth studies for the decay properties of these states.

Previous studies of these resonances were done at the B-factories using the radiative return process, which provides rather low luminosities in the charm threshold region. BESIII studied the process $e^+e^- \to \pi^+\pi^-J/\psi$ at a center of mass energy of 4.26 GeV using 525pb^{-1} data.[36] The Born cross section is measured to be (62.9 $\pm$ 1.9 $\pm$ 3.7) pb, which is consistent with the existing results from the BaBar,[28] CLEO[29] and Belle[30] experiments. In addition, a surprising structure was observed at the mass of (3899 $\pm$ 3.6 $\pm$ 4.9) MeV/c^2 and the width of (46 $\pm$ 10 $\pm$ 20) MeV in the $\pi^\pm J/\psi$ mass spectrum. The ratio R of the production rates is obtained to be $\frac{(\sigma(e^+e^-\to\pi^\pm Z_c(3900)^\mp\to\pi^+\pi^- J/\psi)}{\sigma(e^+e^-\to\pi^+\pi^- J/\psi)} = 21.5 \pm 3.3(stat) \pm 7.5(syst))\%$.

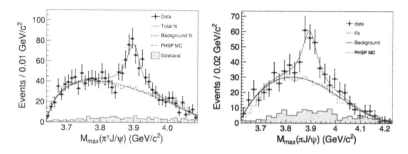

Fig. 5. Fit to the $M_{max}(\pi J/\psi)$ distribution from BESIII (left) and Belle (right). The signal shape is parameterized with same function by using their own mass resolution as described in the text, while background shape is parametrized differently.

At the same time Belle also measured the cross section for $e^+e^- \to \pi^+\pi^- J/\psi$ between 3.8 GeV and 5.5 GeV with 967 fb^{-1} data sample collected by the Belle detector at or near the $\Upsilon(nS)$ (n=1,2,..., 5) resonances.[37] The $Y(4260)$ is observed and its resonant parameters are determined. In the $Y(4260)$ signal region, the peak state $Z(3900)^\pm$ is observed with a mass of $(3894.5 \pm 6.6 \pm 4.5)$ MeV/c^2 and a width of $(63 \pm 24 \pm 26)$ MeV in the $\pi^\pm J/\psi$ mass spectrum with a statistical significance larger than 5.2σ. This new charged state is refer to as the $Z_c(3900)$ or $Z_c(3895)$. Here the ratio of the production rates is obtained from a one-dimentional fit to be $\frac{\mathcal{B}(Y(4260)\to Z(3900)^\pm\pi^\mp)\mathcal{B}(Z(3900)^\pm\to\pi^\pm J/\psi)}{\mathcal{B}(Y(4260)\to\pi^+\pi^- J/\psi)} = (29.0 \pm 8.9)$ %, where the error is statistical only. For this fit the possible interferences between other amplitides are not included. This structure couples to charmonium and has an electric charge, which is suggestive of a state containing four quarks including $c\bar{c}$ and light quarks $u\bar{d}$. The Fig. 5 shows the fit results from BESIII (left) and Belle (right). Unbinned maximum likelihood fit is performed to the distribution of $M(\pi J/\psi)$, using the larger one of the two combinations. The same signal shape for both experiments is parameterized as an S-wave BW function convolved with a Gaussian whose mass resolution is fixed at their own MC-estimated values depending on detectors.

Reference 40 also reports the observation of this charged state at a 6σ significance level in the analysis of 586 pb^{-1} data taken at $\sqrt{s}$=4170 MeV, with the CLEO-c detector at the CESR collider at the Cornell University. The search for Z_c was made in the same decay chain as the BESIII study, $e^+e^- \to \pi^\mp Z_c^\pm$, $Z_c^\pm \to \pi^\pm J/\psi$, but the data was taken at $\sqrt{s}$=4170 MeV, on the peak of the $\psi(4160)$ (2^3D_1) charmonium state. The measured results of the $Z(3900)^\pm$ are M = $(3886 \pm 4(\text{stat}) \pm 2(\text{syst})$ MeV/c^2, $\Gamma = (37 \pm 4(\text{stat}) \pm 8(\text{syst}))$ MeV, and R = $\frac{(\sigma(e^+e^-\to\pi^\pm Z_c(3900)^\mp\to\pi^+\pi^- J/\psi)}{\sigma(e^+e^-\to\pi^+\pi^- J/\psi)} = (32 \pm 8 \pm 10)\%$, These are all in good agreement with the $Z_c(3900)$ signals reported by BESIII and Belle observed in the decay of $Y(4260)$. In addtion, the first evidence for the neutral member state, $Z_c^0(3900)$, of isospin treplet is also reported which decaying into $\pi^0 J/\psi$ at a 3.5σ significance level.

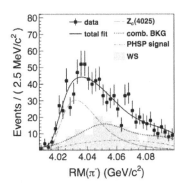

Fig. 6. Unbinned maximum likelihood fit to the π^- recoil mass spectrum in data. The $Z_c^+(4025)$ signal shape is taken as the efficiency-weighted BW shape convoluted with a dectector resolution function. See the text for a detailed description of the other components that were used in the fit.

After this conference, BESIII reported the observation of this charged state at one more decay channel: $e^+e^- \to \pi^\pm(D\bar{D}^*)^\mp$ at the center of mass energy of 4.26 GeV using 525 pb^{-1} data.[41] A distinct charged structure is observed near the threshold of $m(D) + m(\bar{D}^{*\pm})$ in the $(D\bar{D}^*)^\mp$ mass distribution, at $M = (3883.9 \pm 1.5 \pm 4.2)MeV/c^2$, which is 2σ and 1σ, respectively, below those of the $Z_c \to \pi^\pm J/\psi$ peak observed by BESIII and Belle. Here the large Z_c signal yield permitted the establishment of the J^P quantum number of the πZ_c system to be 1^+. Assuming the $Z_c(3885) \to DD^{-*}$ signal and the $Z_c(3900) \to \pi J/\psi$ signal are from the same source, the partial width ratio $\frac{\Gamma(Z_c(3885) \to DD^{-*})}{\Gamma(Z_c(3900) \to \pi^\pm J/\psi)} = 6.2 \pm 1.1 \pm 2.7$ is determined.

More searches for new charged states at higher mass region were followed by exploiting these BESIII scan data sample. A charged charmonium-like structure is observed near the threshold of $m(D^{*+}) + m(\bar{D}^{*0})$ in the $\pi^\pm$ recoil mass spectrum in the process $e^+e^- \to (D^*\bar{D}^*)^\pm \pi^\mp$ at a center of mass energy of 4.26GeV using a 827 pb^{-1} data.[39] Here a partial reconstruction technique is used to identify $(D^*\bar{D}^*)^\pm \pi^\mp$ system, which requires only one $\pi^\mp$ reconstruction from primary decay, the D^+ from $D^{*+} \to D^+\pi^0$, and at least one soft π^0 from either D^{*+} or $\bar{D}^{*0}$. The Fig. 6 shows the $\pi^\mp$ recoil mass spectrum in data and various components in its fit are described in the Ref. 39. From the fit, the mass and width of the $Z_c^+(4025)$ signal are measured to be $(4026.3 \pm 2.6(\text{stat}) \pm 3.7(\text{syst}))$ MeV/c^2 and $(24.8 \pm 5.6(\text{stat}) \pm 7.7(\text{syst}))$ MeV, respectively, with the statistical significance $> 10\sigma$. The ratio $R = \frac{(\sigma(e^+e^- \to Z_c(4025)^\pm \pi^\mp \to (D^*\bar{D}^*)^\pm \pi^\mp)}{\sigma(e^+e^- \to (D^*\bar{D}^*)^\pm \pi^\mp)}$ is determined to be $0.65 \pm 0.09(\text{stat}) \pm 0.06(\text{syst})$.

BESIII has reported searches charged charmonium-like states at the $\pi^\pm h_c$ mass distribution in the process $e^+e^- \to \pi^+\pi^- h_c$ at 13 different CM energies from 3.9 GeV to 4.42 GeV.[38] Here three large statistics CM energies are 4.23 (1090.0), 4.26 (826.8), and 4.36 (544.5 pb^{-1}) GeV. The h_c is reconstructed via E1 transition $h_c \to \gamma\eta_c$, and the η_c is subsequently reconstructed from 16 different exclusive hadronic final states. An unbinned maximum likelihood fit is applied to $\pi^\pm h_c$ mass

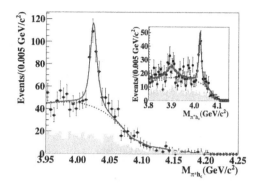

Fig. 7. Sum of the simultaneous fits to the $M_{\pi^\pm h_c}$ distributions at center-of-mass (CM) energies of 4.23 GeV, 4.26 GeV, and 4.36 GeV. The inset shows the sum of the simultaneous fit to the $M_{\pi^\pm h_c}$ distributions at 4.23 GeV and 4.26 GeV with $Z_c(3900)$ and $Z_c(4020)$.

distribution summed over the 16 η_c decay modes. Since each data sets at the CM energy of 4.23, 4.26 and 4.36 GeV shows similar structures, the same signal function with common mass and width is used to fit them simultaneously. Fig. 7 shows huge signal named $Z_c(4020)$ with its statistical significance greater than 8.9σ and fit results give a mass of $(4021.8 \pm 1.0 \pm 2.5)$ MeV/c^2 and a width of $(5.7 \pm 3.4 \pm 1.1)$ MeV. The cross sections are also calculated at each CM energies points. Adding the $Z_c(3900)$ with mass and width fixed to the measurements of Ref. 36 in the fit results in only a 2.1σ significance of it shown in the inset of Fig. 7. The $Z_c(4020)$ agrees within 1.5σ of the $Z_c(4025)$ above which observed at CM energy 4.26 GeV.

7. Summary

The $J^{PC}=1^{--}$ $Y(4260)$ and $\Upsilon(5S)$ have no compelling interpretations. The $Y(4260)$ ($\Upsilon(5S)$) has huge couplings to $\pi^+\pi^- J/\psi$ ($\pi^+\pi^- \Upsilon(nS)$) that were not predicted in any model, and are strong sources of charged Z_c (Z_b) states observed at $\pi^\pm J/\psi$, $\pi^\pm h_c$ ($\pi^\pm \Upsilon(nS)$, $\pi^\pm h_b$) mass, and also with mass near $m(D^{(*)})+m(D^*)$ ($m(B^{(*)})+m(B^*)$).

Numerous non$q\bar{q}$ mesons not specific to QCD have been found, such as a baryonium candidate in $J/\psi \to \gamma p\bar{p}$ at BESII and BESIII and XYZ mesons that contains $c\bar{c}$ and $b\bar{b}$ pairs together with light quarks $u\bar{d}$ or $u\bar{u}$.

QCD motivated spectroscopies predicted by theorists do not seem to exist. No evidence was found for Pentaquarks, and an Hdibaryon with mass near $2m_\Lambda$ is excluded at stringent levels. No hint on D_{s0}^{++} isospin partner state of $D_{s0}^+(2317)$ is observed.

Acknowledgments

This work was supported by the Korean Research Foundation via Grant number 2011-0029457.

References

1. K. Shirotori, *et al.*, *Phys. Rev. Lett.* **109**, 132002 (2012).
2. CLAS Collab. (B. McKinnon *et al.*), *Phys. Rev. Lett.* **96**, 212001 (2006).
3. R. L. Jaffe, *Phys. Rev. Lett.* **38**, 195 (1977).
4. H. Takahashi *et al.*, *Phys. Rev. Lett.* **87**, 212502 (2001).
5. NPLQCD Collab. (S. R. Beane *et al.*), *Phys. Rev. Lett.* **106**, 162001 (2011); HAL QCD Collab. (Takashi Inoue *et al.*), *Phys. Rev. Lett.* **106**, 162002 (2011).
6. ARGUS Collab. (H. Albrecht *et al.*), *Phys. Rev. B* **236**, 102 (1990); CLEO Collab. (D. M. Asner *et al.*), *Phys. Rev. D* **75**, 012009 (2007).
7. Belle Collab. (B. H. Kim *et al.*), *Phys. Rev. Lett.* **110**, 222002 (2013).
8. BES Collab. (J. Z. Bai *et al.*), *Phys. Rev. Lett.* **91**, 022001 (2003).
9. BESIII Collab. (M. Ablikim *et al.*), *Phys. Rev. Lett.* **108**, 112003 (2012).
10. BESIII Collab. (M. Ablikim *et al.*), *Phys. Rev. Lett.* **106**, 072002 (2011).
11. BESIII Collab. (M. Ablikim *et al.*), *Phys. Rev. D* **88**, 091502 (2013).
12. BaBar Collab. (B. Aubert *et al.*), *Phys. Rev. Lett.* **90**, 242001 (2003).
13. CLEO Collab. (D. Besson *et al.*), *Phys. Rev. D* **68**, 032002 (2003).
14. Belle Collab. (P. Krokovny *et al.*), *Phys. Rev. Lett.* **91**, 262002 (2003).
15. BaBar Collab. (B. Aubert *et al.*), *Phys. Rev. Lett.* **93**, 181801 (2004).
16. BaBar Collab. (B. Aubert *et al.*), *Phys. Rev. D* **74**, 032007 (2006).
17. A. Hayashigaki and K. Terasaki, *Prog. Theor. Phys.* **114**, 1191 (2006).
18. Kunihiko Terasaki, *Prog. Theor. Phys.* **116**, 435 (2006).
19. Belle Collab. (S.-K.Choi, S. L. Olsen *et al.*), *Phys. Rev. Lett.* **91**, 262001 (2003).
20. LHCb Collab. (R.Aaij *et al.*), *Phys. Rev. Lett.* **110**, 222001 (2013).
21. BESIII Collab. (M. Ablikim *et al.*), arXiv:1310.4101 (2013).
22. Belle Collab. (S. -K. Choi, S. L. Olsen *et al.*), *Phys. Rev. Lett.* **100**, 142001 (2008).
23. Belle Collab. (R. Mizuk *et al.*), *Phys. Rev. D* **80**, 031104 (2009).
24. BaBar collab. (B. Aubert *et al.*), *Phys. Rev. D* **79**, 112001 (2009).
25. Belle Collab. (R. Mizuk *et al.*), *Phys. Rev. D* **78**, 072004 (2008).
26. Belle Collab. (K. Chilikin *et al.*), *Phys. Rev. D* **88**, 074026 (2013).
27. BaBar Collab. (B. Aubert *et al.*), *Phys. Rev. Lett.* **95**, 142001 (2005).
28. BaBar Collab. (J. P. Lees *et al.*), *Phys. Rev. D* **86**, 051102 (2012).
29. CLEO Collab. (T. E. Coan *et al.*), *Phys. Rev. Lett.* **96**, 162003 (2006).
30. Belle Collab. (C. Z. Yuan *et al.*), *Phys. Rev. Lett.* **99**, 182004 (2007).
31. BES Collab. (J. Z. Bai *et al.*), *Phys. Rev. Lett.* **88**, 101802 (2002).
32. X. H. Mo *et al.*, *Phys. Lett. B.* **640**, 182 (2006).
33. Belle Collab. (K. -F. Chen *et al.*), *Phys. Rev. Lett.* **100**, 112001 (2008).
34. Belle Collab. (A. Bondar *et al.*), *Phys. Rev. Lett.* **108**, 122001 (2012).
35. Belle Collab. (I. Adachi *et al.*), arXiv:1209.6450.
36. BESIII Collab. (M. Ablikim *et al.*), *Phys. Rev. Lett.* **110**, 252001 (2013).
37. Belle Collab. (Z. Q. Liu *et al.*), *Phys. Rev. Lett.* **110**, 252002 (2013).
38. BESIII Collab. (M. Ablikim *et al.*), arXiv:1309.1896.
39. BESIII Collab. (M. Ablikim *et al.*), arXiv:1308.2760.
40. T. Xiao, S. Dobbs, A. Tomaradze and Kamal K. Seth, arXiv:1304.3036.
41. BESIII Collab. (M. Ablikim *et al.*), arXiv:1310.1163.
42. Particle Data Group (J. Beringer *et al.*), *Phys. Rev. D* **86**, 010001 (2012).

Physics in Collision (PIC 2013)
International Journal of Modern Physics: Conference Series
Vol. 31 (2014) 1460294 (7 pages)
© The Author
DOI: 10.1142/S2010194514602944

Third generation superpartners: Results from ATLAS and CMS

J. Poveda

(On behalf of the ATLAS and CMS Collaborations)

Department of Physics, Indiana University
Bloomington, IN 47405-7105, United States
Ximo.Poveda@cern.ch

Published 15 May 2014

These proceedings discuss results from recent searches for third generation SUSY particles by the ATLAS and CMS experiments at the LHC. Analyses performed with 8 TeV data probing direct pair production of bottom and top squarks are presented.

Keywords: LHC; ATLAS; CMS; SUSY; stop; sbottom.

1. Introduction

After the discovery of a Higgs boson[1,2] in 2012, all the particles included in the Standard Model (SM) have been observed experimentally. Despite the many successes of this theory in the last few decades, it has several limitations from the experimental and theoretical perspectives. For instance, the quadratically divergent loop corrections to the Higgs mass would imply a high level of fine-tuning. In addition, the SM in its current formulation cannot explain the abundance of dark matter in the universe and does not provide unification of forces at a high energy scale.

Weak scale supersymmetry (SUSY) is one of the most studied extensions of the SM and, among other features, can solve all the problems mentioned above. SUSY postulates the existence of "superpartners" for all the existing SM particles with spin differing by half a unit. In R-parity conserving SUSY models, the lightest supersymmetric particle (LSP), typically the lightest neutralino ($\tilde{\chi}_1^0$), is stable and therefore a good dark matter candidate.

The experimental search for the superpartners of the SM bottom and top quarks, known as bottom and top squarks or simply as sbottom ($\tilde{b}_1$) and stop ($\tilde{t}_1$), is one of the most active areas in the search for SUSY. These particles can be relatively light, and therefore Naturalness arguments, introduced to avoid fine tuning in the theory, predict the sbottom and stop masses to be a few hundred GeV, which would therefore be produced with sizeable cross sections at colliders, and produce heavy quarks (t and b) in their decays.

2. Search for Third Generation SUSY Particles in ATLAS and CMS

During 2012, the ATLAS[3] and CMS[4] experiments at the LHC[5] acquired more than 20 fb^{-1} of integrated luminosity from proton-proton collisions at $\sqrt{s} = 8$ TeV.

These proceedings will cover recent searches for the direct production of stop and sbottom, most of them performed using the full data set recorded in 2012. In these analyses, the basic elements used to distinguish between stop or sbottom signals and the SM backgrounds are the presence of leptons (produced in chargino, $\tilde{\chi}_1^{\pm}$, or top decays), jets, b-jets (produced in the decays of $\tilde{b}_1$ and top) and missing transverse energy (E_T^{miss}) from the escaping LSPs and neutrinos. Discriminants combining several of these basic objects are also commonly used, such as the transverse mass of a W candidate, defined as $m_T^2 = 2 \cdot p_T^{\ell} \cdot E_T^{\mathrm{miss}} \cdot (1 - \cos(\Delta\phi(\ell, E_T^{\mathrm{miss}})))$. Some analyses also use more complicated methods, such as multivariate techniques or multi-dimensional fits.

Most of the interpretations included in these proceedings are obtained in simplified models,[6] where only a few particles and decays are considered, assuming that the rest of the spectrum is much heavier and decoupled from the process of interest.

3. Results on Sbottom Searches

The main sbottom decay modes include the decay to either the lightest or the second lightest neutralino ($\tilde{\chi}_2^0$) plus a b quark ($\tilde{b}_1 \to b\tilde{\chi}_1^0, b\tilde{\chi}_2^0$), or to charginos via $\tilde{b}_1 \to t\tilde{\chi}_1^{\pm}$.

In scenarios where the sbottom is produced in pairs and decays predominantly via $\tilde{b}_1 \to b\tilde{\chi}_1^0$, signal events have a $b\bar{b}\tilde{\chi}_1^0\tilde{\chi}_1^0$ final state. This is explored with analyses requiring a lepton veto, two b-jets and E_T^{miss}, using event kinematic variables to distinguish the sbottom signal from the $b\bar{b}$ background.[7,8] Figure 1 (left) shows an example of the exclusion limits obtained by ATLAS for this scenario, reaching sbottom masses of approximately 650 GeV for a massless LSP.

For model characterized by the $\tilde{b}_1 \to b\tilde{\chi}_2^0$ decay, Z or h bosons are produced via $\tilde{\chi}_2^0 \to h\tilde{\chi}_1^0$ or $\tilde{\chi}_2^0 \to Z\tilde{\chi}_1^0$, and the additional leptons and b-jets in the $h \to b\bar{b}$ or $Z \to \ell\ell$ decays are used to discriminate signal from background.[9,10] Figure 1 (right) shows example exclusion limits obtained by CMS considering a simplified model defined by the $\tilde{b}_1 \to b\tilde{\chi}_2^0$ with $\tilde{\chi}_2^0 \to Z\tilde{\chi}_1^0$ decay chain, with sensitivity to sbottom masses above 450 GeV.

If the sbottom decays predominantly via $\tilde{b}_1 \to t\tilde{\chi}_1^{\pm}$, the final state includes additional leptons from the top and chargino decays ($\tilde{\chi}_1^{\pm} \to W^{\pm}\tilde{\chi}_1^0$). This signal scenario is explored in ATLAS and CMS with analyses requiring two same-sign leptons[11,12]

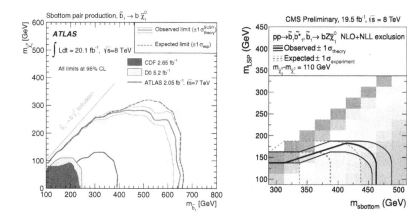

Fig. 1. Expected and observed exclusion limits at 95% CL in the $(m_{\tilde{b}_1}, m_{\tilde{\chi}_1^0})$ mass plane for simplified models with sbottom pair production. On the left,[7] the sbottom decays via $\tilde{b}_1 \to b\tilde{\chi}_1^0$ and, on the right,[10] via $\tilde{b}_1 \to b\tilde{\chi}_2^0$ with $\tilde{\chi}_2^0 \to Z\tilde{\chi}_1^0$.

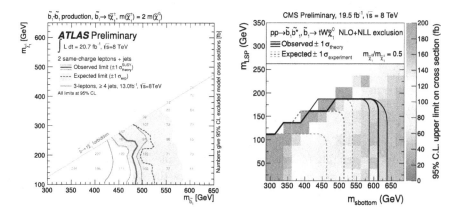

Fig. 2. Expected and observed exclusion limits at 95% CL in the $(m_{\tilde{b}_1}, m_{\tilde{\chi}_1^0})$ mass plane for simplified models with sbottom pair production and $\tilde{b}_1 \to t\tilde{\chi}_1^\pm$ decays obtained with dilepton same-sign[11] (left) and three-lepton[10] (right) analyses.

or three leptons[10, 13] in the final state. Figure 2 shows the interpretations obtained from some of these analyses for the cases where $m_{\tilde{\chi}_1^\pm} = 2m_{\tilde{\chi}_1^0}$, with sensitivities for sbottom masses in the range of 500-600 GeV.

4. Results on Stop Searches

The decay of the top squark depends on the mass splitting between the stop and its possible decay products, leading to very different topologies depending on the mass spectrum.

For a heavy stop, the dominant decays would be $\tilde{t}_1 \to t\tilde{\chi}_1^0$ (kinematically allowed if $m_{\tilde{t}_1} > m_t + m_{\tilde{\chi}_1^0}$) and $\tilde{t}_1 \to b\tilde{\chi}_1^{\pm}$ (allowed if $m_{\tilde{t}_1} > m_{\tilde{\chi}_1^{\pm}} + m_b$). However, if the $\tilde{t}_1 \to t\tilde{\chi}_1^0$ decay is kinematically forbidden ($m_{\tilde{t}_1} < m_t + m_{\tilde{\chi}_1^0}$), the stop would have a three-body decay ($\tilde{t}_1 \to bW\tilde{\chi}_1^0$) via an off-shell top. For a lighter stop, when the decays above are forbidden, four-body decays ($\tilde{t}_1 \to bW^{(*)}\tilde{\chi}_1^0$) via off-shell t and W would occur if $m_{\tilde{t}_1} - m_{\tilde{\chi}_1^0} < m_W + m_b$, or the stop would decay to a charm quark via $\tilde{t}_1 \to c\tilde{\chi}_1^0$ if $m_{\tilde{t}_1} > m_{\tilde{\chi}_1^0} + m_c$.

The searches for a heavy stop decaying predominantly via $\tilde{t}_1 \to t\tilde{\chi}_1^0$ or $\tilde{t}_1 \to b\tilde{\chi}_1^{\pm}$ are designed based on the decay of the W boson in the top or chargino decay modes, leading to topologies with zero, one or two leptons. The semi-leptonic analyses make use of boosted-decision trees[14] at CMS or shape fits in the E_T^{miss} and m_T variables[15] at ATLAS to distinguish a potential stop signal from the $t\bar{t}$ background. Similarly, analyses probing the fully leptonic final state by ATLAS[16, 17] make use of the m_{T2} variable[18, 19] as discriminator between signal and background. Searches for stop are also performed in the fully hadronic channel by both experiments,[20, 21] showing a very good sensitivity at high stop masses.

A dedicated analysis is also performed by ATLAS aiming at the $\tilde{t}_1 \to c\tilde{\chi}_1^0$ decay[22] based on charm-tagging techniques and, for the case of charm jets too soft to be detected, on a monojet approach relying on an initial state radiation high-p_T jet.

Figure 3 summarizes the exclusion limits obtained by ATLAS and CMS as a function of the stop and neutralino masses for simplified models with different stop decays. Sensitivity up to stop masses of around 700 GeV are obtained for a massless neutralino, while for massive neutralinos of 250-300 GeV, stop sensitivity falls to 450-500 GeV.

Despite the good coverage of the stop mass parameter space in Fig. 3, some experimentally challenging regions remain unconstrained. For example, no exclusion is achieved in the $m_{\tilde{t}_1} \approx m_t + m_{\tilde{\chi}_1^0}$ region, where the stop events are kinematically very similar to SM $t\bar{t}$ production. However, this region is explored by ATLAS considering the production of the heavy stop state ($\tilde{t}_2$) decaying via $\tilde{t}_2 \to Z\tilde{t}_1$.[25] As shown in Fig. 4, exclusion limits on $\tilde{t}_2$ masses of approximately 550 GeV are obtained in this very difficult ($m_{\tilde{t}_1}, m_{\tilde{\chi}_1^0}$) configuration.

Other models also feature long decay chains involving stops, such as gauge-mediated SUSY breaking scenarios where the graviton ($\tilde{G}$) is the LSP and the lightest neutralino decays via $\tilde{\chi}_1^0 \to Z\tilde{G}$ or $\tilde{\chi}_1^0 \to h\tilde{G}$. Analyses exploiting the $Z \to \ell\ell$ or $h \to \gamma\gamma$ decays in this kind of model[25–27] lead to sensitivities at stop masses of a few hundred GeV, as illustrated in Fig. 4 (right) by the CMS analysis considering $\tilde{\chi}_1^0 \to h\tilde{G}$ with $h \to \gamma\gamma$.

5. Summary and Conclusions

The search for SUSY is a major part of the LHC physics programme. In particular, naturalness arguments favour the existence of top and bottom squarks with masses of several hundred GeV which can be observed at the LHC energies.

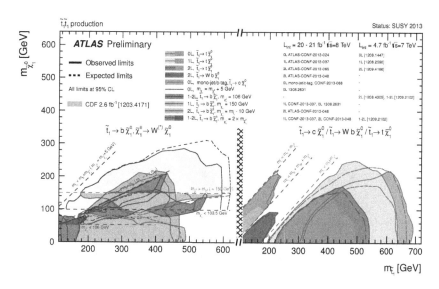

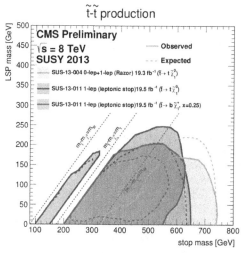

Fig. 3. Summary of the expected and observed exclusion limits at 95% CL in the $(m_{\tilde{t}_1}, m_{\tilde{\chi}_1^0})$ from the ATLAS[23] (top) and CMS[24] (bottom) searches for stop pair production.

A wide program of searches for third generation squarks has been carried out in ATLAS and CMS with the LHC data acquired during 2010-2012. Dedicated analyses optimized for different stop and sbottom decay modes and final state topologies have been conducted. In some cases, complex analysis techniques (boosted-decision trees, multi-dimensional fits, c-jet tagging, etc.) are used to discriminate signals from the background.

No excess has been observed in data above the Standard Model predictions and the current exclusion limits reach stop and sbottom masses at the level of

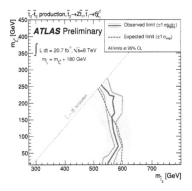 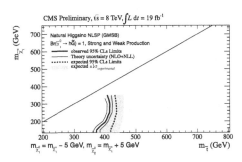

Fig. 4. On the left,[25] expected and observed exclusion limits at 95% CL obtained in the $(m_{\tilde{t}_2},$ $m_{\tilde{\chi}_1^0})$ mass plane for simplified models with $\tilde{t}_2$ pair production and $\tilde{t}_2 \to Z\tilde{t}_1$, $\tilde{t}_1 \to t\tilde{\chi}_1^0$ decays for $m_{\tilde{t}_1} \approx m_t + m_{\tilde{\chi}_1^0}$. On the right,[26] expected and observed exclusion limits at 95% CL considering a "natural" SUSY scenario with gauge-mediated breaking as a function of $m_{\tilde{t}_1}$ and $m_{\tilde{\chi}_1^\pm}$.

600–700 GeV in models with one step decays such as $\tilde{b}_1 \to b\tilde{\chi}_1^0$ or $\tilde{t}_1 \to t\tilde{\chi}_1^0$. Very recent results also address the case of long stop and sbottom decay chains involving Z and Higgs bosons.

References

1. ATLAS Collaboration, *Phys. Lett.* **B716**, 1 (2012).
2. CMS Collaboration, *Phys. Lett.* **B716**, 30 (2012).
3. ATLAS Collaboration, *JINST* **3**, p. S08003 (2008).
4. CMS Collaboration, *JINST* **3**, p. S08004 (2008).
5. L. Evans and P. Bryant (editors), *JINST* **3**, p. S08001 (2008).
6. D. Alves *et al.*, *J. Phys.* **G39**, p. 105005 (2012).
7. ATLAS Collaboration, *arXiv:1308.2631 [hep-ex]* (2013).
8. CMS Collaboration, *Eur.Phys.J.* **C73**, p. 2568 (2013).
9. ATLAS Collaboration, *ATLAS-CONF-2013-061, http://cds.cern.ch/record/1557778*.
10. CMS Collaboration, *CMS-PAS-SUS-13-008, http://cds.cern.ch/record/1547560*.
11. ATLAS Collaboration, *ATLAS-CONF-2013-007, http://cds.cern.ch/record/1522430*.
12. CMS Collaboration, *CMS-PAS-SUS-13-013, http://cds.cern.ch/record/1563301*.
13. CMS Collaboration, *CMS-PAS-SUS-13-002, http://cds.cern.ch/record/1599719*.
14. CMS Collaboration, *arXiv:1308.1586 [hep-ex]* (2013).
15. ATLAS Collaboration, *ATLAS-CONF-2013-037, http://cds.cern.ch/record/1532431*.
16. ATLAS Collaboration, *ATLAS-CONF-2013-048, http://cds.cern.ch/record/1547564* (2013).
17. ATLAS Collaboration, *ATLAS-CONF-2013-065, http://cds.cern.ch/record/1562840*.
18. M. Burns, K. Kong, K. T. Matchev and M. Park, *JHEP* **0903**, p. 143 (2009).
19. W. S. Cho, K. Choi, Y. G. Kim and C. B. Park, *JHEP* **0802**, p. 035 (2008).
20. ATLAS Collaboration, *ATLAS-CONF-2013-024, http://cds.cern.ch/record/1525880*.
21. CMS Collaboration, *CMS-PAS-SUS-13-004, http://cds.cern.ch/record/1596446*.
22. ATLAS Collaboration, *ATLAS-CONF-2013-068, http://cds.cern.ch/record/1562880*.

23. ATLAS Collaboration, https://twiki.cern.ch/twiki/bin/view/AtlasPublic/Combined SummaryPlots.

24. CMS Collaboration, https://twiki.cern.ch/twiki/bin/view/CMSPublic/SUSYSMS SummaryPlots8TeV.

25. ATLAS Collaboration, *ATLAS-CONF-2013-025, http://cds.cern.ch/record/1525881.*

26. CMS Collaboration. CMS-PAS-SUS-13-014, https://cds.cern.ch/record/1598863.

27. CMS Collaboration, *CMS-PAS-SUS-13-002, http://cds.cern.ch/record/1599719.*

Physics in Collision (PIC 2013)
International Journal of Modern Physics: Conference Series
Vol. 31 (2014) 1460295 (5 pages)
© The Author
DOI: 10.1142/S2010194514602956

World Scientific
www.worldscientific.com

Performance and operation experience of the ATLAS Semiconductor Tracker

Zhijun Liang

(On Behalf of the ATLAS Collaboration)

University of Oxford
zhijun.liang@cern.ch

Published 15 May 2014

We report on the operation and performance of the ATLAS Semi-Conductor Tracker (SCT), which has been functioning for 3 years in the high luminosity, high radiation environment of the Large Hadron Collider at CERN. We also report on the few improvements of the SCT foreseen for the high energy run of the LHC. We find 99.3% of the SCT modules are operational, the noise occupancy and hit efficiency exceed the design specifications; the alignment is very close to the ideal to allow on-line track reconstruction and invariant mass determination. We will report on the operation and performance of the detector including an overview of the issues encountered. We observe a significant increase in leakage currents from bulk damage due to non-ionizing radiation and make comparisons with the predictions.

1. Introduction

The ATLAS detector[1] is a multi-purpose apparatus designed to study a wide range of physics processes at the Large Hadron Collider (LHC) at CERN. The recent observation of a Higgs-like particle reported by the ATLAS and CMS collaborations is a milestone in particle physics history.[2,3] The study of the Higgs-like particle rely heavily on the excellent performance of the ATLAS inner detector tracking system. The semiconductor tracker (SCT) is a precision silicon microstrip detector which forms an integral part of this tracking system. The SCT is constructed of 4088 silicon detector modules,[4,5] for a total of 6.3 million strips. Each module operates as a stand-alone unit, mechanically, electrically, optically and thermally. The modules are mounted into two types of structures: one barrel, made of 4 cylinders, and two end-cap systems made of 9 disks each. The SCT silicon micro-strip sensors[6] are processed in the planar p-in-n technology. The signals are processed in the front-end ABCD3T ASICs,[7] which use a binary readout architecture. Data is transferred to the off-detector readout electronics via optical fibres.

2. Detector Operations

The SCT was installed into ATLAS, and was ready for the first LHC proton-proton collisions at a centre of mass energy of 7GeV in March 2010. Since then and until the end of running in February 2013, more than 99% of the 6.3 million readout channels were operational. During operation several enhancements were introduced into the SCT Data Acquisition System (DAQ)[8] in order to avoid potential sources of inefficiency. SCT introduced online monitoring of chip errors in the data and the automatic reconfiguration of the modules with errors. In addition, an automatic global reconfiguration of all SCT module chips every 30 minutes was implemented. There are 90 RODs in the SCT DAQ each of which processes the data for up to 48 modules, if a ROD experiences errors it will exert a BUSY signal which prevents all ATLAS sub-systems from recording data. From 2011 RODs exerting a busy were automatically removed, reconfigured, and reinserted into a run thus having a minimal impact on data taking.

3. Hit Efficiency

A high intrinsic hit efficiency is crucial for the operation of the SCT. The intrinsic efficiencies of the SCT modules are measured by extrapolating well-reconstructed tracks through the detector and counting the number of hits on the track and 'holes' where a hit would be expected but it is not found. Dead channels were removed from the analysis. The efficiency is higher than 99% and the fraction of disabled strips in each layer is less than 0.6% as shown in Fig. 1(a).

To maintain the high hit efficiency and reduce fake hits, dedicated readout timing calibrations were performed during the collisions runs. The in-time hit efficiency as a function of readout delay time has been studied for each SCT module as shown in Fig. 1(b). The readout timing of 4088 modules were optimized and synchronized to 2ns precision level.

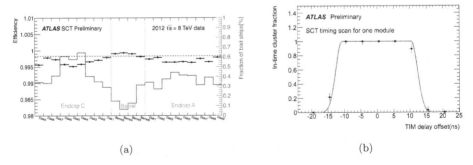

(a)   (b)

Fig. 1. (a) The SCT hit tracking efficiency (black). The blue line and right-hand axis indicates the fraction of disabled strips in each layer. (b) The in-time hit efficiency as a function of readout delay time for one SCT module recorded during a SCT timing calibrations in collisions run.

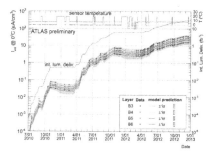

Fig. 2.   SCT barrel leakage currents during 2010, 2011 and 2012, showing correlations with delivered luminosity and temperature, and compared to predictions from Monte Carlo

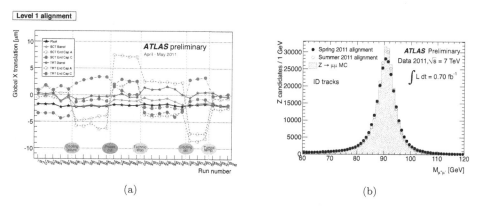

Fig. 3.   (a) Subsystem level, Level One, alignment corrections performed on a run by run basis starting from a common set alignment constants. The corrections shown are for translations in the global x direction in different data taking runs. (b) Invariant mass distribution of $Z \to \mu\mu$ decays, where the mass is reconstructed using track parameters from the Inner Detector track of the combined muons only, using about $702\ pb^{-1}$ of data collected during spring 2011. Ideal alignment performance based on Monte Carlo is compared to observed performance of data processed with spring 2011 alignment and data processed with updated alignment constants.

4. Radiation Damage

Radiation damage was a key consideration during the design phase of the SCT. Irradiation of silicon sensors results in damage in the bulk silicon and the dielectric layers, with main effects so far being the increase in leakage current of the sensor. Therefore the leakage current has been careful measured during the operation. Fig. 2 shows these measurements as a function of time, along with the integrated luminosity delivered by the LHC on the same time scale. The measured values of the fluence and leakage current are in agreement with predictions from a FLUKA based simulation.[9]

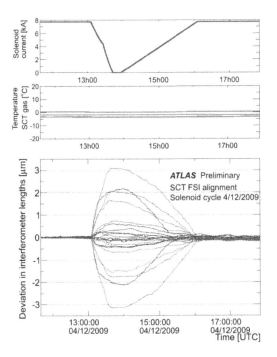

Fig. 4. The effect was observed in the FSI during ramping of the solenoid magnetic field in December 2009. The upper plots show the current in the magnets and the temperature of the gas in the SCT; the lower plot shows the detector movements monitored by different laser lines.

5. Alignment

5.1. *Track based alignment*

Alignment is performed using an algorithm which minimises the the χ^2 between the the measured hit position and the expected position based on on track extrapolation. The alignment is performed at 3 different levels of granularity corresponding to the mechanical layout of the detector: Level 1 corresponds to entire sub-detector barrel and end-caps of Pixel Detector, SCT and TRT. Level 2 deals with silicon barrels and discs, TRT barrel modules and wheels. Level 3 aligns each silicon module and TRT straws having 700,000 degrees of freedom in total. Level 1 alignment corrections performed on a run by run basis starting from a common set alignment constants. As shown in Fig. 3(a) large movements of the detector are measured from track based alignment after hardware incidents. In between these periods, little ($< 1\mu$m) movement is observed indicating that the detector is generally very stable.

After performing three levels of alignment, excellent agreement was found in the residual distributions for both barrel and end cap. The resolution in the $Z \rightarrow \mu\mu$ invariant mass distribution from reconstructed tracks, which is shown in Fig. 3(b), is close to the Monte-Carlo expectation.

5.2. *Laser alignment*

The alignment stability is monitored using a Frequency Scanning Interferometry (FSI) system, which uses data from interferometers attached to the SCT structure to measure changes in length with submicron precision. Fig. 4 shows the effect observed in the FSI during ramping of the solenoid magnetic field in December 2009. Displacements up to 3 μm are observed while the field is ramping but the detector is observed to return to its initial position at the end of the ramping process.

One of the most important result from FSI measurements is that the SCT detector is found to be stable at the μm level over extended periods of time, in agreement with the track-based alignment results.

6. Summary

After three years of operation, the SCT is performing well within its design specification in highly challenging conditions. More than 99% of the detector is still operational. Radiation damage has been observed and is well described by simulation. The detector alignment is very close to the ideal to allow on-line track reconstruction and invariant mass determination.

References

1. The ATLAS Collaboration, *JINST* **3**, S08003 (2008).
2. The ATLAS Collaboration, *Phys. Lett.* **B716**, 1 (2012).
3. The CMS Collaboration, *Phys. Lett.* **B716**, 30 (2012).
4. A. Abdesselam *et al.*, *Nucl. Instrum. Meth.* **A568**, 642 (2006).
5. A. Abdesselam *et al.*, *Nucl. Instrum. Meth.* **A575**, 353 (2007).
6. J. Carter *et al.*, *Nucl. Instrum. Meth.* **A578**, 98 (2007).
7. F. Campabadal *et al.*, *Nucl. Instrum. Meth.* **A552**, 292 (2005).
8. A. Barr *et al.*, *JINST* **3**, P01003 (2008).
9. A. Ferrari, P. R. Sala, A. D. Fasso, J. Ranft, *CERN-2005-010*.

Physics in Collision (PIC 2013)
International Journal of Modern Physics: Conference Series
Vol. 31 (2014) 1460296 (4 pages)
© The Author
DOI: 10.1142/S2010194514602968

The ATLAS Tile Calorimeter performance at LHC

Simon Molander

(On behalf of the ATLAS Collaboration)

Fysikum, Stockholm University, 106 91 Stockholm, Sweden
simon.molander@fysik.su.se

Published 15 May 2014

This paper gives an overview of the performance of the Tile Calorimeter of the ATLAS detector at the Large Hadron Collider. Detector performances with respect to electronic noise and cell response are presented. In addition, an overview of the partially overlapping calibration systems is given.

Keywords: Tile Calorimeter; calibration; performance; electronic noise.

1. Introduction

The Tile Calorimeter[1] is an important sub-detector of the ATLAS experiment[2] at the Large Hadron Collider. It is a scintillating calorimeter consisting of 5182 cells in total, layed out as shown in Fig. 1 for one of the 256 wedges of the detector. Each cell is read out using wavelength shifting fibers of both sides, giving redundancy. These fibers are connected to photo-multiplier tubes (PMT) which in turn are connected to the front end electronics where energy reconstruction is done.

2. Calibration Systems

The Tile Calorimeter is calibrated using multiple partially overlapping systems,[2] giving stability and redundancy for the reconstructed signals. The gain variation for a cell as measured by some of these systems and compared with the integrated luminosity evolution are shown in Fig. 2.

The calibration of the electromagnetic scale was done using response to electrons in beam tests 2001-2004.[3] The precision of the electromagnetic scale in any single cell is 3 %. This scale was then transferred into the ATLAS detector in the cavern and into deeper layers using the cesium system.[4] The cesium system consists of three capsules containing ^{137}Cs sources that are pushed through the detector using a

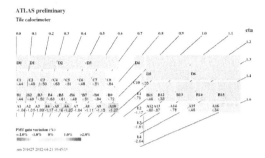

Fig. 1. Mean gain variation as measured by the laser system. Each numbered cell in the graph represents an average of 64 cells. This image also gives an overview of the detector layout, having four layers in the central barrel (to the left in the image), three in the end-cap (to the right) and six gap cells (left part of end-cap). The two intermediate layers in the barrel section (B and C) are read out as one.

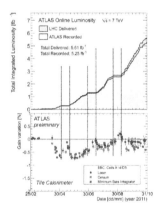

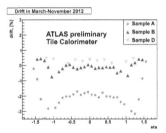

Fig. 3. Calibration drift in percent for for the cesium calibration system between March and November 2012 as a function of η, the different samples represent different radial layers as defined in Fig. 1 and every datapoint represents an average over 64 cells.

Fig. 2. Gain variation of a cell compared to total integrated luminosity. Measurements are done by three different systems.

hydraulic system, thereby injecting signals into the scintillators. These calibrations are done on a monthly basis and can be used to calibrate and to monitor the full optics and photomultiplier response. The precision of the cesium calibration system is 0.3 %. The drift in calibration, mainly caused by variations in PMT gains, for the period between March and November 2012 is shown in Fig. 3 for each layer separately. In order to interpolate between cesium measurements a laser system[2] is used on a weekly basis. The laser system works by injecting light pulses into the photo-multiplier tubes and thereby measures and monitors their gains. The precision of the laser calibration is better than 1 %. The PMT gain variation as measured by the laser system is shown in Fig. 1.

In order to convert the amplitude in ADC counts into a charge in pC, a known charge is injected into the front end electronics. This is known as the charge injection

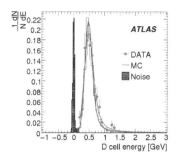

Fig. 4. Noise level compared to signal. The signal is from the response of Tile Calorimeter cells to cosmic muons.

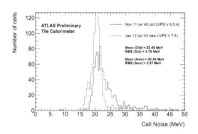

Fig. 5. New and more reliable low voltage power supplies were installed during a maintenance period. They improved the cells' noise as shown by the red line (new) as compared to the blue line (old). Cells with a new power supply have a lower mean noise and a Gaussian distribution.

system (CIS).[2] This is done on a weekly basis, and the constants have been shown to be highly constant and are therefore only updated twice a year. The precision of the CIS calibration is 0.7 %.

3. Electronic noise

The signal-to-noise ratio is a good indicator of detector performance. In addition, knowledge of noise rates is important for reconstruction of constituents of higher level objects such as topological clusters. Figure 4 shows a comparison of signal and noise done by studying cosmic muons.[5] Noise levels were reduced when the low voltage power supplies (LVPS) where replaced during a maintenance period.[6] The result of this replacement can be seen in Fig. 5.

4. Calorimeter Response

The calorimeter energy response can be evaluated by measuring how much energy deposited by a single hadron of a given momentum measured in the inner tracker, expressed as $\frac{E}{p}$ where p is a track momentum. For this evaluation, isolated tracks having energy deposits compatible with minimum ionizing particles in the electromagnetic calorimeter, which is in front of the Tile Calorimeter, are selected. The data is also compared to Monte Carlo simulations. Results are shown as a function of momentum in Fig. 6 and as a function of the angle ϕ in Fig. 7. The largest disagreement between data and Monte Carlo is 12 %, for $11 < p < 12$ GeV.

5. Conclusions

Performance evaluations based on calibration systems, cosmic muons and collision data have been presented. The Tile Calorimeter has redundancy in both read-out and calibration systems. The cell response is monitored and adjusted to the level

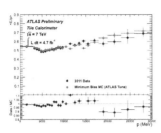

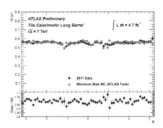

Fig. 6. Calorimeter Response characterised by energy over momentum (E/p) for isolated tracks using proton-proton collision data from 2011 in the Minimum Bias stream. The plot shows the mean of the E/p ratio for hadronic tracks as a function of phi, integrated over the $|\eta| < 0.7$.

Fig. 7. Calorimeter Response characterised by energy over momentum (E/p) for isolated tracks using proton-proton collision data from 2011. The plot shows the mean of the E/p ratio for hadronic tracks as a function of momentum, integrated over ϕ. The lower plot shows the ratio of data to simulation in the same momentum bins.

of $\sim$1 % and in situ energy response validation shows agreement with the hadronic shower simulation prediction within a few percent.

Acknowledgments

In addition to the ATLAS collaboration as a whole, I would like to express my acknowledgements to: My thesis advisor, Sten Hellman, who helped out a lot with technical and practical considerations. Olle Lundberg and Pawel Klimek who helped me with templates and formatting of the poster itself. Eirini Vichou and Tomas Davidek for helping me with improvements.

References

1. G. Aad *et al.*, The ATLAS Experiment at the CERN Large Hadron Collider, *J. Instrum.*, 3:S08003. 437 p., 2008. Also published by CERN Geneva in 2010.

2. E. Berger, R. Blair and J. Dawson, Construction and performance of an iron-scintillator hadron calorimeter with longitudinal tile configuration, Technical Report CERN-LHCC-95-44. LRDB-RD-34, CERN, Geneva, 1995.

3. P, Adragna *et al.* Testbeam Studies of Production Modules of the ATLAS Tile Calorimeter, *Nucl. Instrum. Methods Phys. Res., A*, 606 (ATL-TILECAL-PUB-2009-002. ATL-COM-TILECAL-2009-004. 3):362–394. 73 p., Feb 2009. revised version, includes new author lists.

4. E. A. Starchenko, G. Blanchot *et al.*, Cesium Monitoring System for ATLAS Tile Hadron Calorimeter, Technical Report ATL-TILECAL-2002-003, CERN, Geneva, Feb 2002.

5. G. Aad *et al.*, Readiness of the ATLAS Tile Calorimeter for LHC collisions, *Eur. Phys. J.*, C70:1193–1236, 2010.

6. G. Drake, T. Cundiff *et al.*, An Upgraded Front-End Switching Power Supply Design for the ATLAS TileCAL Detector of the LHC, Technical Report ATL-TILECAL-PROC-2011-015, CERN, Geneva, Nov 2011.

Physics in Collision (PIC 2013)
International Journal of Modern Physics: Conference Series
Vol. 31 (2014) 1460297 (6 pages)
© The Author
DOI: 10.1142/S201019451460297X

The CMS high level trigger

Valentina Gori

Università degli Studi & INFN Firenze,
via G. Sansone 1, Sesto Fiorentino (FI), Italy
valentina.gori@cern.ch

Published 15 May 2014

The CMS experiment has been designed with a 2-level trigger system: the Level 1 Trigger, implemented on custom-designed electronics, and the High Level Trigger (HLT), a streamlined version of the CMS offline reconstruction software running on a computer farm. A software trigger system requires a tradeoff between the complexity of the algorithms running on the available computing power, the sustainable output rate, and the selection efficiency. Here we will present the performance of the main triggers used during the 2012 data taking, ranging from simpler single-object selections to more complex algorithms combining different objects, and applying analysis-level reconstruction and selection. We will discuss the optimisation of the triggers and the specific techniques to cope with the increasing LHC pile-up, reducing its impact on the physics performance.

Keywords: CMS; trigger; HLT; tracking.

1. The CMS trigger

The collision rate at the Large Hadron Collider (LHC) is heavily dominated by large cross section QCD processes, which are not interesting for the physics program of the CMS experiment. The processes we are interested in usually occur at a rate smaller than 10 Hz. Since it is not possible to register all the events and to select them later on, because of a limited bandwith, it becomes mandatory to use a trigger system in order to select events according to physics-driven choices.

The CMS experiment features a two-level trigger architecture. The first level (L1), hardware, operates a first selection of the events to be kept, using muon chambers and calorimeter information. The maximum output rate from L1 is about 100 kHz [1]; this upper limit is given by the CMS data acquisition electronics. The second level, called *High Level Trigger* (HLT), is implemented in software and aims to further reduce the event rate to about 800 Hz on average. Events passing the HLT are then stored on local disk or in CMS Tier-0; about a half of these events were promptly reconstructed (within 48 hours), while the other half have been *parked* and then reconstructed along 2012.

2. The High Level Trigger

The HLT hardware consists of a single processor farm composed of commodity PCs, the event filter farm (EVF).

The main idea is that each HLT trigger path is a sequence of reconstruction and selection steps of increasing complexity. The filtering process uses the full granularity data from the detector, and the selection is based on sophisticated offline-quality reconstruction algorithms. In fact, the HLT algorithms uses a dedicated version of the software framework and the reconstruction code which is used for offline reconstruction and analysis; this online version differs from the offline one only for a different parameter configuration.

HLT starts from the L1 candidate, and then improves the reconstruction and filtering process by exploiting also the tracker information. The starting selection based on the L1 information allows to reduce the rate before tracking reconstruc- tion — a very CPU-expensive process — is performed. In fact, the most challenging aspect is that the CMS high level trigger has to maximize the efficiency while, at the same time, keeping the CPU-time (not only the rate) acceptable.

Events are grouped into a set of non-exclusive streams according to the HLT decisions. In addition to the primary physics stream, "stream A", monitoring and calibration streams are also written. Finally, the event filter farm also collects mon- itoring information and makes it available to the shift crew [2].

3. Timing and Rates in 2012

4. HLT Performance: Object Identification

4.1. *Tracking and vertexing*

Tracking is very important for the reconstruction at the HLT level. A robust and efficient tracking can help reconstruction of particles and can improve their resolu- tion in various ways. For example, it reduces the muon trigger rate by substantially improving the momentum resolution; energy clusters found in the electromagnetic calorimeters can be identified as electrons or photons depending on the presence of a track; the background rejection rate of the lepton triggers can be enhanced further by requiring that leptons should be isolated; it is possible to trigger on jets produced by b-quarks, by counting the numbers of tracks in a jet which have a transverse impact parameter incompatible with the track originating from the beam-line; it is possible to trigger on hadronic τ decays by finding a narrow, isolated jet using tracks in combination with the calorimeter information.

The pixel tracking and other track reconstruction uses about 30% of the total HLT CPU time. This is kept low by only performing track reconstruction when

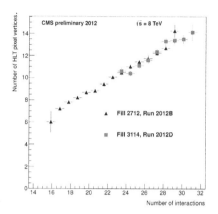

Fig. 1. Number of pixel vertices reconstructed at HLT. The number of interactions is calculated from the bunch luminosity as measured by the forward calorimeters (HF).

necessary (on about 4% of total HLT events) and only after other requirements have been satisfied to reduce the rate at which tracking must be done.

Pixel tracks are used to reconstruct the position of the interaction point. For vertex reconstruction, a simple gap clustering algorithm is used. All tracks are ordered by the z coordinate of their point of closest approach to the beamspot. Then, wherever two neighbouring elements in this ordered set had a gap between them exceeding a distance cut z_{sep}, this point is used to split the tracks on either side of it into separate vertices. In 2012 data taking, where up to 30 interactions per bunch crossing were registered, the number of reconstructed vertices still showed a linear dependence on the number of interactions without saturating (Fig. 1). In this Figure the real number of interactions is measured using the information from the forward calorimeter (HF), which covers the pseudorapidity range $3 < |\eta| < 5$.

4.2. *Muon identification*

The muon high-level triggers at CMS combine information from the muon and the tracker subdetectors to identify muon candidates and determine their transverse momenta, p_T . The algorithm is composed of two main steps: Level-2 (L2), which uses information from the muon system only, and Level-3 (L3), which combines measurements from both tracker and muon subdetectors. In the L2, the reconstruction of a track in the muon spectrometer starts from an initial state, called *seed*, built from patterns of DT and CSC segments. The L3 muon trigger algorithm consists of three main steps: seeding of tracker reconstruction starting from L2 information, track reconstruction in the tracker, and combined fit in the tracker and muon systems. In Fig. 2 the efficiency turn-on curve for an isolated trigger path requiring a single muon with a p_T threshold of 24 GeV is shown. The isolation of L3 muons is evaluated combining information from the silicon tracker and the electromagnetic

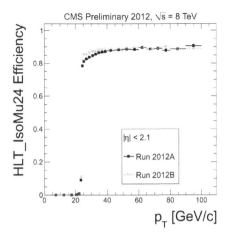

Fig. 2. HLT_IsoMu24 trigger path efficiency calculated with respect to the offline reconstruction.

(ECAL) and hadronic (HCAL) calorimeters. Tracks are reconstructed in the silicon tracker in a geometrical cone of size $\Delta R = \sqrt{\Delta \eta^2 + \Delta \phi^2} = 0.3$ around the L3 muon. In the same cone, ECAL and HCAL deposits are reconstructed. To reduce the dependance of the isolation variable on the pileup of pp collisions, the calorimeter deposits are corrected for the average energy density of the event.

4.3. *Particle flow jets*

At the HLT, jets are reconstructed using the anti-kT clustering algorithm with cone size $R = 0.5$ [3]. The inputs for the jet algorithm can be the calorimeter towers (called CaloJet), or the reconstructed Particle Flow objects (called PFJet). The Particle Flow tecnique allows to use the information from all the detectors and to combine them together to reconstruct the objects [4]. In 2012, most of the jet trigger paths use PFJet. Because of the significant CPU consumption of the Particle Flow algorithm at the HLT, PFJet trigger paths have a pre-selection based on the CaloJet before the particle flow objects will be reconstructed, and PFJets will be formed. The matching between CaloJet and PFJet is also required in single PFJet paths. In Fig. 3 the efficiency turn-on curve of three different trigger paths requiring PFJets with different p_T thresholds are shown.

4.4. *b-Tagging*

The precise identification of b-jets is crucial to reduce the large backgrounds at the LHC. In CMS, using algorithms for b-tagging jets, this background can already be highly suppressed at the HLT, giving lower trigger rates with large efficiency. Algorithms for b-tagging exploit the fact that B hadrons typically have large decay lifetimes and the presence of leptons in the final state compared to those from light

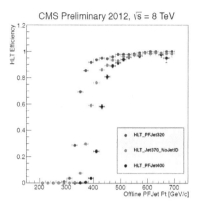

Fig. 3. Turn-on curve measured Vs. offline Particle Flow jet pT. Trigger efficiency measured on an unbiased data sample from Run2012C.

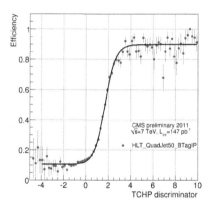

Fig. 4. Turn-on curve of the Track Counting High Purity (TCHP) discriminant efficiency at HLT, with respect to the same variable computed offline.

partons and c quarks. As a consequence, tracks and vertices are largely displaced with respect to the primary vertex. The Track Counting (TC) algorithm uses the impact parameter (IP) significance ($\sigma(IP)/IP$) of the tracks in the jets as a discriminant to distinguish b-jets from other flavours. In Fig. 4 the turn-on curve for the Track Counting discriminant with a High Purity requirement is shown; the online cut for this path is at TCHP= 2. The discriminant is defined as the third highest impact parameter significance for the tracks associated to a jet.

References

1. CMS Collaboration, *CMS The TRIDAS Project, Technical Design Report Vol. 2: Data Acquisition and High-Level Trigger*, CERN LHC 02-26, CMS TDR 6 (2002).

2. CMS Collaboration, *The CMS Trigger System*, CMS PAPER TRG-12-001.
3. Matteo Cacciari, Gavin P. Salam, and Gregory Soyez, *The anti-kt jet clustering algorithm*, *JHEP*, 04:063, 2008.
4. CMS collaboration, *Commissioning of the Particle-Flow reconstruction in Minimum-Bias and Jet Events from pp Collisions at 7 TeV*, CMS-PAS-PFT-10-002, 2010.

Physics in Collision (PIC 2013)
International Journal of Modern Physics: Conference Series
Vol. 31 (2014) 1460298 (5 pages)
© The Authors
DOI: 10.1142/S2010194514602981

MRPC detector for the upgrade of BESIII E-TOF

Rong-Xing Yang, Yong-Jie Sun and Cheng Li*

*Department of Modern Physics, University of Science
and Technology of China, Hefei 230026, Anhui, China*
licheng@ustc.edu.cn

Published 15 May 2014

An end-cap Time-of-Flight (E-TOF) system with higher granularity and intrinsic time resolution of better than 50 ps will extend the K/pion separation (2 sigma) pT range to 1.4 GeV/c and enhance the physics capability of Beijing Spectrometer (BESIII). A R&D work was carried out aiming at upgrading the current BESIII E-TOF with the Multi-gap resistive Plate Chamber (MRPC) detector. The latest best test for the prototype MRPC, together with the custom designed Front End Electronics (FEE) and TDC boards, was performed at the BEPC E3 line in July 2012. The test results show that time resolution of less than 50 ps can be achieved with such a system.

Keywords: Multi-gap resistive plate chamber; BESIII; End-cap TOF.

1. Introduction

The MRPC detector is a good candidate for large area TOF system with very good time resolution, high efficiency and relatively low cost per channel.[1,2] It has been widely used as TOF detector in many experiments, such as LHC/ALICE,[3] RHIC/STAR.[4] The BESIII[5] is a high precision general-purpose detector designed for the high luminosity in the τ-charm energy region at the upgraded Beijing Electron-Positron Collider (BEPCII).[6] A MRPC-based TOF with 50 ps intrinsic time resolution is estimated to extend the BESIII K/π separation (2σ) momentum range to 1.4 GeV/c at BESIII end-cap. A proposal was raised to upgrade the current BESIII ETOF with the MRPC technology and readout electronics, aiming at the 50 ps time resolution.

2. Design

In the conceptual design, each BESIII end-cap ring will be fully covered by 36 trapezium MRPC modules, as shown in Fig. 1. The MRPC module to be used on

*Corresponding author.

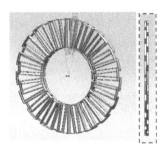

Fig. 1.   The schematic of BESIII MRPC/ETOF.

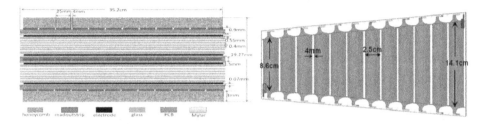

Fig. 2.   Left: The side-view of the MRPC structure. Right: the readout pattern.

Fig. 3.   The setup at BEPC E3 line.

BESIII E-TOF has 12 gas gaps of 0.22 mm thick, arranged in two stacks. Thin floating glass with bulk resistivity $\sim 10^{13} \Omega/\square$ is used as the resistive plate. The signals are picked up by 12 strips with different length and read out from both ends of each strip. The structure of the MRPC is plotted in Fig. 2.

3. Test

The test was performed at BEPC E3 line. The setup of the test is shown in Fig. 3. There MRPC detectors are arranged along the beam line close to each other. The FEE board is custom designed with NINO ASIC chips. The time jitter of the FEE is tested ~ 10 ps. The custom designed far-end electronics is based on HPTDC chips with Leading- and Trailing-edge recording capability. Thus, the TOT (Time-Over-Threshold) of the signal can be used for the slewing correction instead of the traditional T-A correction. The secondary beam contains pions and protons. They can be distinguished clearly by the flight time between scintillators S1 and S2, as shown in Fig. 4.

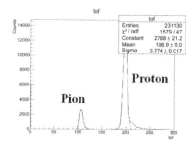

Fig. 4.   The PID with TOF between S1 and S2.

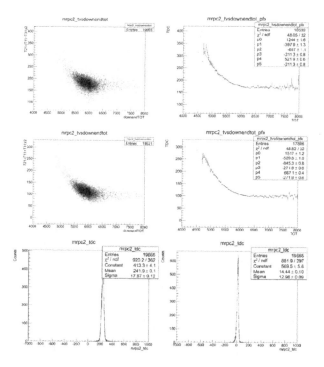

Fig. 5.   The slewing correction and the time spectrum.

The MRPC was first tested with pion samples around 600 MeV which can be considered as MIPs. The slewing correction was made with T-TOT correlation. For each MRPC, the other two modules can be used as reference time (T0). Correcting the time circularly until stable correction relation achieved, the time resolution of each MRPC can be solved. Figure 5 shows the slewing correction and the time spectrum before and after the correction.

The performance of the MRPC is also analyzed with proton samples which is the main contents of the beam. Since proton at such momentum is not MIP, the time resolution is much better, as summarized in Table 1.

Table 1. The time resolution achieved with pion and proton samples at different momentum.

Momentum		500 MeV	600 MeV	800 MeV
Pion sample	MRPC #1	56 PS	47 PS	45 PS
	MRPC #2	51 PS	48 PS	40 PS
	MRPC #3	46 PS	48 PS	45 PS
Proton sample	MRPC #1	29 PS	32 PS	36 PS
	MRPC #2	28 PS	30 PS	35 PS
	MRPC #3	31 PS	31 PS	35 PS

Table 2. The time resolution on different incident positions along a selected strip.

MRPC #2 @600MeV			Strip center → Strip end			
Pion		48 PS		49 PS	49 PS	47 PS
Proton	29 PS	27 PS	26 PS	29 PS	31 PS	36 PS

Table 3. The time resolution under different applied High Voltage.

@600 MeV	±7250 V	±7300 V	±7500 V
Pion	48 PS	48 PS	46 PS
Proton	30 PS	29 PS	30 PS

The performance uniformity along the strip is checked by moving one of the MRPC modules along the strip direction. As shown in Table 2, no position dependence is found. Table 3 shows the time resolution under different applied High Voltage, which indicates the stable performance profited from the long plateau of MRPC.

4. Summary

We successfully design, built and test the MRPC prototype with 12 gas gap structure and double-end readout strips for BESIII ETOF. The proposed performance has been achieved. The time resolution including the custom designed electronics is ~50 ps for MIPs. The detector shows stable performance at different position and working HV. The systematic construction project has been approved and will start soon.

References

1. M. C. S. Williams, E. Cerron-Zeballos *et al.*, *Nucl. Instr. and Meth.* A434 (1999) 362–372.
2. Akindinov, A. Alici *et al.*, *Nucl. Instr. and Meth.* A602 (2009) 709–712.

3. ALICE Collaboration, Addendum to ALICE TDR 8, CERN/LHCC/2002-16, April 2002.
4. STAR TOF Collaboration, Proposal for a large area time of flight system for STAR (2003).
5. M. Ablikim *et al.*, *Nucl. Instr. and Meth.* A614 (2010) 345–399.
6. D. Asner *et al.*, *Ins. J. Mod. Phys.* A24S1 (2009) 499–502.

Physics in Collision (PIC 2013)
International Journal of Modern Physics: Conference Series
Vol. 31 (2014) 1460299 (5 pages)
© The Author
DOI: 10.1142/S2010194514602993

The status of the Double Chooz experiment

Guang Yang
(On behalf of Double Chooa Experiment Collaboration)

Argonne National Laboratory 9700
S. Cass Ave. Lemont, Illinois 60439, USA

Illinois Institute of Technology
3300 S. Federal St, Chicago, Illinois 60616, USA
gyang9@hawk.iit.edu

Published 15 May 2014

Double Chooz is a long baseline neutrino oscillation experiment at Chooz, France. The purpose of this experiment is to measure the non-zero neutrino oscillation parameter θ_{13}, a parameter for changing electron neutrinos into other neutrinos. This experiment uses reactors of the Chooz Nuclear Power Plant as a neutrino source. Double Chooz has published two papers with results showing the measurement of the mixing angle, and 3rd publication is processing.

Keywords: Double Chooz experiment; neutrino oscillation; theta 13.

1. Calculation of Theta 13

Neutrino oscillation comes from a mixture between the flavor and mass eigenstates of neutrinos. As a neutrino propagates through space, the quantum mechanical phases of the three mass states advance at slightly different rates due to the slight differences in the neutrino masses. This results in a changing mixture of mass states as the neutrino travels. So a pure electron neutrino will be a mixture of electron, mu and tau neutrino after traveling some distance.

The formula to calculate the survival probability is

$$P(\bar{\nu}_e \to \bar{\nu}_e) = 1 - \sin^2 2\theta_{13} \sin^2 \left(\frac{1.27 \Delta m_{31}^2 L[m]}{E_v[MeV]} \right). \tag{1}$$

Our plan is to have two detectors: a near detector 400 m away from the reactor core and a far detector 1000 m far away from the reactor core. The difference survival electron neutrino number could be used to calculate the mixing angle theta 13.

Now we only have the far detector undergoing and the near detector will be ready very soon. We are using the comparison of data and Monte Carlo to calculate the survival probability, We will improve a lot when the near detector is ready.

2. Detector

Our far and near detector have the same structure. Our detector consists of outer veto(OV), inner veto(IV), Buffer, γ-Catcher and target. They are concentric cylinders.

Let me introduce them from inside. The target has a capacity of 10.3 m^3 and is built of UV-transparent acrylic with a wall thickness of 8 mm. It is filled with a liquid scintillator. The scintillator is doped with gadolinium Gadolinium has large cross section to capture neutron. The target is the fiducial volume of Double CHooz, where the neutrino interactions are detected by means of the Inverse Beta Decay.

The Target is surrounded by another acrylic vessel, called Gamma Catcher. This vessel has a wall thickness of 12 mm and contains 22.3 m^3 non Gd loaded liquid scintillator. Both the target and Gamma Catcher are fully contained within a stainless steel tank, the buffer, which shields target and gamma catcher from the radioactivity of the PMTs.

The outermost vessel, the so-called Inner Veto, is also a steel tank, but filled with a LAB-based liquid scintillator, which is derectly monitored by 78 8" PMTs. Its main propose is to detect incoming muons and muon secondaries.

The whole setup is additionally shielded against external radioactivity from the rock by 15 cm of steel. On top of it, the detector is covered by the Outer Veto, which consists of plastic scintillator strips and provides further information on the incoming cosmic muons.

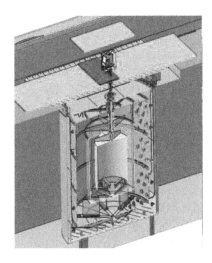

Fig. 1. Far detector Structure.

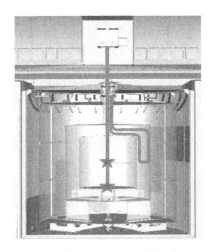

Fig. 2. Calibration system of far detector. Zaxis calibration and Guide Tube calibration systems are presented.

3. Calibration System

The important part to make sure detector works well is calibration. We put sources into the detector to calibrate it. Because the detector could detect very large energy range signals, we need to cover all the energy, we have multiple sources to do that.

First is embedded LEDs inside inner detector and inner veto, it is routinely used to monitor detector stability and PMT gains.

Second is laser, it could calibrate the ability of the detector absorbing UV and green visible light energies.

Third one is calibration radioactivity sources. So far, we have four kinds of radioactivity sources. They are Cs-137, Co-60, Ge-68 and Cf-252. They could cover a very wide energy range. There are three deployment systems to put the radioactivity sources into the target and gamma catcher. We use Zaxis deployment system to calibrate the target and use the Guide Tube to calibrate the gamma catcher. The Fig. 2 shows the idea of first 2 deployment systems. We also have Articulated Arm to calibrate the target, the data of that is being used for analysis work.

4. Energy Calibration

The energy calibration is the most important part of oscillation analysis. We have four factors to get the visible energy of Inverse Beta Decay signal,

$$E_{vis} = PE^m(\rho, z, t) \times f_u^m(\rho, z) \times f_s^m(t) \times f_{MeV}^m. \tag{2}$$

The first factor is the calibration of the linearity of charge/Phtonelectron. We test that for every singal channel, so it is as a function of space and time. From Fig. 3 we could see for the example channel, as the charge incresing, the gain keeps stable. The second factor is the spatical uniformity correction, the Fig. 4 shows the spatial map, it is as a function of space. Third correction is the time stability correction. Figure 5 shows the detector response as a function of time in unit of day. Last factor is how we transfer the PE to MeV. We use the radioactivity sources to finish it. Figure 6 shows the example of Co-60 at center and bottom of target.

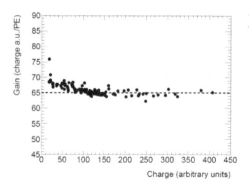

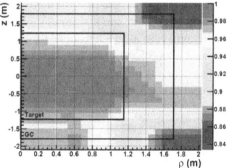

Fig. 3.   The gain stability calibration.   Fig. 4.   The spatial uniformity calibration.

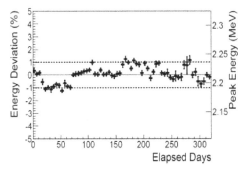

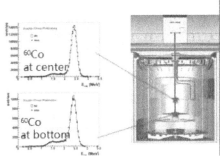

Fig. 5.   Time stability calibration.

Fig. 6.   PE->MeV calibration.

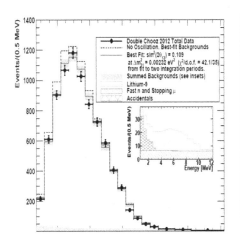

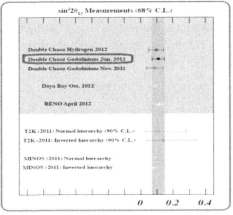

Fig. 7.   MC to data oscillation analysis.

Fig. 8.   Comparison of different experiments.

5. Background and Neutrino Candidate

We define three kinds of backgrounds as follows:

Accidentals: Measured from off-time windows.

Cosmogenic Li9: Measured from At(u) & spatial muon.

Fast-n & Stopping muons: measured from high-energy spectrum

After removing all background, for Gd analysis, we use the candidates that satisfy:

Prompt signal Evis = [0.7, 12.2] MeV

Delayed signal Evis = [6.0, 12.0] MeV

Delayed Coincidence At = [2, 100] μ sec.

6. Oscillation Analysis

We input the candidates into the final fit, comparing data to MC, the best fit of Fig. 7 gives the Gd oscillation analysis result:

$$\sin^2 2\theta_{13} = 0.109 \pm 0.030 \text{ (stat.)} \pm 0.025 \text{(syst.)}.$$

On 2nd publication we analyze 8249 neutrino candidates. Doubled data set since then now. New analysis results will be published soon.

References

1. Y. Abe *et al.*, *Phys. Rev. D* **86**, 052008 (2012), arXiv: 1207.6632.
2. Double Chooz Experiment Official Website: doublechooz.in2p3.fr.

Physics in Collision (PIC 2013)
International Journal of Modern Physics: Conference Series
Vol. 31 (2014) 1460300 (5 pages)
© The Author
DOI: 10.1142/S2010194514603007

Overview of the Jiangmen Underground Neutrino Observatory (JUNO)

Yu-Feng Li

Institute of High Energy Physics,
Chinese Academy of Sciences, Beijing 100049, China
liyufeng@ihep.ac.cn

Published 15 May 2014

The medium baseline reactor antineutrino experiment, Jiangmen Underground Neutrino Observatory (JUNO), which is being planed to be built at Jiangmen in South China, can determine the neutrino mass hierarchy and improve the precision of three oscillation parameters by one order of magnitude. The sensitivity potential on these measurements is reviewed and design concepts of the central detector are illustrated. Finally, we emphasize on the technical challenges we meet and the corresponding R&D efforts.

Keywords: Neutrino oscillation; mass hierarchy; reactors; JUNO.

PACS Numbers: 14.60.Pq, 29.40.Mc, 28.50.Hw, 13.15.+g

1. Introduction

After the discovery of non-zero θ_{13} in latest reactor[1–5] and accelerator[6,7] neutrino oscillation experiments, the neutrino mass hierarchy (i.e., the sign of Δm_{31}^2 or Δm_{32}^2) and lepton CP violation are the remaining oscillation parameters to be measured in the near future. The methods of determining the neutrino mass hierarchy (MH) include the matter-induced oscillations in the long-baseline accelerator neutrino experiments[8–10] and atmospheric neutrino experiments,[11,12] and the vacuum oscillations in the medium baseline reactor antineutrino experiments.[13–16]

The Jiangmen Underground Neutrino Observatory (JUNO) is a multipurpose liquid scintillator (LS) neutrino experiment, whose primary goal[15] is to determine the neutrino mass hierarchy using reactor antineutrino oscillations. The layout of JUNO is shown in Fig. 1, where the candidate site is located at Jiangmen in South China, and 53 km away from the Taishan and Yangjiang reactor complexes. The overburden for the experimental hall is required to be larger than 700 meters in order to reduce the muon-induced backgrounds.

2. Physics Potential

Because the relative size of two fast oscillation components is different ($|\Delta m_{31}^2| > |\Delta m_{32}^2|$ or $|\Delta m_{31}^2| < |\Delta m_{32}^2|$), the interference between the two oscillation frequencies in the reactor antineutrino energy spectrum gives us discrimination ability of two different MHs (normal or inverted). The discrimination power is maximized when the Δm_{21}^2 oscillation is maximal (see Fig. 1 of Ref. [15]).

To calculate the sensitivity of MH determination at JUNO, we assume the following nominal setups. A LS detector of 20 kton is placed 53 km away from the Taishan and Yangjian reactor complexes. The detailed distance and power distribution of reactor cores summarized in Table 1 of Ref. [15] is used to include the reduction effect of baseline difference. In the simulation, we use nominal running time of six years, 300 effective days per year, and a detector energy resolution $3\%/\sqrt{E(\text{MeV})}$ as a benchmark. A normal MH is assumed to be the true one while the conclusion won't be changed for the other assumption. The relevant oscillation parameters are taken from the latest global analysis.[17] To illustrate the effect of energy non-linearity and the power of self-calibration, we assume a residual non-linearity curve parametrized in Fig. 3 of Ref. [15] and a testing polynomial non-linearity function with 100% uncertainties for the coefficients. Taking into account all above factors in the least squares method, we can get the MH sensitivity as shown in Fig. 2, where the discriminator is defined as

$$\Delta\chi_{\text{MH}}^2 = |\chi_{\text{min}}^2(\text{Normal}) - \chi_{\text{min}}^2(\text{Inverted})|, \tag{1}$$

and Δm_{ee}^2 and $\Delta m_{\mu\mu}^2$ are the effective mass-squared differences[18] in the electron and muon neutrino disappearance experiments, respectively. From the figure we can learn that a confidence level of $\Delta\chi^2 \simeq 11$ is achieved for the reactor-only analysis, and it will increase to $\Delta\chi^2 \simeq 19$ by using a prior measurement of $\Delta m_{\mu\mu}^2$ (1%).

Fig. 1. Layout of the JUNO experimental design. Current site makes use of the Taishan and Yangjiang reactor complexes, where the previous site is not considered because the third reactor complex (Lufeng) is being planed.

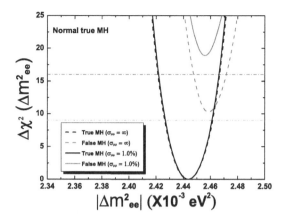

Fig. 2. MH sensitivity of JUNO using reactor antineutrino oscillations. The vertical difference of the curves for the true and false MHs is the discriminator defined in Eq. (1). The solid and dashed lines are for the analyses with and without the prior measurement of $\Delta m^2_{\mu\mu}$.

Other important goals of JUNO include the precision measurement of oscillation parameters and unitarity test, observation of supernova neutrinos, geo-neutrinos, solar neutrinos and atmospheric neutrinos, and so on. Using reactor antineutrino oscillations, we can measure three of the oscillation parameters (i.e. $\sin^2 \theta_{12}$, Δm^2_{21} and $|\Delta m^2_{31}|$) better than 1%.

3. Design Concepts

The design of the central detectors is still open for different options. One basic option is shown in Fig. 3, where the concept of three separated layers is used for better radioactivity protection and muon tagging. The inner acylic tank contains 20 kton linear alkylbenzene (LAB) based LS as the antineutrino targets. 15,000 20-inch photomultiplier tubes (PMTs) are installed in the internal surface of the outer stainless steel tank. 6 kton mineral oil is filled between the inner and outer tanks as buffer of radioactivities. 10 kton high-purity water is filled outside the stainless steel tank. It serves as a water Cherenkov detector after being mounted with PMTs. Other design concept contains the balloon option, single tank option, PMTs module option and mixtures among them. The energy resolution, radioactivity level and technical challenges are the main concerns of different options.

To obtain an unprecedent energy resolution level of $3\%/\sqrt{E(\mathrm{MeV})}$ (or 1,200 photon electrons per MeV) is a big challenge for a LS detector of 20 kton. Much better performance for PMTs and LS is required. R&D efforts to overcome the above challenges are being developed within the JUNO working groups. A new type[19] of low-cost high-efficiency PMTs is being designed, which uses the micro channel

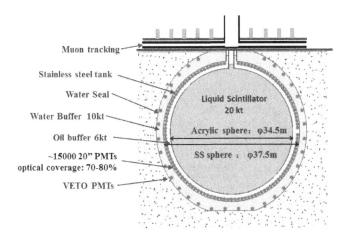

Fig. 3.   A basic design of the JUNO central detector (see the context for details).

plate (MCP) as the dynode and receives both the transmission and reflection light using the reflection photocathode. A coverage level of 80% can be realized with a careful consideration of the PMTs spacing and arrangement. Moreover, highly transparent LS with longer attenuation length (>30 m) is also being developed. Both the method of LS purification by using Al_2O_3 and the distillation facility and the method to increase the light yield are considered.

4. Conclusion

JUNO is designed to determine the neutrino MH ($3\sigma - 4\sigma$ for six years) and measure three of the oscillation parameters better than 1% using reactor antineutrino oscillations. It can also detect the neutrino sources from astrophysics and geophysics. It has strong physics potential, meanwhile contains significant technical challenges. This program is supported from the Chinese Academy of Sciences and planed to be in operation in 2020.

References

1. Daya Bay Collaboration, (F. P. An *et al.*), *Phys. Rev. Lett.* **108**, 171803 (2012).
2. Daya Bay Collaboration, (F. P. An *et al.*), *Chin. Phys. C* **37**, 011001 (2013).
3. Daya Bay Collaboration, (F. P. An *et al.*), *Phys. Rev. Lett.* **112**, 061801 (2014).
4. Double Chooz Collaboration, (Y. Abe *et al.*), *Phys. Rev. Lett.* **108**, 131801 (2012).
5. RENO Collaboration, (J. K. Ahn *et al.*), *Phys. Rev. Lett.* **108**, 191802 (2012).
6. T2K Collaboration, (K. Abe *et al.*), *Phys. Rev. Lett.* **107**, 041801 (2011).
7. MINOS Collaboration, (P. Adamson *et al.*), *Phys. Rev. Lett.* **107**, 181802 (2011).
8. HyperK Collaboration, (K. Abe *et al.*), arXiv:1109.3262.
9. LBNE Collaboration, (T. Akiri *et al.*), arXiv:1110.6249.
10. S. Bertolucci *et al.*, arXiv:1208.0512.
11. A. Samanta, *Phys. Lett. B* **673**, 37 (2009).
12. D. J. Koskinen, *Mod. Phys. Lett. A* **26**, 2899 (2011).

13. L. Zhan, Y. Wang, J. Cao and L. Wen, *Phys. Rev. D* **78**, 111103 (2008).
14. L. Zhan, Y. Wang, J. Cao and L. Wen, *Phys. Rev. D* **79**, 073007 (2009).
15. Y. F. Li, J. Cao, Y. Wang and L. Zhan, *Phys.Rev. D* **88**, 013008 (2013).
16. S. B. Kim, Proposal for RENO-50: detector design and goals, International Workshop on "RENO-50" toward Neutrino Mass Hierarchy, Seoul, June 13-14, (2013).
17. G. L. Fogli *et al.*, *Phys. Rev. D* **86**, 013012 (2012).
18. H. Nunokawa, S. Parke and R. Z. Funchal, *Phys. Rev. D* **72**, 013009 (2005).
19. Y. Wang *et al.*, *Nucl. Inst. Meth. A* **695**, 113 (2012).

Physics in Collision (PIC 2013)
International Journal of Modern Physics: Conference Series
Vol. 31 (2014) 1460301 (5 pages)
© The Author
DOI: 10.1142/S2010194514603019

World Scientific
www.worldscientific.com

Highest energy particle physics with the Pierre Auger Observatory

Analisa Mariazzi

IFLP, Universidad Nacional de La Plata and CONICET,
La Plata(1900), Argentina
mariazzi@fisica.unlp.edu.ar

For the Pierre Auger Collaboration*
Observatorio Pierre Auger, Av. San Martin Norte 304,
5613 Malargüe, Argentina
auger_spokespersons@fnal.gov

Published 15 May 2014

Astroparticles offer a new path for research in the field of particle physics, allowing investigations at energies above those accesible with accelerators. Ultra-high energy cosmic rays can be studied via the observation of the showers they generate in the atmosphere. The Pierre Auger Observatory is a hybrid detector for ultra-high energy cosmic rays, combining two complementary measurement techniques used by previous experiments, to get the best possible measurements of these air showers. Shower observations enable one to not only estimate the energy, direction and most probable mass of the primary cosmic particles but also to obtain some information about the properties of their hadronic interactions. Results that are most relevant in the context of determining hadronic interaction characteristics at ultra-high energies will be presented.

Keywords: Cosmic rays; extensive air showers; hadronic interactions.

PACS Numbers: 96.50.S−, 96.50.sd, 96.50.sb

1. Introduction

Cosmic rays from astrophysical sources provide a natural beam of ultra high energy particles that can be used to probe particle interactions at the highest energies.

The interpretation of cosmic ray measurements requires modeling of hadronic interactions in an energy range beyond that which can be studied in accelerator experiments. The knowledge of the relevant properties of hadronic interactions in

*Full author list: http://www.auger.org/archive/authors_2013_05.html

this energy range is therefore of central importance for the interpretation of the cosmic ray data. Nevertheless, it is in principle possible to obtain information about hadronic interactions from the cosmic ray observations, but dealing with the fact that this natural cosmic ray beam has an unknown energy spectrum and an unknown mass composition. The mass composition of the primary particles must be estimated from the same data set. Solving the ambiguity between composition and hadronic interaction modeling is a key problem for ultra-high energy cosmic ray observations.

The Pierre Auger Observatory, the worlds largest cosmic ray observatory, is located near Malargüe, in the Province of Mendoza, Argentina. It was designed to investigate the origin and the nature of ultra-high energy cosmic rays by taking advantage of two available techniques to detect extensive air showers initiated by ultra-high energy cosmic rays: a surface detector (SD) array and a fluorescence detector (FD). The SD consists of an array of about 1600 water-Cherenkov surface detectors deployed over a triangular grid of 1.5 km spacing and covering an area of 3000 km^2. The SD is overlooked by 27 fluorescence telescopes, grouped in four sites, making up the fluorescence detector. The FD observes the longitudinal development of the shower in the atmosphere by detecting the fluorescence light emitted by excited nitrogen molecules and Cherenkov light induced by shower particles in air. The FD provides a calorimetric measurement of the primary particle energy.

These two detection methods are complementary, so that combining them in hybrid mode will help resolve mass composition and hadronic interaction information.

2. Proton-Air Cross Section from Air Showers

The tail of the X_{max} (depth at which the shower reaches its maximum size) distributions, is shown to be highly sensitive to cross section. Considering only the most deeply penetrating air showers enhances the proportion of protons in a sample as the average depth of shower maximum is higher in the atmosphere for heavier primaries. An unbiased sample of deep X_{max} events in the energy interval between 10^{18} to $10^{18.5}$ eV was used to measure the cross section. In Fig. 1 (left) the result

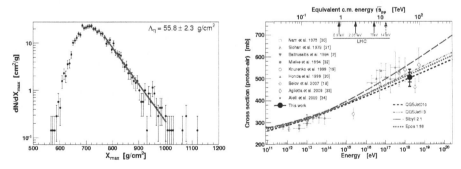

Fig. 1. (left)-Fit to the tail of the X_{max} distribution. (right)- Resulting inelastic proton-air cross-section compared to other measurements and model predictions.

of an unbinned maximum likelihood fit of an exponential function to the the tail of the X_{max} distribution is shown. To properly account for shower fluctuations and detector effects, the exponential tail is compared to Monte Carlo predictions. Any disagreement between data and predictions can then be attributed to a modified value of the proton-air cross section.[1] The result, after averaging the values of the cross section for different hadronic interaction models, yields $\sigma_{p\,air} = 505$ (± 22 stat) ($+20 - 15$ syst) mb at a center-of-mass energy of 57 ± 6 TeV. The systematic uncertainty is dominated by mass composition(mainly the Helium fraction) and hadronic interaction models. The result, shown in Fig. 1 (right), favors a moderately slow rise of the cross-section towards higher energies, as observed at the LHC.[5]

3. Longitudinal Profile and Ground Signal Missmatch: Muon Content

Discrepancies were found in showers measured in hybrid mode, when measurements of the longitudinal shower profile were compared to the lateral particle distributions at ground level.[2] For each shower, Monte Carlo simulated events with similar energies were generated selecting those matching the measured longitudinal profile. When the predicted lateral distributions of the signal are compared to the data recorded by the SD, the Monte Carlo predictions are found to be systematically below the observed signals, regardless of the hadronic model being used, and for all composition mixes that fit the X_{max} distribution of the data sample (see Fig. 2).

Independent methods of extracting muon content from very inclined showers[3] or relying on the different signal shape produced by muons in the water-Cherenkov detectors,[4] together with this analysis of hybrid events, indicate that there is a deficit of muons in the simulations when current hadronic interaction models are used, unless a pure iron primary composition is assumed. A pure iron composition would be in contradiction to the X_{max} data when interpreted using the same models, leading to the conclusion that shower models do not correctly describe the muonic

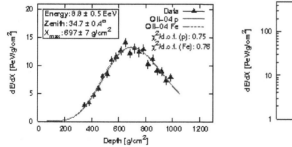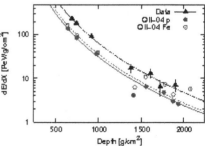

Fig. 2. (left)-Measured longitudinal profile of a typical air shower with two of its matching simulated air showers, for a proton and an iron primary, simulated using QGSJETII-04. (right)- The observed simulated ground signals for the same event.

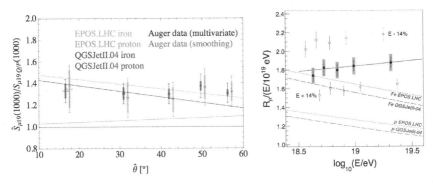

Fig. 3. (left) The measured muon signal at 1000 m from the shower axis vs. zenith angle, with respect to QGSJETII-04 proton at 10^{19} eV as baseline, obtained by different methods. (right) Number of muons from inclined showers compared to model predictions, as a function of energy.

ground signals. The muon content can be obtained from a SD signal time structure as muon signals stand above the smooth electromagnetic component. Results using the multivariate and smoothing methods are shown in Fig 3 (left). Above a zenith angle of 62°, muons dominate the recorded signals at ground level, as the electromagnetic component is absorbed in the atmosphere. A direct measurement of the muon content from very inclined showers is shown in Fig. 3 (right).

4. Conclusions

The Pierre Auger Observatory can be used to test the properties of hadronic interactions by comparing simulations with data for independent observables.

The proton-air inelastic cross section was extracted from measurements of the longitudinal development of air showers at a centre of mass energy per nucleon of 57 TeV, well above energies of accelerator experiments.

Deviations were found if the FD longitudinal profile and SD signals are compared in hybrid events. Independent methods using the SD indicate that this is due to a significant muon deficit in the predictions. A realistic treatment of the mass composition does not remove the muon discrepancy. Multiple methods reach the same conclusion: current hadronic interaction models do not accurately describe the muon signal.

References

1. Pierre Auger Collab.(P. Abreu *et al.*), *Phys. Rev. Lett.* **109**, 062002 (2012).
2. G.Farrar for the Pierre Auger Collab.(P. Abreu *et al.*), The muon content of hybrid events recorded at the Pierre Auger Observatory, in *Proceedings of the 33rd International Cosmic Ray Conference*, arXiv:1307.5059, p. 44.
3. I. Valino for the Pierre Auger Collab.(P. Abreu *et al.*), A measurement of the muon number on showers using inclined events recorded at the Pierre Auger Observatory, in *Proceedings of the 33rd International Cosmic Ray Conference*, arXiv:1307.5059, p. 52.

4. B. Kegl for the Pierre Auger Collab.(P. Abreu *et al.*), Measurement of the muon signal using the temporal and spectral structure of the signals in surface detectors of the Pierre Auger Observatory, in *Proceedings of the 33rd International Cosmic Ray Conference*, arXiv:1307.5059, p. 48.

5. TOTEM Collab.(T. Csörgö *et al.*), *Prog. Theor. Phys. Suppl.* **193**, 180 (2012).

Physics in Collision (PIC 2013)
International Journal of Modern Physics: Conference Series
Vol. 31 (2014) 1460302 (4 pages)
© The Author
DOI: 10.1142/S2010194514603020

Search for muon to electron conversion at J-PARC: COMET experiment

Y. Yuan

(On behalf of the COMET collaboration)

Experimental Physics Center, Institute of High Energy Physics,
19B YuquanLu, Shijingshan District, Beijing, 100049, P.R.China
yuany@ihep.ac.cn

Published 15 May 2014

The COherent Muon to Electron Transition (COMET) experiment aims to search for the charged-lepton-flavor-violating process through measure muon to electron conversion in a muonic atom to very high sensitivity of 2.6×10^{-17}. A two-stage approach in order to realize the experiment has been taken and the first-stage(COMET Phase-I) has been approved by KEK.

Keywords: COMET; CLFV; $\mu - e$ conversion; Phase-I.

1. Introduction

Charged lepton flavor violation(CLFV) is strictly suppressed in minimal Standard Model(SM) and has never been observed. Simple extension of SM with tiny neutrino masses predicted a very small rate of CLFV that can't be experimental reached. However, many new physics models that go beyond the SM predict a branching ratio that may be observed by experiment, $10^{-13} - 10^{-15}$ from supersymmetric theories for example. Therefor, any observation of CLFV would indicate a clear signal of physics beyond the SM with massive neutrinos

The J-PARC E21 experiment which is called COherent Muon to Electron Transition(COMET) is an experiment aims to search for a CLFV process of neutrinoless muon-to-electron conversion($\mu - e$ conversion) in a muonic atom: $\mu^- + N(A, Z) \rightarrow e^- + N(A, Z)$. The current upper limit of the $\mu - e$ conversion process was determined by SINDRUM-II[1] at PSI which is 7×10^{-13} using a gold target. The anticipated sensitivity goal of the COMET experiment is a factor of 10 000 better than that to a single-event sensitivity(SES) of 2.6×10^{-17} (or $< 6 \times 10^{-17}$ @ 90% confidence level upper limit).

In recently, a two-stage approach in order to realize the experiment has been taken. The original full-sized COMET experiment(now called COMET Phase-II) be treated as second-stage. The first-stage COMET Phase-I aims at an intermediate SES of 3×10^{-15} and intended to take measurements of potential background sources for COMET Phase-II. COMET Phase-I had been approved in January, 2013 by KEK. Design and R&D is well underway, construction has already begun.

2. $\mu^- N \to e^- N$ Conversion Experiment

In physics beyond the Standard Model, the lepton flavor violation process of neutrinoless muon capture, such as $\mu^- + N(A, Z) \to e^- + N(A, Z)$ is expected to happen when a negative muon is stopped in material and form a muonic atom. This process is called $\mu - e$ conversion in a muonic atom.

The event signature of coherent $\mu - e$ conversion in a muonic atom is a monoenergetic single electron, approximately 105 MeV in aluminium. It makes $\mu - e$ conversion a very attractive process because the energy is far above the end-point energy of the muon decay spectrum(~52.8 MeV). Futhermore, the search for this process can improve sensitivity by using a high muon rate without suffering from accidental background events.

3. COMET Phase-II and Phase-I

Layout of the experimental setup of COMET Phase-II is shown in Fig 1. An 8 GeV, pulsed proton beam hit target and produce pions. The backward low momentum pions are then collected. The first C-shaped solenoid will transfer and select momentum of pions and muons. The low momentum muons will stop in the stopping target

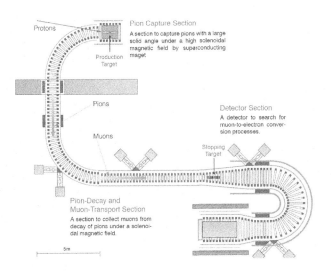

Fig. 1.   Layout of COMET phase-II.

and form muonic atoms. Another C-shaped electron spectrometer will transport and select high energy electrons of momenta above 70 MeV/c . The energy and momentum of the electrons are measured in the detector which consists of a straw tube tracker and an electron calorimeter.

To improve the sensitivity some important features have been considered:

- *Highly intense muon source:* A proton beam of high beam power as well as high efficiency pion capture system proposed by muon collider and neutrino factory R&D which has been experimentally demonstrated by MuSIC[2] are adopted.
- *Pulsed proton beam:* Muons in muonic atoms have lifetimes of the order of 1μsec, a pulsed beam with short width would allow removal of prompt beam background events by allowing measurements to be performed in a delayed time window.
- *Muon transport system with curved solenoids:* The centers of the helical motion of charged particles drift perpendicular to the plane in curved solenoids proportional to their momentum. This effect can be used to eliminate beam particles of high momenta which would produce electron background events.
- *Spectrometer with curved solenoids:* The same principle of momentum selection is used in electron spectrometer to allow selection of electrons and suppress the detection hit rates.

The proton beam line and pion production system of COMET Phase-I is identical to Phase-II, but the muon beamline will only bends to the end of the first 90 degree. The stopping target will be moved to just after the end of the beam pipe. For the $\mu - e$ conversion search, a cylindrical drift chamber will be used.

4. Proton Beam

The main proton pulses occur every 1.1 μsec with a beam width of 100 ns. After a short time muon beam hits the stopping target and a prompt background is emitted. Also at this time, the muonic atoms form and start to decay. Data will only be recorded in a certain time window in order to reduce background and maintain a good sensitivity to the $\mu - e$ conversion process.

The proton beam that will be used is the J-PARC Main Ring (MR) beam and slow extracted. Only half of the buckets will be filled and a novel injection method using single bunch kick injection will be used to fullfill the proton extinction requirement. Early tests have shown an extinction factor of 3×10^{-11} can be expected.

5. Pion Production

Pion capture system consists of a proton target, a surrounding tungsten radiation shield, a superconducting solenoid magnet with 5 T magnetic field, and a matching section connected to the transport solenoid system with a 3 T field. COMET will run at a beam power of 3.2 kW for Phase-I which will be upgraded for Phase-II to 56 kW. The pion capture system needs to be designed to operate at the higher beam power of Phase-II.

The pion production target for Phase-I will be a radiation cooled 60 cm long piece of graphite with a radius of 2 cm. For Phase-II the plan is to use a tungsten target since this will give us a higher pion yield and need to be helium cooled.

6. Muon Transport and Stopping

In Phase-I, a collimator will be installed at the end of the first 90 degrees in order to remove pions and high momentum muons since there will be a larger contamination in the beam than in Phase-II.

The baseline design of the muon stopping target for both Phase-I and Phase-II is 17 aluminium disks. A double cone geometry as used in SINDRUM has also been evaluated.

7. Cylindrical Drift Chamber

The requirements for the cylindrical drift chamber are that it must have a gas gain of greater than 105, a position resolution in the x and y directions of less than 150 μm, a position resolution in the z direction of less than 2mm.

The current design of the cylindrical drift chamber is to have a 1.5m long chamber covering the radius 50 cm to 80 cm. The drift gas used will be a 90:10 mix of helium and butane. There will also be Cherenkov hodoscopes at the front and back of the drift chamber to act as triggers . Finally, there will also be a 1 T magnetic field applied to the whole section.

8. Phase-II Detector

The design of the straws in the straw tube tracker will be based on those developed at JINR for the NA62 experiment with some modifications. A prototype straw tube tracker will be build for testing before the design is finalised.

Currently, there are two candidate crystals, GSO and LYSO that are under consideration. A beam test has been performed at J-PARC to test both crystals and decide which will be used in the final detector. Additional beam tests at Tohoku University and BINP to further test and design the electron calorimeter.

9. Summary

The COMET experiment is going to seach for the CLFV 4 order of magnitude better than curren upper limit. The first-stage approach which is called COMET Phase-I has been approved. The fund is secured and the construction has begun. It is expected to take measurement in 2016 and reach a single event sensitivity of 3×10^{-15} or better.

References

1. SINDRUM II Collab. (W. Burtl *et al.*), *Eur. Phys. J. C* **47**, 337 (2006).
2. MuSIC Collab. *The MuSIC project under the center of excellence of sub atomic physics*(2010) (Available at: http://133.1.141.121/~sato/music/doc/MuSIC.pdf).

Physics in Collision (PIC 2013)
International Journal of Modern Physics: Conference Series
Vol. 31 (2014) 1460303 (4 pages)
© The Author
DOI: 10.1142/S2010194514603032

Search for the Higgs Boson decaying into tau pairs

Jakob Salfeld-Nebgen

Deutsches Elektronen-Synchrotron,
Notkestrasse 85, 22607 Hamburg, Germany
jakob.salfeld@cern.ch

Published 15 May 2014

A search for the Standard Model Higgs Boson decaying into τ pairs is performed using events recorded by the CMS experiment at the LHC in 2011 and 2012. An excess of events is observed over a broad range of Higgs mass hypotheses, with a maximum local significance of 2.93 standard deviations at $m_H = 120$ GeV. The excess is compatible with the presence of a standard-model Higgs boson of mass 125 GeV/c^2.

1. Introduction

We present a search for the Standard Model Higgs Boson decaying into tau-pairs in the invariant mass region 110-145 GeV/c^2 performed using events recorded by the CMS experiment at the LHC in 2011 and 2012 corresponding to an integrated luminosity of 4.9 fb^{-1} at a centre-of-mass energy of 7 TeV and 19.4 fb^{-1} at 8 TeV 1. Since July 4th 2012 an experimental observation of a new Boson of a mass around 125 GeV and compatible with the Standard Model Higgs Boson hypothesis is established both by the Atlas and CMS collaborations Ref. 2, 3, 4. The observation of the Higgs-like Boson is predominantly based on its couplings to gauge bosons. To seek direct evidence for its fermionic couplings, the search for the Standard Model Higgs Boson decaying into tau-pairs is a distinguished channel to be analyzed with the CMS detector.

2. Analysis Strategy

The search is simultaneously performed in several final states of the decaying di-tau system, henceforth denoted as: $\tau_h\tau_h$, $e\tau_h$, $\mu\tau_h$, $e\mu$ and $\mu\mu$, where τ_h declares the hadronically decaying tau final states. For each final state the events are then divided into several disjoint event categories in order to enhance the overall sensetivity by exploiting the distinct event topologies of the Higgs production processes.

Major Higgs production mechanisms considered are processes via gluon fusion, vector boson fusion and the vector boson associated production. The dedicated $H \to \tau\tau$ analyses exploiting the associated Higgs productions (see Ref. 5) are not discussed in detail in this review, however are included in the results in Fig. 3 and 2.

The most sensitive category is the Vector Boson Fusion (VBF) category where the event topology of Standard Model Higgs production via Vector Boson Fusion is exploited which is characterized by two high p_T jets in the forward regions of the detector. In addition to the event and lepton selection criteria events in the VBF category are required to have two jets with $p_T > 30$ GeV/c and $|\eta| > 4.7$, having an invariant mass of $m_{jj} > 500$ GeV/c^2, with a seperation in the pseudo-rapdity $\Delta\eta_{jj} > 3.5$ and no addtional jet with $p_T > 30$ GeV/c in the eta gap between the two corresponding jets.

In addition a 1-Jet category is defined. Events with at least one jet with $p_T > 30$ GeV/c and $|\eta| > 4.7$ are selected in this category, exluding events of the VBF category. The Standard Model Higgs production via gluon fusion dominates the signal contribution in the 1-Jet category.

Further the 0-Jet category is defined for events with no jet with $p_T > 30$ GeV/c and $|\eta| > 4.7$. This category is largely dominated by background processes and is used to constrain the backgrounds in the VBF and 1-Jet categories.

For all categories a b-tagged jet with $p_T > 20$ GeV/c veto is applied to decouple the analysis from the MSSM Higgs search in the di-tau channel.

The background compositions are channel and category dependent with the $Z\gamma* \to \tau\tau$ processes representing typically the most dominant contributions over $t\bar{t}$, W+Jets, $Z \to \ell\ell$, QCD multijet-events and di-boson productions. To further reduce the various backgrounds additional topological event selection criteria depend on the final state are defined.

For the $e\mu$ final state the $t\bar{t}$ background is significantly reduced with selection criteria on $D_\zeta = p_\zeta - 0.85 \cdot p_\zeta^{vis} > -20$ GeV/c, where ζ is the bisector of the two leptons, $p_\zeta = \vec{p}_{T,1} \cdot \hat{\zeta} + \vec{p}_{T,2} \cdot \hat{\zeta} + \vec{E}_T^{miss} \cdot \hat{\zeta}$ and $p_\zeta^{vis} = \vec{p}_{T,1} \cdot \hat{\zeta} + \vec{p}_{T,2} \cdot \hat{\zeta}$. D_ζ is a

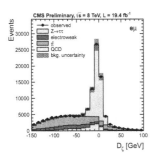 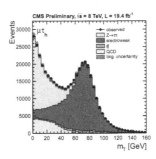 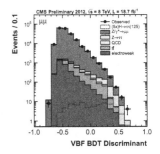

Fig. 1. Left: Distribution of the P_ζ variable in the $e\mu$ channel. Middle: Distribution of the m_T variable in the $\mu\tau_h$ channel. Right: Distribution of the BDT discriminant in $\mu\mu$ channel as an example in the VBF category.

measure of how collinear the missing transverse energy vector is with the ransverse momentum of the di-lepton system (see Fig. 1 (left)).

For the $e\tau_h$ and $\mu\tau_h$ final states the W+Jets background is reduced by selection criteria on the transverse mass variable $m_T = \sqrt{2p_T E_T^{miss}(1 - cos(\Delta\phi))}$ (see Fig. 1 (middle)).

For the $\mu\mu$ final state the overwhelming $Z/\gamma* \to \mu\mu$ background is reduced by selection criteria on dedicated multivariate Boosted Decision Trees in each category. (see Fig. 1 (right)).

3. Results

The observed and expected 95% CL exclusion limit on the Standard Model Higgs Boson production rate is derived by a combined fit on the reconstructed di-tau mass $m_{\tau\tau}$ in each category for each channel and shown in Fig. 2 (right). For the $\mu\mu$ final state the 2 dimensional distribution of $m_{\tau\tau}$ versus the invariant mass of the two

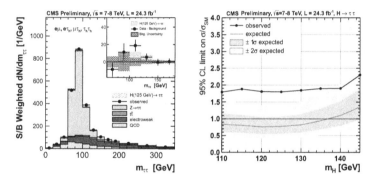

Fig. 2. Left: Combined $m_{\tau\tau}$ distribution of $\tau_h\tau_h$, $e\tau_h$, $\mu\tau_h$ and $e\mu$ final states weighted by the ratio of the expected signal and background yields in each category. Right: Observed and expected 95% CL exclusion limit on the Standard Model Higgs Boson production rate

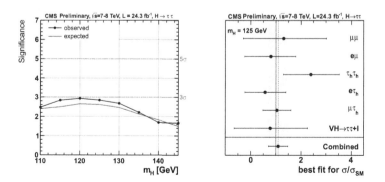

Fig. 3. Observed and expected p-values as a function of m_H. Right: Best-fit signal strength at $m_H = 125$ GeV/c relative to the standard model expectation.

muons is used. Figure 2 (left) shows the combined $m_{\tau\tau}$ distribution obtained after summing up the $m_{\tau\tau}$ distribution of each category of the $\tau_h\tau_h$, $e\tau_h$, $\mu\tau_h$ and $e\mu$ final states weighted by the ratio of the expected signal and background yields.

An excess of the observed limit when compared to the background-only hypothesis is found and the local p-value and signifiacance of this excess as a function of the Higgs mass is shown in Fig. 3 (left). A maximum significance of 2.93 standard deviations is observed for a Higgs mass of $m_H = 120$ GeV/c^2 and a significance of 2.85σ at $m_H = 125$ GeV/c^2 compared to the Standard Model expectation of 2.63σ. The signal strength for m_H is measured to be 1.1 ± 0.4 times the Standard Model Higgs production rate (see Fig. 3 (right))and thus the excess is compatible with the hypothesis of the presence of the Standard Model Higgs Boson.

References

1. CMS Collaboration, *Search for the Standard-Model Higgs boson decaying to tau pairs in proton-proton collisions at sqrt(s) = 7 and 8 TeV*, CMS-PAS-HIG-13-004, 2–13.
2. CMS Collaboration, *Observation of a new boson at a mass of 125 GeV with the CMS experiment at the LHC*, Phys. Lett. B **716** (2012) 30–61, doi:10.1016/j.physletb.2012.08.021, arXiv:1207.7235.
3. Atlas Collaboration, *Observation of a new particle in the search for the Standard Model Higgs boson with the ATLAS detector at the LHC*, Phys. Lett. B **716** (2012) 1–29, doi:10.1016/j.physletb.2012.08.020, arXiv:1207.7214.
4. CMS Collaboration, *Combination of standard model Higgs boson searches and measurements of the properties of the new boson with a mass near 125 GeV*, CMS-PAS-HIG-13-005, 2013.
5. CMS Collaboration, *Search for the standard model Higgs boson decaying to tau pairs produced in association with a W or Z boson*, CMS-PAS-HIG-12-053, 2013.

Physics in Collision (PIC 2013)
International Journal of Modern Physics: Conference Series
Vol. 31 (2014) 1460304 (5 pages)
© The Author
DOI: 10.1142/S2010194514603044

Test of QCD at large Q^2 with exclusive hadronic processes

Chengping Shen

(For the Belle Collaboration)

School of Physics and Nuclear Energy Engineering,
Beihang University, Beijing, 100191, China
shencp@ihep.ac.cn

Published 15 May 2014

My report consists of two parts: (1). Using samples of 102 million $\Upsilon(1S)$ and 158 million $\Upsilon(2S)$ events at Belle, we study 17 exclusive hadronic decays of these two bottomonium resonances to some Vector-Pseudoscalar (VP), Vector-Tensor (VT) and Axial-vector-Pseudoscalar (AP) processes and their final states. Branching fractions are measured for all the processes. The ratios of the branching fractions of $\Upsilon(2S)$ and $\Upsilon(1S)$ decays into the same final state are used to test a perturbative QCD (pQCD) prediction for OZI-suppressed bottomonium decays. (2). Using data samples of 89 fb^{-1}, 703 fb^{-1}, and 121 fb^{-1} collected at center-of-mass (CMS) energies 10.52, 10.58, and 10.876 GeV, respectively, we measure the cross sections of $e^+e^- \to \omega\pi^0$, $K^*(892)\bar{K}$, and $K_2^*(1430)\bar{K}$. The energy dependence of the cross sections is presented.

Keywords: $\Upsilon(1S)$; $\Upsilon(2S)$; hadronic decays; cross sections.

PACS Numbers: 13.25.Gv, 14.40.Pq, 13.66.Bc, 13.40.Gp

1. Measurement of $\Upsilon(1S)$ and $\Upsilon(2S)$ Decays into VP Final States

We know for the OZI (Okubo-Zweig-Iizuka) suppressed decays of the J/ψ and $\psi(2S)$ to hadrons proceed via the annihilation of the quark-antiquark pair into three gluons or a photon, pQCD provides a relation for the ratios of branching fractions ($\mathcal{B}$) for J/ψ and $\psi(2S)$ decays

$$Q_\psi = \frac{\mathcal{B}_{\psi(2S) \to \text{hadrons}}}{\mathcal{B}_{J/\psi \to \text{hadrons}}} = \frac{\mathcal{B}_{\psi(2S) \to e^+e^-}}{\mathcal{B}_{J/\psi \to e^+e^-}} \approx 12\%, \tag{1}$$

which is referred to as the "12% rule" and is expected to apply with reasonable accuracy to both inclusive and exclusive decays. However, it is found to be severely violated for $\rho\pi$ and other VP and VT final states.

A similar rule can be derived for OZI-suppressed bottomonium decays:

$$Q_\Upsilon = \frac{\mathcal{B}_{\Upsilon(2S)\to\text{hadrons}}}{\mathcal{B}_{\Upsilon(1S)\to\text{hadrons}}} = \frac{\mathcal{B}_{\Upsilon(2S)\to e^+e^-}}{\mathcal{B}_{\Upsilon(1S)\to e^+e^-}} = 0.77 \pm 0.07. \qquad (2)$$

Recently, using 102 million $\Upsilon(1S)$ and 158 million $\Upsilon(2S)$ events Belle studied exclusive hadronic decays of these two bottomonium resonances to the three-body final states $\phi K^+ K^-$, $\omega\pi^+\pi^-$ and $K^{*0}(892)K^-\pi^+$,[1] and to the two-body VT states $(\phi f_2'(1525)$, $\omega f_2(1270)$, $\rho a_2(1320)$ and $K^{*0}(892)\bar{K}_2^{*0}(1430))$ and AP $(K_1(1270)^+K^-$, $K_1(1400)^+K^-$ and $b_1(1235)^+\pi^-)$ states.[2] Signals are observed for the first time in the $\Upsilon(1S) \to \phi K^+ K^-$, $\omega\pi^+\pi^-$, $K^{*0}K^-\pi^+$, $K^{*0}K_2^{*0}$ and $\Upsilon(2S) \to \phi K^+ K^-$, $K^{*0}K^-\pi^+$ decay modes. Besides $K^{*0}K_2^{*0}$, no other two-body processes are observed in all investigated final states. For the processes $\phi K^+ K^-$, $K^{*0}K^-\pi^+$, and $K^{*0}\bar{K}_2^{*0}(1430)$, the Q_Υ ratios are consistent with the expected value, while for $\omega\pi^+\pi^-$, the measured Q_Υ ratio is 2.6σ below the pQCD expectation. The results for the other modes are inconclusive due to low statistical significance.

We also used the same data samples of $\Upsilon(1S)$ and $\Upsilon(2S)$ to study exclusive hadronic decays to the $K_S^0 K^+\pi^-$, $\pi^+\pi^-\pi^0\pi^0$, and $\pi^+\pi^-\pi^0$, and two-body VP $(K^*(892)^0\bar{K}^0$, $K^*(892)^-K^+$, $\omega\pi^0$, and $\rho\pi)$ final states.[3] After event selections, Fig. 2 shows the $K^+\pi^-$ and $K_S^0\pi^-$ invariant mass distributions for the $K_S^0 K^+\pi^-$ final state, the $\pi^+\pi^-\pi^0$ invariant mass distribution for the $\pi^+\pi^-\pi^0\pi^0$ final state, and the $\pi\pi$ invariant mass distribution for the $\pi^+\pi^-\pi^0$ final state. An unbinned simultaneous maximum likelihood fit was done to these mass spectra. The results of the fits are shown in Fig. 2 and listed in Table 1.

Table 1. Results for the $\Upsilon(1S)$ and $\Upsilon(2S)$ decays, where N_{sig} is the number of signal events from the fits, $N_{\text{sig}}^{\text{UL}}$ is the upper limit on the number of signal events, Σ is the statistical significance (σ), $\mathcal{B}$ is the branching fraction (in units of 10^{-6}), $\mathcal{B}^{\text{UL}}$ is the 90% C.L. upper limit on the branching fraction.

Channel	$\Upsilon(1S)$			$\Upsilon(2S)$		
	$N_{\text{sig}}/N_{\text{sig}}^{\text{UL}}$	Σ	$\mathcal{B}/\mathcal{B}^{\text{UL}}$	$N^{\text{sig}}/N_{\text{sig}}^{\text{UL}}$	Σ	$\mathcal{B}/\mathcal{B}^{\text{UL}}$
$K_S^0 K^+\pi^-$	37.2 ± 7.6	6.2	$1.59 \pm 0.33 \pm 0.18$	39.5 ± 10.3	4.0	$1.14 \pm 0.30 \pm 0.13$
$\pi^+\pi^-\pi^0\pi^0$	143.2 ± 22.4	7.1	$12.8 \pm 2.01 \pm 2.27$	260.7 ± 37.2	7.4	$13.0 \pm 1.86 \pm 2.08$
$\pi^+\pi^-\pi^0$	25.5 ± 8.6	3.4	$2.14 \pm 0.72 \pm 0.34$	15	—	0.80
$K^{*0}\bar{K}^0$	16.1 ± 4.7	4.4	$2.92 \pm 0.85 \pm 0.37$	30	2.7	4.22
$K^{*-}K^+$	6.3	1.3	1.11	13	2.0	1.45
$\omega\pi^0$	6.8	1.6	3.90	4.6	0.1	1.63
$\rho\pi$	22	2.2	3.68	14	—	1.16

Signals are observed for the first time in the $\Upsilon(1S) \to K_S^0 K^+\pi^-$, $\pi^+\pi^-\pi^0\pi^0$ and $\Upsilon(2S) \to \pi^+\pi^-\pi^0\pi^0$ decay modes. There is an indication for large isospin-violation between the branching fractions for the charged and neutral $K^*(892)\bar{K}$ for both $\Upsilon(1S)$ and $\Upsilon(2S)$ decays, as in $\psi(2S)$ decays, which indicates that the electromagnetic process plays an important role in these decays. For the processes $K_S^0 K^+\pi^-$ and $\pi^+\pi^-\pi^0\pi^0$, the Q_Υ ratios are consistent with the expected value; for

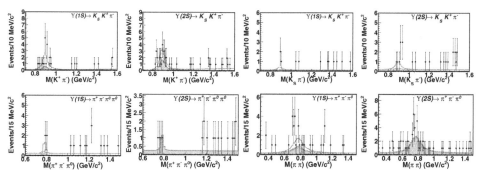

Fig. 1. The fits to the $K^+\pi^-$, $K_S^0\pi^-$, $\pi^+\pi^-\pi^0$ and $\pi\pi$ mass distributions for the $K^*(892)^0$, $K^*(892)^-$, ω and ρ vector meson candidates from $K_S^0 K^+\pi^-$, $\pi^+\pi^-\pi^0\pi^0$ and $\pi^+\pi^-\pi^0$ events from $\Upsilon(1S)$ and $\Upsilon(2S)$ decays. The solid histograms show the results of the simultaneous fits, the dotted curves show the total background estimates, and the shaded histograms are the normalized continuum contributions.

$\pi^+\pi^-\pi^0$, the Q_Υ ratio is a little lower than the pQCD prediction. The results for the other modes are inconclusive due to low statistical significance.

2. Measurement of $e^+e^- \rightarrow \omega\pi^0$, $K^*(892)\bar{K}$ and $K_2^*(1430)\bar{K}$

Different models predict different energy dependence of the cross sections for the process $e^+e^- \rightarrow$ VP. If SU(3) flavor symmetry is perfect, one expects the cross sections of $\omega\pi^0 : K^*(892)^0\bar{K}^0 : K^*(892)^-K^+$ production equal 9:8:2. However, this relation was found to be violated severely at $\sqrt{s} = 3.67$ GeV and 3.773 GeV by the CLEO experiment with the ratio $R_{VP} = \frac{\sigma_B(e^+e^- \rightarrow K^*(892)^0\bar{K}^0)}{\sigma_B(e^+e^- \rightarrow K^*(892)^-K^+)}$ greater than 9 and 33 at $\sqrt{s} = 3.67$ GeV and 3.773 GeV, respectively, at the 90% C.L. By taking into account SU(3)$_f$ symmetry breaking, a pQCD calculation predicts $R_{VP} = 6.0$. In the quark model, one may naively expect $R_{TP} = \frac{\sigma_B(e^+e^- \rightarrow K_2^*(1430)^0\bar{K}^0)}{\sigma_B(e^+e^- \rightarrow K_2^*(1430)^-K^+)} = R_{VP}$.

The cross sections of $e^+e^- \rightarrow \omega\pi^0$, $K^*(892)\bar{K}$, and $K_2^*(1430)\bar{K}$ are measured,[4] based on data samples of 89 fb^{-1}, 703 fb^{-1}, and 121 fb^{-1} collected at $\sqrt{s} =10.52$, 10.58 ($\Upsilon(4S)$ peak), and 10.876 GeV ($\Upsilon(5S)$ peak), respectively. After event selections, Fig. 2 shows the $\pi^+\pi^-\pi^0$, $K^+\pi^-$, and $K_S^0\pi^-$ invariant mass distributions for the $\pi^+\pi^-\pi^0\pi^0$ and $K_S^0 K^+\pi^-$ final states. Unbinned maximum likelihood fitted results are listed in Table 2 together with the calculated Born cross sections. Assuming $1/s^n$ dependence, the fit gives $n = 3.83 \pm 0.07$ and 3.75 ± 0.12 for $e^+e^- \rightarrow K^*(892)^0\bar{K}^0$ and $\omega\pi^0$ cross sections distributions. With the calculated Born cross sections, we obtain $R_{VP} > 4.3, 20.0, 5.4$, and $R_{TP} < 1.1, 0.4, 0.6$, for $\sqrt{s} = 10.52$, 10.58, and 10.876 GeV, respectively, at the 90% C.L. Both the ratios are different from the predictions from exact or broken SU(3) symmetry models.

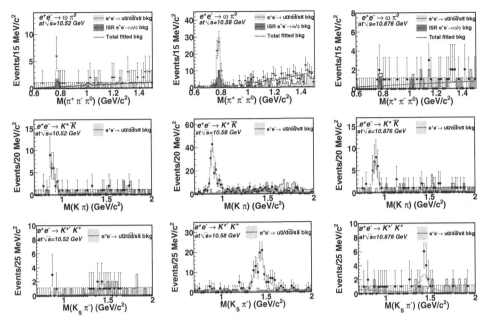

Fig. 2. The fits to the $\pi^+\pi^-\pi^0$ (top row), $K^+\pi^-$ (middle row) and $K_S^0\pi^-$ (bottom row) invariant mass distributions for the ω, $K^*(892)$, and $K_2^*(1430)$ meson candidates from $e^+e^- \to \pi^+\pi^-\pi^0\pi^0$ and $K_S^0K^+\pi^-$ events from the $\sqrt{s} = 10.52$ GeV, 10.58 GeV, and 10.876 GeV data samples. The solid lines show the results of the fits described in the text, the dotted curves show the total background estimates, the dark shaded histograms are from the normalized ISR backgrounds $e^+e^- \to \gamma_{\mathrm{ISR}}\omega/\phi \to \gamma_{\mathrm{ISR}}\pi^+\pi^-\pi^0$ and the light shaded histograms are from the normalized $e^+e^- \to u\bar{u}/d\bar{d}/s\bar{s}$ backgrounds.

Table 2. Results for the Born cross sections, where N_{sig} is the number of fitted signal events, $N_{\mathrm{sig}}^{\mathrm{UL}}$ is the upper limit on the number of signal events, Σ is the signal significance, σ_B is the Born cross section, σ_B^{UL} is the upper limit on the Born cross section. All the upper limits are given at the 90% C.L. The first uncertainty in σ_B is statistical, and the second systematic.

Channel	$\sqrt{s}$ (GeV)	N_{sig}	$N_{\mathrm{sig}}^{\mathrm{UL}}$	$\Sigma\ (\sigma)$	σ_B (fb)	σ_B^{UL} (fb)
$\omega\pi^0$	10.52	$4.1^{+3.3}_{-2.6}$	9.9	1.6	$4.53^{+3.64}_{-2.88} \pm 0.50$	11
	10.58	$38.8^{+8.3}_{-7.6}$	—	6.7	$6.01^{+1.29}_{-1.18} \pm 0.57$	—
	10.876	$-0.7^{+2.9}_{-2.1}$	7.0	—	$-0.68^{+2.71}_{-1.97} \pm 0.20$	6.5
$K^*(892)^0\bar{K}^0$	10.52	$34.6^{+6.9}_{-6.1}$	—	7.4	$10.77^{+2.15}_{-1.90} \pm 0.77$	—
	10.58	187 ± 17	—	>10	$7.48 \pm 0.67 \pm 0.51$	—
	10.876	$34.6^{+7.5}_{-6.7}$	—	7.2	$7.58^{+1.64}_{-1.47} \pm 0.63$	—
$K^*(892)^-K^+$	10.52	$4.6^{+3.6}_{-2.7}$	9.3	1.4	$1.14^{+0.90}_{-0.67} \pm 0.15$	2.3
	10.58	$5.9^{+4.7}_{-3.8}$	14	1.5	$0.18^{+0.14}_{-0.12} \pm 0.02$	0.4
	10.876	$1.6^{+3.9}_{-3.0}$	8.5	0.3	$0.28^{+0.68}_{-0.52} \pm 0.10$	1.5
$K_2^*(1430)^0\bar{K}^0$	10.52	$1.3^{+4.3}_{-3.9}$	6.8	0.3	$0.76^{+2.53}_{-2.26} \pm 0.14$	4.0
	10.58	21^{+11}_{-10}	40	2.1	$1.65^{+0.86}_{-0.78} \pm 0.27$	3.1
	10.876	$1.0^{+4.5}_{-3.7}$	8.9	0.2	$0.38^{+1.79}_{-1.47} \pm 0.07$	3.5
$K_2^*(1430)^-K^+$	10.52	$12.0^{+6.2}_{-5.8}$	21	2.1	$6.06^{+3.13}_{-2.93} \pm 1.34$	11
	10.58	129 ± 15	—	>10	$8.36 \pm 0.95 \pm 0.62$	—
	10.876	$17.6^{+5.3}_{-4.6}$	—	4.5	$6.20^{+1.86}_{-1.63} \pm 0.64$	—

Acknowledgments

This work is supported partly by the Fundamental Research Funds for the Central Universities of China (303236).

References

1. Charge-conjugate decays are implicitly assumed throughout the report.
2. Belle Collab. (C. P. Shen *et al.*), *Phys. Rev. D* **86**, 031102(R) (2012).
3. Belle Collab. (C. P. Shen *et al.*), *Phys. Rev. D* **88**, 011102(R) (2013).
4. Belle Collab. (C. P. Shen *et al.*), *Phys. Rev. D* **88**, 052019 (2013).

Physics in Collision (PIC 2013)
International Journal of Modern Physics: Conference Series
Vol. 31 (2014) 1460305 (5 pages)
© The Author
DOI: 10.1142/S2010194514603056

World Scientific
www.worldscientific.com

Measurement of strong phase in $D^0 \to K\pi$ decay at BESIII

Xiao-Kang Zhou

(On Behalf of the BESIII Collaboration)

University of Chinese Academy of Sicences (UCAS)
Beijing 100049, China
zhouxiaokang08@mails.ucas.ac.cn

Published 15 May 2014

Based on $2.92\,\text{fb}^{-1}$ of e^+e^- collision data collected with the BESIII detector at $\sqrt{s} = 3.773$ GeV, we measured the asymmetry $\mathcal{A}_{CP \to K\pi}$ of the branching fractions of $D^{CP\pm} \to K^-\pi^+$ ($D^{CP\pm}$ are the CP-odd and CP-even eigenstates) to be $(12.77 \pm 1.31^{+0.33}_{-0.31})\%$. $\mathcal{A}_{CP \to K\pi}$ is used to extract the strong phase difference $\delta_{K\pi}$ between the doubly Cabibbo-suppressed process $\overline{D}^0 \to K^-\pi^+$ and Cabibbo-favored $D^0 \to K^-\pi^+$. By taking inputs of other parameters in world measurements, we obtain $\cos\delta_{K\pi} = 1.03 \pm 0.12 \pm 0.04 \pm 0.01$. This is the most accurate result of $\cos\delta_{K\pi}$ to date and can improve the world constrains on the mixing parameters and on γ/ϕ_3 in the CKM matrix.

Keywords: BESIII; D^0-$\overline{D}^0$ mixing; strong phase difference.

1. Introduction

The strong phase difference $\delta_{K\pi}$ between the doubly Cabibbo-suppressed (DCS) decay $\overline{D}^0 \to K^-\pi^+$ and the corresponding Cabibbo-favored (CF) $D^0 \to K^-\pi^+$ is denoted as

$$\frac{\langle K^-\pi^+|\overline{D}^0\rangle}{\langle K^-\pi^+|D^0\rangle} = \frac{\overline{A}_f}{A_f} = -re^{-i\delta_{K\pi}},$$

which plays an important role in measurements of D^0-$\overline{D}^0$ mixing parameters and helps to improve accuracy of the γ/ϕ_3 angle measurement in CKM matrix.

In the limit of CP conservation, $\delta_{K\pi}$ is the same in the final states of $K^-\pi^+$ and $K^+\pi^-$. Here, the notation of $D^0 \to K^-\pi^+$ is used for CF decay and $\overline{D}^0 \to K^-\pi^+$ for DCS, and its charge conjugation mode is always implied to be included.

Using quantum-correlated technique, $\delta_{K\pi}$ can be measured in the mass-threshold production process $e^+e^- \to D^0\overline{D}^0$.[1-4] In this process, the initial $J^{PC} = 1^{--}$

requires the D^0 and $\overline{D}^0$ are in a CP-odd quantum-coherent state. Therefore, at any time, the two D mesons have opposite CP-eigenstates until one of them decays. When omitting the high orders, we obtain the $\mathcal{A}_{CP\to K\pi}$ is the asymmetry between CP-odd and CP-even states decaying to $K^-\pi^+$ [2–5]

$$\mathcal{A}_{CP\to K\pi} \equiv \frac{\Gamma_{D^{CP-}\to K^-\pi^+} - \Gamma_{D^{CP+}\to K^-\pi^+}}{\Gamma_{D^{CP-}\to K^-\pi^+} + \Gamma_{D^{CP+}\to K^-\pi^+}}$$

and the relationship between $\delta_{K\pi}$ and $\mathcal{A}_{CP\to K\pi}$

$$2r\cos\delta_{K\pi} + y = (1 + R_{\text{WS}})\mathcal{A}_{CP\to K\pi},$$

when y is the mixing parameter, R_{WS} is the decay rate ratio of the wrong sign process $\overline{D}^0 \to K^-\pi^+$ and the right sign process $D^0 \to K^-\pi^+$. Therefore, with result of $\mathcal{A}_{CP\to K\pi}$ and external inputs of y, r^2 and R_{WS}, we can obtain $\cos\delta_{K\pi}$.

CLEO-c used $818\,\text{pb}^{-1}$ of $\psi(3770)$ data and used a global fit method to get the $\cos\delta = 1.15^{+0.19+0.00}_{-0.17-0.08}$ (with the external parameters) and $\cos\delta = 0.81^{+0.22+0.07}_{-0.18-0.05}$ (without the external parameters), where the uncertainties are statistical and systematic.[5]

2. Determine $\cos\delta_{K\pi}$ in Experiment

To determine these branching fractions, we follow the 'DTag' technique firstly introduced by the MARK-III collaboration.[6] We select $D \to CP$ without reference to the other particle as single tag (ST) events, and reconstruct D to CP and another D to $K\pi$ as double tag (DT) events. The branching fraction for D decays can be obtained from the fraction of DT events in STs with no need of the total number of $D\overline{D}$ events produced.

The total dataset analyzed is about $2.92\,\text{fb}^{-1}$. 5 $CP+$ modes and 3 $CP-$ modes are reconstructed from combinations of $\pi^\pm$, $K^\pm$, π^0, η, and K^0_S candidates with $\pi^0 \to \gamma\gamma$, $\eta \to \gamma\gamma$, $K^0_S \to \pi^+\pi^-$ and $\omega \to \pi^+\pi^-\pi^0$, as list in Table 1. Two kinematic variables are defined; the beam constrained mass M_{BC} and the energy difference ΔE

$$M_{\text{BC}} \equiv \sqrt{E_0^2/c^4 - |\vec{p}_{\text{D}}|^2/c^2},$$

$$\Delta E \equiv E_{\text{D}} - E_0,$$

Table 1. D decay modes.

Type	Mode
Flavored	$K^-\pi^+, K^+\pi^-$
$CP+$	$K^+K^-, \pi^+\pi^-, K^0_S\pi^0\pi^0, \pi^0\pi^0, \rho^0\pi^0$
$CP-$	$K^0_S\pi^0, K^0_S\eta, K^0_S\omega$

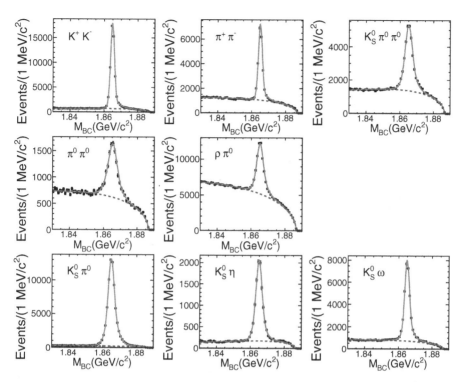

Fig. 1. ST M_{BC} distributions of the $D \to CP$ decay final states and fits to the distributions. Data are shown in points with error bars. The solid lines show the total fits and the dashed lines show the background shapes.

where $\vec{p}_D$ and E_D are the total momentum and energy of the D candidate, and E_0 is the beam energy. D signal candidates produce a peak in M_{BC} at the D mass and in ΔE at zero. To obtain the ST yields, we fit the M_{BC} distribution with the mode-dependent requirements on ΔE. We accept only one candidate per mode per event; when multiple candidates are present, we choose the one with the least $|\Delta E|$. The M_{BC} distribution, shown in Fig. 1, which has a peak at the mass of the D meson and a smooth background cut-off at the point of the beam energy, are fitted by a signal shape derived from MC simulation, convoluted with a smearing Gaussian function, and a background function modeled with the ARGUS function.[7] The Gaussian function is supposed to compensate for the difference in the signal shape between data and MC simulation.

To select DT signals, we require the ΔE of both D mesons, restrict the mass window $1.86\,\mathrm{GeV}/c^2 < M_{BC}(D \to K\pi) < 1.875\,\mathrm{GeV}/c^2$, and extract the signal yields of DTs by fitting the distributions of $M_{BC}(D \to CP)$. The $M_{BC}(D \to CP)$ signal is described by the signal MC shape convoluted with a Gaussian function, and the background is modeled with the ARGUS function.

We obtain $\mathcal{A}_{CP \to K\pi} = (12.77 \pm 1.31^{+0.33}_{-0.31})\%$, where the first uncertainty is statistical and the second is systematic. When quote the external inputs of of

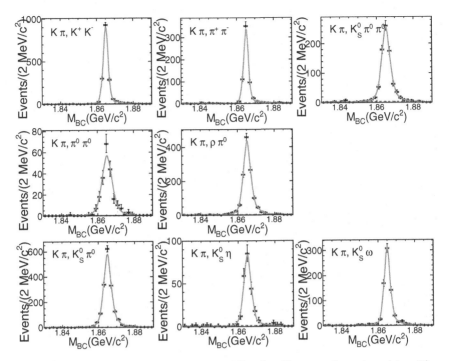

Fig. 2. DT M_{BC} distributions and the corresponding fits. Data are shown in points with error bars. The solid lines show the total fits and the dashed lines show the background shapes.

$R_{\mathrm{D}} = 3.47 \pm 0.06\text{‰}$, $y = 6.6 \pm 0.9\text{‰}$ from HFAG,[8] and $R_{\mathrm{WS}} = 3.80 \pm 0.05\text{‰}$ from PDG,[9] we obtain $\cos\delta_{K\pi} = 1.03 \pm 0.12 \pm 0.04 \pm 0.01$, where the first uncertainty is statistical, the second uncertainty is systematic, and the third uncertainty is due to the errors introduced by the external inputs.

3. Summary

We employ a double tagging technique to analyze a $2.92\,\mathrm{fb}^{-1}$ quantum-correlated data of $e^+e^- \to D^0\overline{D}^0$ at the $\psi(3770)$ peak. We measure the asymmetry $\mathcal{A}_{CP\to K\pi} = (12.77 \pm 1.31^{+0.33}_{-0.31})\%$. Using the inputs of r^2, y from HFAG and R_{WS} from PDG, we get $\cos\delta_{K\pi} = 1.03 \pm 0.12 \pm 0.04 \pm 0.01$. The first uncertainty is statistical, the second is systematic, and the third is due to the external inputs. This is the most precise measurement of the $D \to K\pi$ strong phase difference to date. It helps to constrain the measurements of the D^0-$\overline{D}^0$ mixing parameters and the γ/ϕ_3 angles in the CKM matrix.

References

1. D. M. Asner and W. M. Sun, Phys. Rev. D **73** (2006) 034024 [Erratum-ibid. D **77** (2008) 019901]; X.-D. Cheng, K.-L. He, H.-B. Li, Y.-F. Wang and M.-Z. Yang, Phys. Rev. D **75** (2007) 094019; D. Atwood and A. A. Petrov, Phys. Rev. D **71** (2005) 054032.

2. Z.-Z. Xing, Phys. Rev. D **55** (1997) 196.
3. H.-B. Li and M.-Z. Yang, Phys. Rev. D **74** (2006) 094016.
4. M. Gronau, Y. Grossman and J. L. Rosner, Phys. Lett. B **508** (2001) 37.
5. D. M. Asner *et al.* [CLEO Collaboration], Phys. Rev. D **86** (2012) 112001.
6. R. M. Baltrusaitis *et al.* [MARK-III Collaboration], Phys. Rev. Lett. **56** (1986) 2140;
   J. Adler *et al.* [MARK-III Collaboration], Phys. Rev. Lett. **60** (1988) 89.
7. H. Albrecht *et al.* [ARGUS Collaboration], Phys. Lett. B **241** (1990) 278.
8. Heavy Flavor Averaging Group: http://www.slac.stanford.edu/xorg/hfag/charm.
9. J. Beringer *et al.* [Particle Data Group], Phys. Rev. D **86** (2012) 010001.

Physics in Collision (PIC 2013)
International Journal of Modern Physics: Conference Series
Vol. 31 (2014) 1460306 (5 pages)
© The Author
DOI: 10.1142/S2010194514603068

Partial wave analysis at BESIII

Beijiang Liu

(For BESIII collaboration)

Institute of High Energy Physics, Chinese Academy of Sciences,
19B Yuanquanlu, Shijingshan District, Beijing, 100049, China
liubj@ihep.ac.cn

Published 15 May 2014

The BESIII experiment in Beijing takes data in τ-charm domain since 2009. For the moment the world largest samples of J/ψ, $\psi(3686)$, $\psi(3770)$ and $\psi(4040)$ data have been collected. Hadron spectroscopy is a unique way to access QCD, which is one of the most important physics goals of BESIII. Experimental search of new forms of hadrons and subsequent investigation of their properties would provide validation of and valuable input to the quantitative understanding of QCD. The key to success lies in high levels of precision during the measurement and high statistics in the recorded data set complemented with sophisticated analysis methods. Partial wave analysis (PWA) is a powerful tool to study the hadron spectroscopy, that allows one to extract the resonance's spin-parity, mass, width and decay properties with high sensitivity and accuracy. In this poster, we present the working PWA framework of BESIII – GPUPWA and the recent results of PWA of $J/\psi \rightarrow \gamma\eta\eta$. GPUPWA is a PWA framework for high statistics partial wave analyses harnessing the GPU parallel computing.

Keywords: Hadron spectroscopy; partial wave analysis; BESIII.

PACS Numbers: 13.25.Gv, 13.40.Hq, 14.40.Be

1. Introduction

Hadron spectroscopy is a unique way to access QCD. This theory predicts that there should exist new forms of matter, such as glueballs (pure-gluon objects) and hybrids ($q\bar{q}$ states with explicit gluon).[1-5] Experimental search of these predictions and subsequent investigation of their properties would provide validation of and valuable input to the quantitative understanding of QCD. From the experimental side, the basic task is to systematically map out all the resonances with the determination of their properties like mass, width, spin-parity as well as partial decays widths.

BESIII (Beijing Spectrometer)is a new state-of-the-art 4π detector at the upgraded BEPCII (Beijing Electron and Positron Collider) that operated in the τ-charm threshold energy region.[6] Since 2009, it has collected the worlds largest data samples of J/ψ, $\psi(3686)$, $\psi(3770)$ and $\psi(4040)$ decays. These data are being used to make a variety of interesting and unique studies of light hadron spectroscopy precision charmonium physics and high-statistics measurements of D meson decays.[7]

The study of radiative decays of J/ψ is considered most suggestive in the glueball search. After photon emission, the $c\bar{c}$ annihilation can go through $C-even$ gg states, and hence may have a strong coupling to the low-lying glueballs.[1–5]

2. Partial Wave Analysis Method

Extracting resonance properties from experimental data is however far from straightforward; resonances tend to be broad and plentiful, leading to intricate interference patterns. In such an environment, simple fitting of mass spectra is usually not sufficient and a partial wave analysis (PWA) is required to disentangle interference effects and to extract resonance properties. In the cases discussed here, the full kinematic information is used and fitted to a model of the amplitude in a partial wave decomposition. The partial wave amplitude is constructed with an angular part and a dynamical part. The model parameters are determined by an unbinned likelihood fit to the data, while the event-wise efficiency correction is included. In a typical PWA (we use the radiative decay $J/\psi \rightarrow \gamma\eta\eta$[8] as an example), the quasi two-body decay amplitudes (isobar model) in the sequential decay process $J/\psi \rightarrow \gamma X, X \rightarrow \eta\eta$ are constructed using covariant tensor amplitudes described in Ref. 9.

The probability to observe the event characterized by the measurement ξ is

$$P(\xi) = \frac{\omega(\xi)\epsilon(\xi)}{\int d\xi\omega(\xi)\epsilon(\xi)}, \tag{1}$$

where $\epsilon(\xi)$ is the detection efficiency and $\omega(\xi) \equiv \frac{d\sigma}{d\Phi}$ is the differential cross section, and $d\Phi$ is the standard element of phase space.

$$\frac{d\sigma}{d\Phi} = |\sum_{j}^{wave} \Lambda_j A_j|^2, \tag{2}$$

where A_j is the partial wave amplitude with coupling strength determined by a complex coefficient Λ_j. The normalization integral is performed numerically by Monte Carlo techniques. The likelihood for a particular model is

$$\mathcal{L} = \prod_{i=1}^{N_{data}} P(\xi_i). \tag{3}$$

A series of likelihood fits are performed for parameter estimation and model evaluation. In the log likelihood calculation, the likelihood value of background events are given negative weights, and are removed from data since the log likelihood value of data is the sum of signal and background.

As this involves the computation of the amplitude for every event in every iteration of a fit, this becomes computationally very expensive for large data samples. As events are independent and the amplitude calculation does not vary from event to event, this task is trivially parallelizable. This and the floating point intensity predestine PWA for implementation on graphics processing units (GPUs). GPUPWA[10][13] has been developed as the working framework of BESIII, harnessing GPU parallel acceleration. The framework now provides facilities for amplitude calculation, minimization and plotting and is widely used for analyses at BESIII.

GPUPWA has been developed as the working framework of BESIII, harnessing GPU parallel acceleration. GPUPWA is now developed with *OpenCL*[11] as described in Ref. 12. The framework now provides facilities for amplitude calculation, minimization and plotting and is widely used for analyses at BES III. It continues to be developed and is available at Ref. 13.

3. PWA Results of $J/\psi \to \gamma\eta\eta$

For a J/ψ radiative decay to two pseudoscalar mesons, it offers a very clean laboratory to search for scalar and tensor glueballs because only intermediate states with $J^{PC} = even^{++}$ are possible. An early study of $J/\psi \to \gamma\eta\eta$ was made by the Crystal Ball Collaboration with the first observation of $f_0(1710)$, but the study suffered from low statistics. The results of partial wave analysis (PWA) on $J/\psi \to \gamma\eta\eta$ (Fig. 1) are presented based on a sample of 2.25×10^8 J/ψ events collected with BESIII.[8]

4. Summary and Outlook

A full partial wave analysis was performed to disentangle the structures present in $J/\psi \to \gamma\eta\eta$ decays. The scalar contributions are mainly from $f_0(1500)$, $f_0(1710)$ and $f_0(2100)$, while no evident contributions from $f_0(1370)$ and $f_0(1790)$ are seen. Recently, the production rate of the pure gauge scalar glueball in J/r radiative decays predicted by the lattice QCD[14] was found to be compatible with the production rate of J/ψ radiative decays to $f_0(1710)$; this suggests that $f_0(1710)$ has a larger overlap with the glueball compared to other glueball candidates (eg. $f_0(1500)$). In this analysis, the production rate of $f_0(1710)$ and $f_0(2100)$ are both about one order of magnitude larger than that of the $f_0(1500)$, which are both consistent with, at least not contrary to, lattice QCD predictions.[14]

Now five years from our first collisions, BESIII has established a broad and successful program in charm physics. Recently, in 2012, even larger samples have been accumulated at the J/ψ and $\psi(3686)$; total samples are now about 1.2 billion and 0.35 billion decays, respectively. Furthermore, our 2013 dataset includes more data near 4260 MeV, and also a large sample at the $Y(4360)$. With the excellent performance of the accelerator and detector, more interesting results are expected.

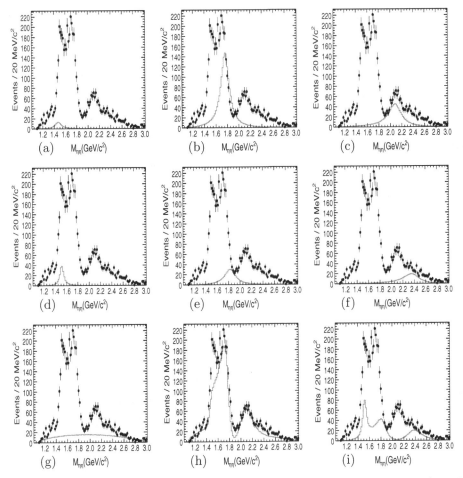

Fig. 1. Contribution of the components. (a) $f_0(1500)$, (b) $f_0(1710)$, (c) $f_0(2100)$, (d) $f_2'(1525)$, (e) $f_2(1810)$, (f) $f_2(2340)$, (g) 0^{++} phase space, (h) total 0^{++} component, and (i) total 2^{++} component. The dots with error bars are data with background subtracted, and the solid histograms are the projection of the PWA results.[8]

References

1. F. E. Close, Rept. Prog. Phys. **51**, 833 (1988).
2. S. Godfrey and J. Napolitano, Rev. Mod. Phys. **71**, 1411 (1999).
3. C. Amsler and N. A. Tornqvist, Phys. Rept. **389**, 61 (2004).
4. E. Klempt and A. Zaitsev, Phys. Rept. **454**, 1 (2007).
5. V. Crede and C. A. Meyer, Prog. Part. Nucl. Phys. **63**, 74 (2009).
6. M. Ablikim *et al.* (BESIII Collaboration), Nucl. Instrum. Meth. A **614**, 345 (2010).
7. Edited by Kuang-Ta Chao and Yi-Fang Wang, Int. J. Mod. Phys. A **24**, S1 (2009)
8. M. Ablikim *et al.* [BESIII Collaboration], Phys. Rev. D. 87, **092009** (2013)
9. B. S. Zou and D. V. Bugg, Eur. Phys. J. A **16**, 537 (2003).
10. N. Berger, L. Beijiang and W. Jike, J. Phys. Conf. Ser. **219**, 042031 (2010).

11. http://www.khronos.org/opencl/
12. Berger N 2011 *AIP Conf. Proc.* **1374** 553–556
13. GPUPWA package, available at http://sourceforge.net/projects/gpupwa/
14. L.-C. Gui, Y. Chen, G. Li, C. Liu, Y.-B. Liu, J.-P. Ma, Y.-B. Yang and J.-B. Zhang, Phys. Rev. Lett. **110**, 021601 (2013)

Physics in Collision (PIC 2013)
International Journal of Modern Physics: Conference Series
Vol. 31 (2014) 1460307 (4 pages)
© The Authors
DOI: 10.1142/S201019451460307X

Observation of $e^+e^- \to \gamma X(3872)$, $X(3872) \to \pi^+\pi^- J/\psi$

Qing Gao[*,‡] and Ke Li[*,†]

(on behalf of the BESIII Collaboration)

*Institute of High Energy Physics
Chinese Academy of Sciences, Beijing 100049, China
‡gaoq@ihep.ac.cn

†Shandong University, Jinan 250100, China
like@ihep.ac.cn

Published 15 May 2014

Using data samples collected with the BESIII detector operating at the BEPCII storage ring at central-of-mass(CM) energies from 4.009 to 4.420 GeV, the $e^+e^- \to \gamma X(3872)$ process is observed with a statistical significance of more than 5σ. The measured mass is in agreement with previous measurements. The products of cross section of $e^+e^- \to \gamma X(3872)$ and the branching fraction $\mathcal{B}(X(3872) \to \pi^+\pi^- J/\psi)$ at CM energies 4.009, 4.229, 4.260, and 4.360 GeV is reported. The results support the possibility that $Y(4260) \to \gamma X(3872)$.

Keywords: X(3872).

PACS Numbers: 14.40.Rt, 13.20.Gd, 13.66.Bc, 13.40.Hq, 14.40.Pq

1. Introduction

The X(3872) was discovered[1] by Belle in $B^\pm \to K^\pm\pi^+\pi^- J/\psi$ in 2003. Since its discovery, $X(3872)$ has simulated great interest in its nature. The LHCb experiment determined[2] the quantum numbers of the $X(3872)$ to be $J^{PC} = 1^{++}$, and CDF also found that the $\pi^+\pi^-$ system was dominated by the $\rho^0(770)$ resonance.[3]

Since the mass is near $D\bar{D}^*$ threshold, the $X(3872)$ was interpreted as a good candidate for a hadronic molecule or a tetraquark state. Currently, the $X(3872)$ has only been observed in B meson decays and hadron collisions. BESIII can hunt for it in excited 1^{--} E1 transitions, using the process $e^+e^- \to \gamma X(3872)$.

2. Observation of the $X(3872)$

The process of $e^+e^- \to \gamma X(3872) \to \gamma\pi^+\pi^- J/\psi$, $J/\psi \to l^+l^-$ ($l^+l^- = e^+e^-$ or $\mu^+\mu^-$) is observed with a statistical significance of more than 5σ for the first time with data collected with the BESIII detector operating at the BEPCII storage ring[4] at e^+e^- center-of-mass (CM) energies from $\sqrt{s} = 4.009$ GeV to 4.420 GeV[5] with an integral luminosity of 3301.0 ± 33.1 pb^{-1}.

Figure 1 shows the $\pi^+\pi^- J/\psi$ invariant mass distribution for all data samples. Where $M(\pi^+\pi^- J/\psi) = M(\pi^+\pi^- l^+l^-) - M(l^+l^-) + m(J/\psi)$ is used to reduce the resolution effect of the lepton pairs, and $m(J/\psi)$ is the nominal mass of J/ψ.[6] There is a huge $e^+e^- \to \gamma_{ISR}\psi(2S)$ peak which is used to calibrate and to validate the analysis. In addition, X(3872) can also be clearly seen. Remaining backgrounds mainly come from $e^+e^- \to \gamma_{ISR}\pi^+\pi^- J/\psi$, $\eta' J/\psi$ and $\pi^+\pi^-\pi^+\pi^- (\pi^0/\gamma)$ processes. But none of them form peaks around the $X(3872)$ signal region.

The mass of X(3872) is determined by fitting the $M(\pi^+\pi^- J/\psi)$ distribution (shown in Fig. 2), which use a MC simulated signal histogram convolved with a Gaussian function representing the difference in the mass resolution between data and MC simulation as the signal shape, and a linear function for the background. The fit result is $M(X(3872)) = (3871.9 \pm 0.7)$ MeV/c^2, $\sigma = 1.14$ MeV/c^2, $N^{obs} = 20.1 \pm 4.5$. The statistical significance of $X(3872)$ is 6.3σ.

Figure 3 shows the $\pi^+\pi^-$ invariant mass distribution for the selected $X(3872)$ candidates, which is dominated by the $\rho^0(770)$ resonance and consistent with the CDF observation.[3]

The product of the Born-order cross section times the branching ratio $X(3872) \to \pi^+\pi^- J/\psi$ is calculated using $\sigma^B(e^+e^- \to \gamma X(3872)) \times \mathcal{B}(X(3872) \to \pi^+\pi^- J/\psi) = \frac{N^{obs}}{\mathcal{L}_{int}(1+\delta)\epsilon\mathcal{B}}$, where N^{obs} is the number of observed events obtained

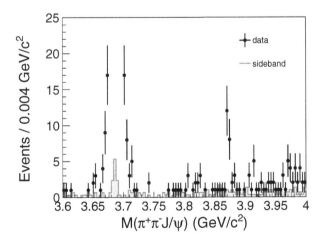

Fig. 1. The invariant mass distribution of $M(\pi^+\pi^- J/\psi)$ for all data samples. Dots with error bars are data, green shaded histograms are normalized J/ψ sideband events.

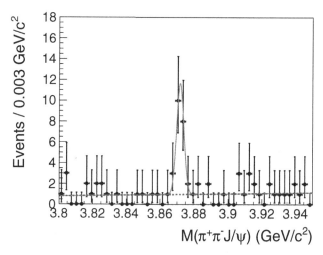

Fig. 2. Fit of the $M(\pi^+\pi^- J/\psi)$ distribution with a MC simulated histogram convolved with a Gaussian function for signal and a linear background function. Dots with error bars are data, the red curve shows the total fit result, while the blue dashed curve shows the background contribution.

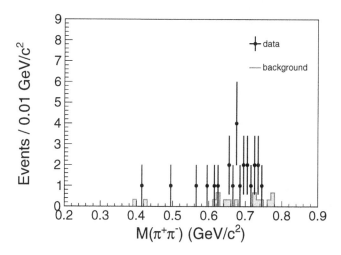

Fig. 3. The $M(\pi^+\pi^-)$ distribution for $X(3872) \to \pi^+\pi^- J/\psi$ candidate events. Dots with error bars are data, and the green shaded histogram is normalized background events in $X(3872)$ sideband region.

from the fit to the $M(\pi^+\pi^- J/\psi)$ distribution, $\mathcal{L}_{int}$ is integrated luminosity, ϵ is the detection efficiency, $\mathcal{B}$ is the branching ratio of $J/\psi \to l^+l^-$ and $1+\delta$ is the radiative correction factor. The $(1+\delta)$ factor, detection efficiency, number of $X(3872)$ signal events, and $\sigma^B(e^+e^- \to \gamma X(3872)) \times \mathcal{B}(X(3872) \to \pi^+\pi^- J/\psi)$ at $\sqrt{s} = 4.009$ GeV, 4.229 GeV, 4.260 GeV and 4.360 GeV are listed in Table. 1. For 4.009 and 4.360

Table 1. The number of $X(3872)$ events, radiative correction factor, detection efficiency, measured Born cross section $\sigma^B(e^+e^- \to \gamma X(3872))$ times $\mathcal{B}(X(3872) \to \pi^+\pi^- J/\psi)$ ($\sigma^B \times \mathcal{B}$, where the first errors are statistical and the second are systematic) at different energy points. The upper limits are given at the 90% C.L..

Energy (MeV)	$\epsilon(\%)$	$1+\delta$	N^{obs}	$\sigma^B \times B$ pb^{-1}
4009	25.5	0.861	<1.4	<0.12
4229	31.5	0.799	9.6±3.1	0.29±0.10±0.02
4260	30.5	0.814	8.7±3.0	0.36±0.13±0.03
4360	21.1	1.023	<5.1	<0.39

GeV data, since the $X(3872)$ signal is not significant, upper limits for production yield at the 90% C.L. are given.

3. Summary

The process of $e^+e^- \to \gamma X(3872)$ is observed for the first time. The measured mass of the $X(3872)$ is $M(X(3872)) = (3871.9 \pm 0.7 \pm 0.2)$ MeV/c^2, which agrees well with previous measurements.[6] The production rate $\sigma^B(e^+e^- \to \gamma X(3872)) \times \mathcal{B}(X(3872) \to \pi^+\pi^- J/\psi)$ is measured to be $(0.29\pm0.10\pm0.02)$ pb at $\sqrt{s} = 4.229$ GeV, $(0.36\pm0.13\pm0.03)$ pb at $\sqrt{s} = 4.260$ GeV, < 0.12 pb at $\sqrt{s} = 4.009$ GeV, and < 0.39 pb at $\sqrt{s} = 4.360$ GeV at the 90% C.L., respectively. Where the first errors are statistical and the second are systematic. The observation suggests that the $X(3872)$ might be from the radiative transition of the $Y(4260)$.

References

1. S. K. Choi *et al.* (Belle Collaboration), Phys. Rev. Lett. **91**, 262001 (2003).
2. R. Aaij *et al.* (LHCb Collaboration), Eur. Phys. J. C **72**, 1972 (2012); arXiv:1302.6269.
3. A. Abulencia *et al.* (CDF Collaboration), Phys. Rev. Lett. **96**, 102002 (2006).
4. M. Ablikim *et al.* (BESIII Collaboration), Nucl. Instrum. Methods Phys. Res. Sect. A **614**, 345 (2010).
5. M. Ablikim *et al.* (BESIII Collaboration), arXiv:1309.1896.
6. J. Beringer *et al.* (Particle Data Group), Phys. Rev. D **86**, 010001 (2012).

Physics in Collision (PIC 2013)
International Journal of Modern Physics: Conference Series
Vol. 31 (2014) 1460308 (6 pages)
© The Author
DOI: 10.1142/S2010194514603081

Rare kaon decays with NA48/2 and NA62 at CERN

Patrizia Cenci*

INFN Perugia, Via A. Pascoli, Perugia, 06123, Italy
patrizia.cenci@pg.infn.it

Published 15 May 2014

The Kaon physics program at CERN will be shortly presented, addressing the most recent results from the NA48/2 and NA62 experiments at the CERN SPS and future prospects.

Keywords: Decays of K mesons; standard model; chiral perturbation theory.

PACS Numbers: 13.25.Es, 13.20.Eb, 12.15.−y, 12.60.−i

1. Introduction

In 2003-04, the NA48/2 experiment has collected, at the CERN SPS, the largest world sample of charged kaon decays, with the main goal of searching for direct CP violation.[1] At an early stage, in 2007-08, the NA62 experiment[2] collected a large minimum bias data sample exploiting the same detector with modified data taking conditions. The main goal was the measurement of the ratio of the rates of the leptonic kaon decays $K^\pm \to e\nu_e$ (K_{e2}) and $K^\pm \to \mu\nu_\mu$ ($K_{\mu2}$).[3] The large statistics accumulated by both experiments has allowed the studies of a variety of rare kaon decay modes.

2. Latest Results from the NA48/2 Experiment

The beam line of the NA48/2 experiment has been designed to deliver simultaneous, narrow momentum band, K^+ and K^- beams produced by 400 GeV/c primary

*On behalf of the NA48/2 and NA62 Collaborations at CERN (The NA48/2 Collaboration: Cambridge, CERN, Dubna, Chicago, Edinburgh, Ferrara, Firenze, Mainz, Northwestern University, Perugia, Pisa, Saclay, Siegen, Torino, Wien. The NA62 Collaboration: Birmingham, CERN, Dubna JINR, Fairfax, Ferrara, Firenze, Frascati LNF, Mainz, Merced, Moscow INR, Napoli, Perugia, Pisa, Roma I, Roma Tor Vergata, Saclay IRFU, San Luis Potosi, Stanford, Sofia, Torino, TRIUMF).

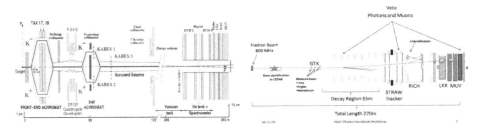

Fig. 1. Schematic drawing of the NA48/2 experimental layout (left) and of the NA62 detector (right).

protons extracted from the CERN SPS impinging on a beryllium target. The experimental layout is displayed in Fig. 1 (left). The momentum of the charged particles from $K^{\pm}$ decays was measured by a magnetic spectrometer consisting of four drift chambers (DCH1-DCH4) and a dipole magnet. The spectrometer was located in a tank filled with helium at atmospheric pressure and followed by a scintillator trigger hodoscope. A liquid Krypton calorimeter (LKr) was used to measure the energy of electrons and photons. A hadron calorimeter (HAC) and a muon veto system (MUV), essential to distinguish muons from pions, were located further downstream. The detector description is available in Ref. 4.

2.1. *The $K^{\pm} \to \pi^{\pm}\gamma\gamma$ Decays*

Measurements of radiative non-leptonic kaon decays provide crucial tests of the Chiral Perturbation Theory (ChPT), describing weak low energy processes. In ChPT the main contributions to the $K^{\pm} \to \pi^{\pm}\gamma\gamma$ decay at the lowest order $O(p^4)$ depend on an unknown parameter $\hat{c}$. Higher order corrections including $O(p^6)$ contribution have been found to modify the decay spectrum and rate significantly. A sample of more than 300 $K^{\pm} \to \pi^{\pm}\gamma\gamma$ rare decays with a background contamination below 10% has been collected during a 3-day NA48/2 run (2004) and a 3-month NA62 run (2007) at low intensity with minimum bias trigger configuration. The results[5-6] are obtained at increased precision and set stronger constrains on ChPT predictions. The observed decay spectrum agrees with the ChPT description and the preliminary combination of the measured ChPT parameters are $\hat{c}_4 = 1.56 \pm 0.23$ and $\hat{c}_6 = 2.00 \pm 0.26$. The preliminary combined branching ratio, calculated in the full kinematic range assuming the $O(p^6)$ description, is $\mathrm{BR}_6 = (1.01 \pm 0.06) \times 10^{-6}$. The results agree with the only measurement published so far and represent a significant improvement with respect to the earlier data. The publication of the final combined result is in preparation. The invariant mass distribution of $\pi^{\pm}\gamma\gamma$ candidates (2007 data), with MC expectation for signal and background, is displayed in Fig. 2 (left).

2.2. *Semileptonic kaon decays*

Semileptonic kaon decays offer the most precise determination of the CKM matrix element $|V_{us}|$. The experimental precision is however limited by the knowledge of

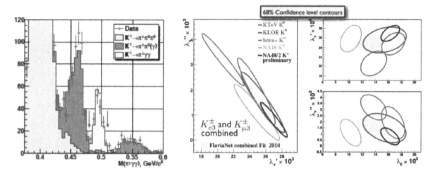

Fig. 2. The $\pi^{\pm}\gamma\gamma$ invariant mass distribution (2007 data) with MC expectation for signal and background (left). Preliminary combined quadratic fit results for $K_{\ell 3}^{\pm}$ decays (right). The ellipses are 68% C.L. contours.

the form factors (FF) of this decay, since they enter both the phase space integral and the detector acceptances. The NA48/2 experiment presents new measurements of the FF of $K^{\pm}$ semileptonic decays, based on 4.3 million $K^{\pm} \rightarrow \pi^0 e\nu_e$ ($K_{e3}^{\pm}$) and 3.5 million $K^{\pm} \rightarrow \pi^0 \mu\nu_e$ ($K_{\mu 3}^{\pm}$) decays collected in 2004, both with negligible background. The hadronic matrix element of the $K_{\ell 3}^{\pm}(\ell = e$ or $\mu)$ decays is described by two dimensionless FF which depend on the squared four-momentum transferred to the $\ell - \nu$ system. They can be expressed in terms of vector $f_+(t)$ and scalar $f_0(t)$ exchange contributions, parametrized either as a Taylor expansion ("quadratic parametrization"), quantified with λ coefficients, or by assuming the dominance of vector (V) or scalar (S) resonance exchange ("pole parametrization"), where pole masses are the only free parameters.

Figure 2 (right) shows the preliminary comparison of combined quadratic fit results for $K_{\ell 3}^{\pm}$ decays from NA48/2 and other experiments.[7] The ellipses represent 68% Confidence Level (C.L.) contours. The NA48/2 results agree with other experiments and match the precision of the world average on the vector and scalar FF, allowing an improved precision on the $K_{\ell 3}^{\pm}$ form factor contribution to the $|V_{us}|$ uncertainty. The combined $K_{e3}^{\pm}$–$K_{\mu 3}^{\pm}$ results from NA48/2 are the first high precision FF measurement with both K^+ and K^- mesons. The comparison of both channels sets tight constraints on lepton flavor violation and other possible new physics.

2.3. The $K^{\pm} \rightarrow \pi^+\pi^- e\nu_e$ and the $K^{\pm} \rightarrow \pi^0\pi^0 e\nu_e$ decays

The NA48/2 collaboration has analyzed 1.13 million charged kaon decays $K^{\pm} \rightarrow \pi^+\pi^- e\nu_e$ (K_{e4}^{+-}) leading to detailed FF studies and an improved determination of the BR at percent level precision.[8] The hadronic FF in the S- and P-wave and their variation with energy are obtained concurrently with the phase difference between the S- and P-wave states of the $\pi^+\pi^-$ system. The latter measurement allows a precise determination of a_0^0 and a_0^2, the $I = 0$ and $I = 2$ S-wave $\pi^+\pi^-$ scattering lengths.

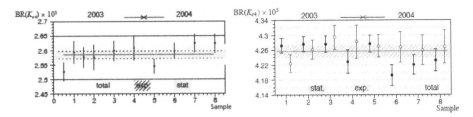

Fig. 3.   Measurements of BR(K_{e4}^{00}) (left) and BR(K_{e4}^{+-}) (right, both charges) made with different data samples.

A combination of this result with another NA48/2 measurement, obtained in the study of $K^\pm \to \pi^\pm \pi^+ \pi^-$ decays, brings a further improved determination of a_0^0 and the first precise experimental measurement of a_0^2. These measurements deliver new inputs to low energy QCD calculations and are crucial tests of ChPT and lattice QCD predictions.

New preliminary results on FF and BR measurements are available for the neutral channel $K^\pm \to \pi^0 \pi^0 e \nu_e$ (K_{e4}^{00}).[9] The NA48/2 experiment collected about 66,000 K_{e4}^{00} decays, increasing the world available statistics by several orders of magnitude. The low background contamination, at 1% level, and the very good π^0 reconstruction allow the first accurate measurements of FF and of the decay BR. The achieved precision makes possible the observation of small effects such as a deficit of events at low $\pi^0\pi^0$ invariant mass, explained by charge exchange rescattering effects in the $\pi\pi$ system below $2m_{\pi+}$ threshold. Figure 3 displays the measurements of both neutral and charged BR based on statistically independent samples collected in 2003-04. The results are: BR(K_{e4}^{00}) = $(2.585 \pm 0.010_{stat} \pm 0.010_{syst} \pm 0.032_{ext}) \times 10^{-5}$ (preliminary) and BR(K_{e4}^{+-}) = $(4.257 \pm 0.004_{stat} \pm 0.016_{syst} \pm 0.031_{ext}) \times 10^{-5}$. The systematic errors include uncertainties on acceptance, resolution, particle identification, beam geometry, trigger efficiencies and radiative corrections. The external error stems from the normalization mode BR uncertainties and is now the dominant error. The hatched band shows the experimental error ($\sigma_{exp} = (\sigma_{stat} \oplus \sigma_{syst})$). The total uncertainty (shaded band) includes the external error.

3. Search for Lepton Flavor Violation and Rare Decays at the NA62 Experiment

The primary goal of NA62[2], the last generation experiment for kaon physics at the CERN SPS, is to measure BR($K^+ \to \pi^+ \nu \bar{\nu}$) with 10% precision. The decay $K^+ \to \pi^+ \nu \bar{\nu}$ is highly suppressed in the Standard Model (SM), while its rate can be predicted with minimal theoretical uncertainty (BR$_{SM}$ = $(0.781 \pm 0.080) \times 10^{-10}$).[10] This BR is a sensitive probe of the flavor sector of the SM. However, its smallness and the challenging experimental signature make it very difficult to measure. The only measurement[11] is compatible with the SM within a large uncertainty. The challenging aspect which drives the design of the NA62 experiment is the suppression of kaon decays with branching fractions of many orders of magnitude higher than the signal This is achieved by collecting about 100 signal events with 10% background in two years of data taking, which requires the observation of at least 10^{13} K^+ decays in the experiment's fiducial volume. To achieve the needed level of background rejection, NA62 relies on: high-resolution timing, to support a high-rate environment; kinematic rejection, involving cutting on the squared missing mass of the system of incident kaon and observed charged particle under pion assumption; particle identification of kaons, pions, muons, electrons and photons; hermetic vetoing of photons out to large angles from the decay region and of muons within the acceptance; redundancy of information. A suitable experimental apparatus[12] has been conceived to fulfill these purposes. The layout, shown in Fig. 1 (right), has the following main features.

- An intense charged hadron beam of secondary particles at 75 GeV/c momentum and 800 MHz average rate, with a 6% K^+ component, will be exploited.
- The K^+ component in the beam is identified with respect to the other beam particles by an upgraded differential Čerenkov detector (KTAG).
- The coordinates and the momenta of individual beam particles are registered before entering the decay region by 3 silicon pixel tracking detectors (GTK).
- A large-acceptance magnetic spectrometer with low mass tracking detectors in vacuum (STRAW tracker) is required to detect and measure the coordinates and momenta of charged particles originating from the decay region.
- A ring-imaging Čerenkov (RICH) detector, consisting of a long vessel filled with Neon gas, identifies pions with respect to muons.
- A set of photon-veto detectors provides hermetic coverage from zero out to large angles (50 mr) from the decay region. It consists of the existing, high-resolution, electromagnetic calorimeter (LKr), supplemented, at small and forward angles, by intermediate ring (IRC) and small-angle (SAC) Shashlyk calorimeters, and, at large angles, by a series of annular lead-glass photon-veto (LAV) detectors.
- Muon-veto detectors (MUV) are made of a two-part hadron calorimeter followed by iron and a transversally-segmented scintillator hodoscope. This system supplements and provides redundancy with respect to the RICH in muon rejection.
- Scintillator counters (CHANTI) surrounding the last GTK station veto charged particles upstream of the decay region.
- A segmented charged-particle hodoscope (CHOD) made of scintillator counters covers the detector acceptance between the RICH and the LKr calorimeter.
- All the detector components are inter-connected with a high-performance trigger and data-acquisition (TDAQ) integrated system.

Part of the experimental apparatus was commissioned during a technical run in 2012; installation continues and data taking is expected to begin in late 2014.

The features of the NA62 experiment allow a rich physics program, in addition to the main goal. A lepton universality test has been performed using data collected at an early stage of the experiment, in 2007-08. A precise measurement of the helicity-suppressed ratio of the leptonic kaon decay rates, $R_K = \Gamma(K_{e2})/\Gamma(K_{\mu 2})$, has been obtained[3]. A world-record sample of about 150,000 reconstructed K_{e2} candidates with 11% background contamination was analyzed. The result $R_K = (2.488 \pm 0.007_{stat} \pm 0.007_{syst}) \times 10^{-5}$ agrees with the SM expectation and is the most precise measurement to date. The record accuracy of 0.4% constrains the parameter space of new physics models with extended Higgs sector, a fourth generation of quarks and leptons or sterile neutrinos.

References

1. NA48/2 Collab. (R. Batley *et al.*), *Eur. Phys. J C* **52**, 875 (2007).
2. NA62 Collab. (G. Anelli *et al.*), CERN-SPSC-2005-013, CERN-SPSC-P-326, (2005).
3. NA62 Collab. (C. Lazzeroni *et al.*), *Phys. Lett. B* **719**, 326 (2013).
4. NA48/2 Collab. (V. Fanti *et al.*), *Nucl. Instrum. Method A* **574**, 433 (2007).
5. NA48/2 Collab. (J.R.Batley *et al.*), CERN-PH-EP-2013-197, arXiv:1310.5499 [hep-ex].
6. NA48/2 and NA62 Collab. (E. Goudzovsky *et al.*), *Rare kaon decays*, in Proc. of the European Physical Society Conference on High Energy Physics EPS-HEP2013, PoS(EPS-HEP 2013) 426.
7. NA62 Collab. (M. Raggi *et al.*), *High precision measurement of the form factors of the semi leptonic decays at NA48/2*, in Proc. of the 2013 Kaon Physics International Conference, PoS(KAON13) 017.
8. NA48/2 Collab. (R. Batley *et al.*), *Eur. Phys. J C* **70**, 635 (2010), *ibidem Phys. Lett. B* **715**, 105 (2012).
9. NA62 Collab. (B. Bloch-Devaux *et al.*), *Study of the $K^{\pm} \to \pi^0 \pi^0 e \nu_e$ decay with NA48/2 at CERN*, in Proc. of the 2013 Kaon Physics International Conference, PoS(KAON13) 028.
10. J. Brod, M. Gorbahn and E. Stamou, *Phys. Rev. D* **83**, 034030 (2011).
11. E949 Collab. (A. V. Artamonov *et al.*), *Phys. Rev. D* **79**, 092004 (2009).
12. NA62 Technical Design, https://cdsweb.cern.ch/record/14049857, NA62-10-07, (2010).

Physics in Collision (PIC 2013)
International Journal of Modern Physics: Conference Series
Vol. 31 (2014) 1460309 (5 pages)
© The Authors
DOI: 10.1142/S2010194514603093

A new inner drift chamber for BESIII MDC

M. Y. Dong*, Z. H. Qin, X. Y. Ma, J. Zhang, J. Dong, W. Xie, Q. L. Xiu,
X. D. Ju, R. G. Liu and Q. Ouyang

Experiment Physics Center, Institute of High Energy Physics
Chinese Academy of Sciences, Beijing, 100049, China
*dongmy@ihep.ac.cn

Published 15 May 2014

Due to the beam related background, the inner chamber of BESIII MDC has aging effect after 5 years running. The gains of the inner chamber cells decrease obviously, and the max gain decrease is about 26% for the first layer cells. A new inner drift chamber with eight stereo sense wire layers as a backup for MDC is under construction, which is almost the same as the current one but using stepped endplates to shorten the wire length beyond the effective solid angle. This new structure will be of benefit to reducing the counting rate of single cell. The manufacture of each component is going smoothly, and the new inner drift chamber will be finished by the end of April 2014.

Keywords: Inner drift chamber; aging effect; MDC; BESIII.

1. Introduction

The Beijing spectrometer III (BESIII) is a high precision detector for the Beijing electron positron collider II (BEPCII), a high luminosity, multi-bunch $e^+ e^-$ collider running at the tau-charm energy region.[1] As its center tracking detector, BESIII main drift chamber (MDC) is a low-mass cylindrical chamber with small-cell geometry, using helium-based gas and operating in a 1T magnetic field. MDC consists of an inner chamber and an outer chamber, which are joined together at the endplates, sharing a common gas volume.[2] The inner chamber was designed to be replaced in case of radiation damage.

After it has been running five years, the inner chamber of BESIII MDC is suffering from the aging problems because of the huge beam related background. The gains of the inner chamber have an obvious decrease, and the gain of the inner most layer decrease a maximum of about 26%. In addition, in order to reduce the dark currents of the sense wires to protect the detector, the operation high voltages of the first four layers have to be set to 96%, 97%, 98% and 99% of the normal value respectively, which results in the decrease of the effective gain and leads to the performance of the inner chamber become bad. Besides the study on upgrading the inner chamber with some new technologies, such as cylindrical gas electron multiplier (GEM), CMOS pixel sensor, building a backup of inner chamber is to be accomplished firstly. In this paper, we report the design and construction of the new inner drift chamber.

2. The New Inner Drift Chamber

The new inner drift chamber is composed of an inner carbon fiber cylinder and two multi-stepped endplates, including eight stereo sense wire layers. The length of the new inner chamber is 1092 mm, and the radial extent is from 59 mm to 183.5 mm, which is almost the same as the current one. In order to reduce the background counting rate, new multi-stepped endplates are adopted to shorten the wire length beyond the effective detection solid angle, as shown in Fig. 1. This new structure will be of benefit to reducing the dark current of single cell. The wire length of the new inner chamber compared with the current one is shown in Table 1, and the expected reduction of the background counting rate is in the range from 31% to 4% for each layer.

2.1. *Major components*

New feed-through models have been tested on two prototypes with 234 wires. The test results show that the feed-throughs have a good insulated performance, and

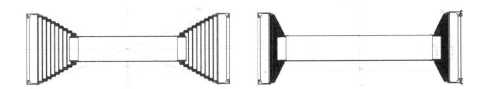

Fig. 1.   The structure of the new inner drift chamber (left) and the current one (right).

Table 1.   The wire length of the new inner chamber compared with the current one.

No. of wire layer	Current wire length (cm)	New wire length (cm)	Background rate reduction expected
1	78.0	53.8	31%
2	79.2	58.0	27%
3	80.4	62.2	23%
4	81.6	66.4	19%
5	82.8	70.6	15%
6	84.0	74.8	11%
7	85.2	79.0	7%
8	86.4	83.2	4%

a long term test for crimping shows no wire broken or tension lost happening for more than three months. Total 12000 feed-throughs have been made and delivered for test.

New endplates are under construction in the manufactory. There are nine steps in each aluminum endplate, and about 2000 $\varphi 3.2$ mm holes with the tolerance of less than 25 μm in radius will be drilled on the plate. The assembly tolerance of two endplate is less than 50 μm.

The new inner carbon fiber cylinder is also being manufactured. The thickness of the inner cylinder is 1mm to reduce the material budget. The tolerance in diameter and length is less than 80 μm, and the deflection in Z direction under stress of 500 kg is less than 50 μm. The aluminum foils will be pasted on both surfaces of the cylinder for shielding.

2.2. *Wire stringing test*

A three-stepped endplate prototype with six layers was built for wire stringing test, due to the increased difficulty caused by the new endplate structure. The prototype has three steps just same as the new inner chamber structure in geometry. Total 120 stereo wires were arranged with the similar cell size. Uniform distribution of wire tension test result proved that wire stringing on the new structure was not a critical problem.

3. Simulation

Monte Carlo simulation has been done with the samples of π^+ track. The simulation results are shown in Figs. 2 and 3. Tracking efficiency changes with momentum and theta, momentum resolution look quite similar for current MDC and MDC with new inner chamber. Because the positions of the feed-throughs are not changed in φ direction, shortening the wire length leads to a larger stereo angle than the current one. As a result, the spatial resolution in Z direction for the new chamber improves a little bit.

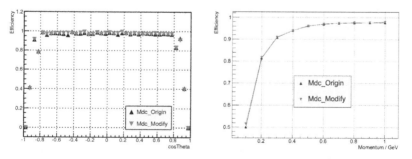

Fig. 2. Tracking efficiency vs. cosθ (left) and momentum (right) compared between current MDC and MDC with new inner chamber.

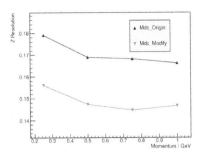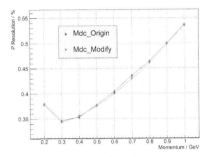

Fig. 3.   Spatial resolution in Z direction (left) and momentum resolution (right) compared between current MDC and MDC with new inner chamber.

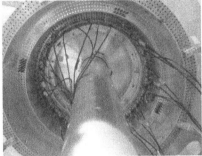

Fig. 4.   The inner chamber removal prototype (left) and wire tension test during the removal operation (right).

4. Study on Inner Chamber Removal

Because MDC has no outer cylinder for the inner chamber and no inner cylinder for the outer chamber, the inner chamber and the outer chamber are joined together at the endplates. The replacement of the inner chamber is not easy. A test prototype was constructed to simulate the whole replacement procedure, which include the removal of the sealing glue at the endplates, pulling out the inner chamber, and monitoring the wire tension during the operation. Figure 4 shows the experiments of inner chamber removal. From the experiments, we know that the glue can be removed without problems, but very carefully operation is needed to avoid damaging the feed-throughs and the wires. The inner chamber can be removed out with the suitable designed toolings, but stress on inner chamber due to some deformation or external force need to be well considered, so a real-time wire tension measurement is very necessary. Wire tension probably changes a little during the operating but will come back to the normal value after then.

5. Conclusion

A new inner drift chamber with multi-stepped endplates to shorten the wire length out of effective detection solid angle is designed and under construction, which will reduce about 31% to 4% of the background counting rate for each layer. Although this new structure can not relieve aging effect of each cell, it can reduce the wire dark current, which will be of benefit not only to the resolution in Z direction but also to chamber safe working. The manufactures of the major components are going smoothly. The construction of the new inner drift chamber will be accomplished by the end of April 2014.

A test prototype was built for the inner chamber removal study. We have some experiences from the removal experiments. The inner chamber can be removed out with the suitable toolings safely with very careful operation.

Acknowledgments

We would like to thank Chinese Academy of Sciences for financial support.

References

1. BESIII Collaboration, Preliminary design report of the BESIII detector, 2004.
2. C. Chen *et al.*, The BESIII drift chamber, *IEEE Nuclear Science, Symposium Conference Record*, 1844 (2007).

Physics in Collision (PIC 2013)
International Journal of Modern Physics: Conference Series
Vol. 31 (2014) 1460310 (4 pages)
© The Author
DOI: 10.1142/S201019451460310X

Cosmic muon induced backgrounds
in the Daya Bay reactor neutrino experiment

Dengjie Li

*Department of Modern Physics, University of Science
and Technology of China, Hefei 230026, P. R. China*
lidengjie@ihep.ac.cn

Published 15 May 2014

Muon induced neutrons and long-lived radioactive isotopes are important background sources for low-energy underground experiments. We study the produced processes and properties of cosmic muon induced backgrounds, show the muon veto system used for rejecting these backgrounds and the methods to estimate residual backgrounds in the Daya Bay Reactor Neutrino Experiment.

Keywords: Muon spallation; neutron; radioactive isotopes; background; neutrino.

PACS Numbers: 25.30.Mr, 29.40.−n

1. Introduction

Muons are secondary products of collisions between high-energy cosmic rays and atmosphere nuclei. The muon flux at sea level is roughly 160 Hz/m^2. Muon induced neutrons and long-lived radioactive isotopes are important background sources for low-energy underground experiments. There are three important muon induced backgrounds in the Daya Bay experiment, accidentals, fast neutron and $^8He/^9Li$.

2. Muon Induced Backgrounds

Following are the processes that muon induces backgrounds. Muon induced neutron can be produced by muon spallation that muon interactions with nuclei via a virtual photon producing a nuclear disintegration, muon elastic scattering with neutrons bound in nuclei, photo-nuclear reactions associated with electromagnetic showers generated by muons, and secondary neutron production following any of the above processes. Muon induced isotopes can be produced by muon spallation and their secondary shower particles.[1]

3. Muon Veto System

Locating the detectors at sites with adequate overburden is the only way to reduce the muon flux and the associated background to a tolerable level. Being supplemented with a good muon identifier outside the detector, we can tag the muons going through or near the detector modules and reject backgrounds efficiently. Figure 1 is the muon veto system in Daya Bay Reactor Neutrino Experiment.[2]

This water buffer is used as a cherenkov detector to detect muons. Thus neutrons produced by muons in the detector module, the water buffer or surrounding rock will be effectively attenuated by the 2.5 m water buffer. Together with RPC outside the water buffer, the combined muon tag efficiency is 99.5%, with an uncertainty smaller than 0.25%. Veto time windows after tagged muon are [0 μs, 600 μs] for water pool tagged muon, [0 ms, 1 ms] for AD tagged muon, [0 s, 1 s] for AD tagged shower muon, respectively.

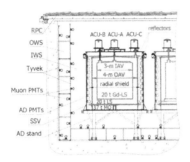

Fig. 1.   Muon veto system.

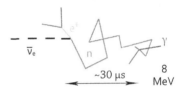

Fig. 2.   Physical process of inverse beta decay.

4. Three Backgrouds In DayaBay

4.1. *Inverse beta decay*

The reaction employed to detect the $\bar{\nu}_e$ from reactor is the inverse beta decay.

$$\bar{\nu}_e + p \rightarrow e^+ + n$$

Figure 2 is the physical process of inverse beta decay. The reaction has a distinct signature : a prompt positron followed by a neutron-capture. Selections of inverse beta decay events(IBD) are

  (i) Energy of prompt signal : [0.7 MeV, 12.0 MeV],
 (ii) Energy of delayed signal : [6.0 MeV, 12.0 MeV],
(iii) Delta time between delayed and prompt signal: [1 μs, 200 μs],
(iv) Multiple cut : no other >0.7 MeV triggers in 200 μs time window before prompt signal or after delayed signal.

4.2. *Accidentals*

A single neutron capture signal has some probability to fall accidentally within the time window of a preceding signal due to natural radioactivity (U, Th, K, Co, Rn, Kr), producing an accidental background. Some other long-lived cosmogenic isotopes, such as $^{12}B/^{12}N$, can fake the delayed 'neutron' signal if they have beta decay energy in the 6 MeV $\sim$ 12 MeV range.

The expected accidental background can be calculated as:

$$N_{accBkg} = \Sigma N^i_{n-like\ singles} \cdot (1 - e^{R^i_{e+-like\ triggers} \cdot 200\mu s}). \qquad (1)$$

where, $N^i_{n-like\ singles}$ is rate of the triggers in prompt signal energy region that calculated every 4 hours, $R^i_{e+-like\ triggers}$ is number of the triggers in delayed signal energy region that counted every 4 hours.

4.3. *Fast neutron*

A fast neutron produced by cosmic muons in the surrounding rock or the detector can produce a signal mimicking the inverse beta decay reaction in the detector. Figure 3 is the physical process of fast neutron, the recoil proton generates the prompt signal and the capture of the thermalized neutron provides the delayed signal.The energy spectrum of prompt signal of only inner water pool tagged fast neutron between 0 MeV and 100 MeV is linear distribution.

Thus we use linear extrapolation to estimate fast neutron background (see Fig. 4). First,relax the 12 MeV prompt energy criterion in the IBD selection to 100 MeV, then take the mean of zero and first order polynomial fit values in [12 MeV, 100 MeV] range to extrapolate the backgrounds into the [0.7 MeV, 12 MeV] region.

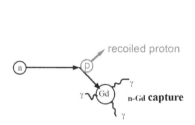

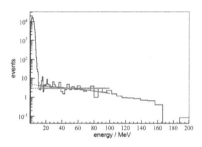

Fig. 3.   Physical process of fast neutron.   Fig. 4.   Fitting method of fast neutron.

4.4. $^{8}He/^{9}Li$

The $^{8}He/^{9}Li$ isotopes produced by cosmic muons have substantial beta-neutron decay branching fractions (see Fig. 5). The beta energy of the beta-neutron cascade overlaps the energy range of positron signals from neutrino events, mimicking the prompt signal, and the neutron emission forms the delayed signal.

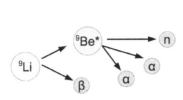

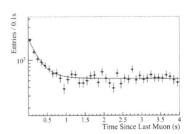

Fig. 5.   Physical process of ^9Li.   Fig. 6.   Fitting method of $^8He/^9Li$.

$$f(t) = N_{Li+He}(R \cdot \lambda_{Li} \cdot e^{-\lambda_{Li} \cdot t} + (1 - R) \cdot \lambda_{He} \cdot e^{-\lambda_{He} \cdot t}) + N_{ibd} \cdot R_\mu \cdot e^{-R_\mu \cdot t}. \quad (2)$$

Fit the time since last muon distribution with Eq.2 to measure $^8He/^9Li$ background.[3] Figure 6 shows an example of fitting for $^8He/^9Li$ backgrounds.

5. Results

The muon induced backgrounds in six antineutrino detectors(AD) after muon rejection is listed in following Table.[4]

Table 1.   Result of backgrounds in Daya Bay Reactor Neutrino Experiment

	AD1	AD2	AD3	AD4	AD5	AD6
IBD rate(/day)	662.47±3.00	670.87±3.01	613.53±2.69	77.57±0.85	76.62±0.85	74.97±0.84
Accidentals(/day)	9.73±0.10	9.61±0.10	7.55±0.08	3.05±0.04	3.04±0.04	2.93±0.03
Fast neutron(/day)	0.77±0.24	0.77±0.20	0.58±0.33	0.05±0.02	0.05±0.02	0.05±0.02
^{8}He/^{9}Li(/day)	2.9±1.5		2.0±1.1		0.22±0.12	

Acknowledgments

Sincerely thanks to every contributor to this study at Daya Bay Collaboration. Thanks to everyone who helped me.

References

1. Y. F. Wang, V. Balic, G. Gratta *et al.*, *Phys. Rev. D* **62**, 013012 (2001).
2. DAYA-BAY Collab. (F. P. An *et al.*), *Phys. Rev. Lett.* **108**, 171803 (2012).
3. L. J. Wen, J. Cao, K. B. Luk *et al*, *Nuclear Instruments and Methods in Physics Research A* **564**, 471 (2006).
4. DAYA-BAY Collab. (F. P. An *et al.*), *Chinese Phys. C* **37**, 011001 (2013).

Physics in Collision (PIC 2013)
International Journal of Modern Physics: Conference Series
Vol. 31 (2014) 1460311 (4 pages)
© The Author
DOI: 10.1142/S2010194514603111

Observation of the $X(1840)$ at BESIII

Jingqing Zhang
(on Behalf of the BESIII Collaboration)

Institute of High Energy Physics
Chinese Academy of Sciences, Beijing 100049, China
zhangjq@ihep.ac.cn

Published 15 May 2014

Observation of the $X(1840)$ in the $3(\pi^+\pi^-)$ invariant mass in $J/\psi \to \gamma 3(\pi^+\pi^-)$ at BESIII is reviewed. With a sample of 225.3×10^6 J/ψ events collected with the BESIII detector at BEPCII, the $X(1840)$ is observed with a statistical significance of 7.6σ. The mass, width and product branching fraction of the $X(1840)$ are determined. The decay $\eta' \to 3(\pi^+\pi^-)$ is searched for, and the upper limit of the branching fraction is set at the 90% confidence level.

PACS Numbers: 13.20.Gd, 12.39.Mk

1. Introduction

Within the standard model framework, the strong interaction is described by Quantum Chromodynamics (QCD), which suggests the existence of the unconventional hadrons, such as glueballs, hybrid states and multiquark states. The establishment of such states remains one of the main interests in experimental particle physics.

Decays of the J/ψ particle are ideal for the study of the hadron spectroscopy and the searching for the unconventional hadrons. In the decays of the J/ψ particle, several observations in the mass region 1.8 GeV/c^2 - 1.9 GeV/c^2 have been presented in different experiments,[1–8] such as the $X(p\bar{p})$,[1–3] $X(1835)$,[4,5] $X(1810)$[6,7] and $X(1870)$.[8]

2. Observation of the $X(1840)$

Recently, using a sample of 225.3×10^6 J/ψ events[9] collected with BESIII detector[10] at BEPCII,[11] the decay of $J/\psi \to \gamma 3(\pi^+\pi^-)$ was analyzed,[12] and the $X(1840)$ was observed in the $3(\pi^+\pi^-)$ mass spectrum with a statistical significance of 7.6σ.

The $3(\pi^+\pi^-)$ invariant mass spectrum is shown in Fig. 1, where the $X(1840)$ can be clearly seen. The parameters of the $X(1840)$ are extracted by an unbinned maximum likelihood fit. In the fit, the background is described by two contributions: the contribution from $J/\psi \to \pi^0 3(\pi^+\pi^-)$ and the contribution from other sources. The contribution from $J/\psi \to \pi^0 3(\pi^+\pi^-)$ is determined from MC simulation and fixed in the fit (shown by the dash-dotted line in Fig. 2). The other contribution is described by a third-order polynomial. The signal is described by a Breit-Wigner function modified with the effects of the detection efficiency, the detector resolution, and the phase space factor. The fit result is shown in Fig. 2. The mass and width of the $X(1840)$ are $M = 1842.2 \pm 4.2^{+7.1}_{-2.6}$ MeV/c^2 and

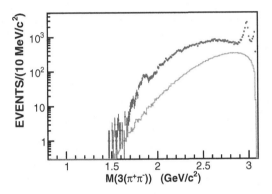

Fig. 1.   Distribution of the invariant mass of $3(\pi^+\pi^-)$. The dots with error bars are data; the histogram is phase space events with an arbitrary normalization.

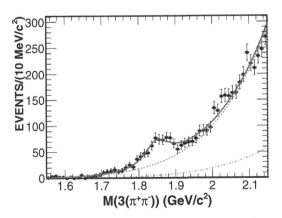

Fig. 2.   The fit of mass spectrum of $3(\pi^+\pi^-)$. The dots with error bars are data; the solid line is the fit result. The dashed line represents all the backgrounds, including the background events from $J/\psi \to \pi^0 3(\pi^+ pi^-)$ (dash-dotted line, fixed in the fit) and a third-order polynomial representing other backgrounds.

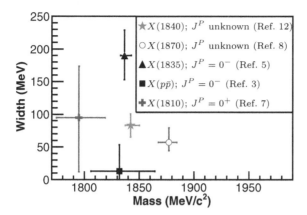

Fig. 3. Comparisons of observations at BESIII. The error bars include statistical, systematic, and, where applicable, model uncertainties.

$\Gamma = 83 \pm 14 \pm 11$ MeV, respectively; the product branching fraction of the $X(1840)$ is $\mathcal{B}(J/\psi \to \gamma X(1840)) \times \mathcal{B}(X(1840) \to 3(\pi^+\pi^-)) = (2.44 \pm 0.36^{+0.60}_{-0.74}) \times 10^{-5}$. In these results, the first errors are statistical and the second errors are systematic.

Figure 3 shows the comparisons of the $X(1840)$ with other observations at BESIII.[12] The comparisons indicate that at present one can not distinguish whether the $X(1840)$ is a new state or the signal of a $3(\pi^+\pi^-)$ decay mode of an existing state.

3. Search for the Decay of $\eta' \to 3(\pi^+\pi^-)$

With the same data sample, the decay of $\eta' \to 3(\pi^+\pi^-)$ was searched for.[12] The mass spectrum of the $3(\pi^+\pi^-)$ is shown in Fig. 1, where no events are observed in the η' mass region. With the Feldman-Cousins frequentist approach,[13] the upper limit of the branching fraction is set to be $\mathcal{B}(\eta' \to 3(\pi^+\pi^-)) < 3.1 \times 10^{-5}$ at the 90% confidence level, where the systematic uncertainty is taken into account.

4. Summary

With a sample of 225.3×10^6 J/ψ events collected at BESIII, the decay of $J/\psi \to \gamma 3(\pi^+\pi^-)$ was analyzed.[12] The $X(1840)$ was observed in the $3(\pi^+\pi^-)$ invariant mass spectrum. The mass, width and product branching fraction of the $X(1840)$ are $M = 1842.2 \pm 4.2^{+7.1}_{-2.6}$ MeV/c^2, $\Gamma = 83 \pm 14 \pm 11$ MeV and $\mathcal{B}(J/\psi \to \gamma X(1840)) \times \mathcal{B}(X(1840) \to 3(\pi^+\pi^-)) = (2.44 \pm 0.36^{+0.60}_{-0.74}) \times 10^{-5}$, respectively. The decay $\eta' \to 3(\pi^+\pi^-)$ was searched for. No events were observed in the η' mass region and the upper limit of the branching fraction was set to be $\mathcal{B}(\eta' \to 3(\pi^+\pi^-)) < 3.1 \times 10^{-5}$ at the 90% confidence level.

References

1. J. Z. Bai *et al.* [BES Collaboration], *Phys. Rev. Lett.* **91**, 022001 (2003).
2. J. P. Alexander *et al.* [CLEO Collaboration], *Phys. Rev. D* **82**, 092002 (2010).
3. M. Ablikim *et al.* [BESIII Collaboration], *Phys. Rev. Lett.* **108**, 112003 (2012).
4. M. Ablikim *et al.* [BES Collaboration], *Phys. Rev. Lett.* **95**, 262001 (2005).
5. M. Ablikim *et al.* [BESIII Collaboration], *Phys. Rev. Lett.* **106**, 072002 (2011).
6. M. Ablikim *et al.* [BES Collaboration], *Phys. Rev. Lett.* **96**, 162002 (2006).
7. M. Ablikim *et al.* [BESIII Collaboration], *Phys. Rev. D* **87**, 032008 (2013).
8. M. Ablikim *et al.* [BESIII Collaboration], *Phys. Rev. Lett.* **107**, 182001 (2011).
9. M. Ablikim *et al.* [BESIII Collaboration], *Chin. Phys. C* **36**, 915 (2012).
10. M. Ablikim *et al.* [BESIII Collaboration], *Nucl. Instrum. Meth. A* **614**, 345 (2010).
11. J. Z. Bai *et al.* [BES Collaboration], *Nucl. Instrum. Meth. A* **458**, 627 (2001).
12. M. Ablikim *et al.* [BESIII Collaboration], arXiv:1305.5333 [hep-ex].
13. G. J. Feldman and R. D. Cousins, Phys. Rev. D **57**, 3873 (1998).

Physics in Collision (PIC 2013)
International Journal of Modern Physics: Conference Series
Vol. 31 (2014) 1460312 (4 pages)
© The Authors
DOI: 10.1142/S2010194514603123

Energy non-linearity studies at Daya Bay

Masheng Yang* and Yaping Cheng[†]

(On behalf of the Daya Bay Collaboration)

Institute of High Energy Physics, Beijing, China
*yangms@ihep.ac.cn
†chengyp@ihep.ac.cn

Published 15 May 2014

The Daya Bay Reactor Neutrino Experiment has measured a non-zero value of the neutrino mixing angle θ_{13} with a significance of 7.7 standard deviations by a rate-only analysis.[1] The distortion of neutrino energy spectrum carries additional oscillation information and can improve the sensitivity of θ_{13} as well as measure neutrino mass splitting Δm^2_{ee}. A rate plus shape analysis is performed and the results have been published.[2] Understanding detector energy non-linearity response is crucial for the rate plus shape analysis. In this contribution, we present a brief description of energy non-linearity studies at Daya Bay.

Keywords: Neutrino experiment; oscillation; energy response; non-linearity; spectral analysis.

1. General introduction

It is well established that the flavor of a neutrino oscillates with time. Neutrino oscillations can be described by the three mixing angles(θ_{12}, θ_{23}, and θ_{13},), a phase of the Pontecorvo-Maki-Nakagawa-Sakata matrix, and the neutrino mass squared differences.[3,4] To date, the Daya Bay experiments has made the most precise measurement of the neutrino mixing angle θ_{13}.[1,5] The Daya Bay Experiment has three underground Experiment Halls(EH) and totally 8 anti-neutrino detectors(ADs). The AD contains a structure of three layers, with Gadolinium loaded Liquid Scintillator(GdLS) in the center, LS in the middle as the gamma catcher, and oil in the outer layer to shield the radioactive components like PMTs. The $\bar{\nu}_e$ from the reactor interacts with the detector via the inverse beta decay(IBD), and a positron together with a neutron come out after the interaction, namely, $\bar{\nu}_e + p \rightarrow e^+ + n$. With the positron kinetic energy deposited followed by its annihilation, this event presents as a prompt signal. The neutron is captured by the Gd nucleus, and several gammas with total energy about 8 MeV are emmited. This event forms the delayed signal. IBD candidates are selected through time-correlation method. The prompt energy from the positron gives an estimate of the incident $\bar{\nu}_e$ energy through: $E_\nu \simeq E_{e+} + 0.8 MeV$.

2. Energy Response Process

The Daya Bay detector energy response can be understood as follows. A particle with its true energy deposit its energy in the AD. For a e^-, E_{true} is the kinetic energy; for a positron, E_{true} is the sum of the kinetic energy and the energy from annihilation. After that, the LS translates the deposit energy into visible energy E_{vis}. The visible photons are detected by photomultiplier tubes(PMT). After the calibration and reconstruction, E_{vis} is converted to be the reconstructed energy E_{rec}. The energy response is not linear due to scintillator and electronics effects. The non-linearity, E_{rec}/E_{true}, can be separated into two parts: the electronics non-linearity, E_{rec}/E_{vis}, and the scintillator non-linearity, E_{vis}/E_{true}. Non-linearity study is significant to deduce the true energy of positron from the reconstructed energy.

3. Non-Linearity from the Electronics

The electronics non-linearity is due to the interplay between slow component of LS light emitting > 100 ns later after the first light(see Fig. 1) and the front end electronics system. Later hits formed by slow component may not be included in the hit collection, thus would make the collection efficiency decrease with increasing E_{vis}. We used several models to parameterize the electronics non-linearity, for example, $\frac{E_{vis}}{E_{rec}} = (1 - \alpha e^{\frac{E_{rec}}{\tau}})$.

4. Non-Linearity from Liquid Scintillator

The scintillator nonlinearity is particle and energy-dependent, and it is related to the intrinsic scintillator quenching and Cherenkov light emission. Different models are

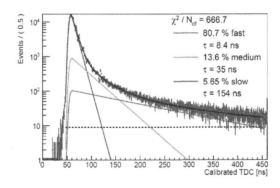

Fig. 1. Fast and slow components of LS photons, fitted by exponential functions.

used to constrain the LS non-linearity of electron. One is a model using the Birk's law[6] and Cherenkov radiation theory, $E_{vis}/E_{true} = f_q(E_{true}; K_B) + K_C f_c(E_{true})$, with the 1st term for quenching effect using the Birk's law, and the 2ns term for Cherenkov radiation. Another one is an empirical model with 4 parameters, $E_{vis}/E_{true} = (p_0 + p_3 E_{true})/(1 + p_1 e^{-p_2 E_{true}})$. The gamma non-linearity and the electron non-linearity in the LS are connected by the energy conversion processes, gamma-rays interact with matter mainly in three ways, namely, Computon scattering,Photoelectric, and pair production, all these finally result in e^+ or e^-. With a Geant4 simulation, the gamma to e^+/e^- converting probability function can be obtained. The non-linearity of gamma can be deduced from the secondary electron's non-linearity(see Fig. 2).

5. Available Data to Constrain the Non-Linearity

During a special calibration period in summer 2012, we used gamma sources and neutron sources deployed at the center of detector to do the nonlinearity study. Gamma sources such as $^{137}Cs, ^{54}Mn, ^{40}K$, and neutron sources such as $^{241}Am-^9Be$

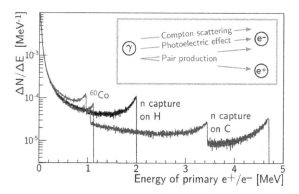

Fig. 2. The gamma to electron conversion probability density function via Geant4 for different gammas.

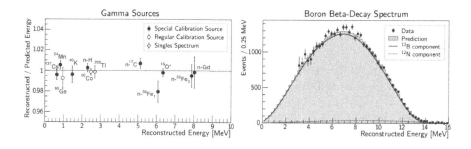

Fig. 3. The comparison of predicted vs measure gamma peaks(left) and ^{12}B spectrum (right).

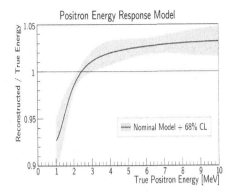

Fig. 4.   Positron non-linearity result and its uncertainty contour.

and $Pu-^{13}C$ were used. Neutrons emitted from the neutron radiation sources can be captured on hydrogen or Gd. A single 2.2 MeV gamma is emitted when captured on hydrogen, providing one more gamma peak for nonlinearity study. The calibration data were fitted with the models described above, and consistency is obtained among different models. Energy spectrum from cosmic muon induced $\gamma +^{12}B$ spectrum, is also used to test the consistency of the nonlinearity. The comparisons of the predicted using the best fit nonlinearity model vs measured gamma peaks and ^{12}B spectrum are shown in Fig. 3.

6. Summary

After comparison between different models and comparison between models and data, the energy non-linearity for positron is show as follow(see Fig. 4), the shadow area represents the nonlinearity uncertainty both systematic and statistic combined within 1 sigma significance level, and the relative uncertainty of the non-linearity for positron is about 1.5%.

Acknowledgments

We'd like to thank the PIC organization committee for this opportunity to present the nonlinearity studies at Daya Bay. We would also like to thank everyone who helped us on this contribution.

References

1. Daya Bay Collab. (F. P. An *et al.*), *Chin.Phys.C.* **108**, 011001 (2013).
2. Spectral measurement. at Daya Bay (Daya Bay Collab.) *Phys. Rev. Lett.* **112**, 061801 (2014).
3. B. Pontecorvo, *Sov. Phys. JETP* **6**, 429 (1957) and **26**, 984 (1968)
4. Z. Maki, M .Nakagawa and S. Sakata, *Prog. Theor. Phys.* **28**, 870 (1962)
5. Daya Bay Collab. (F. P. An *et al.*), *Phys. Rev. Lett.* **108**, 171803 (2012).
6. J. B. Birks, *Phys. Lett. B* **218**, 365(1989)

Printed in the United States
By Bookmasters